D1199218

The Skin and Eye

The Skin and Eye

A Dermatologic Correlation of
Diseases of the Periorbital Region

By GUNTER W. KORTING, M.D.
Professor and Director of the Department
of Dermatology, University Hospital, Mainz

Translated and Adapted by

WILLIAM CURTH, M.D.
Assistant Clinical Professor of Dermatology,
New York Medical College

HELEN O. CURTH, M.D.
Special Lecturer in Dermatology,
College of Physicians and Surgeons, Columbia University

FREDERICK F. URBACH, M.D.
Chairman of the Department of Dermatology,
Temple University School of Medicine

DANIEL M. ALBERT, M.D.
Associate Professor of Ophthalmology,
Yale University School of Medicine

With 390 Figures in Color

W. B. SAUNDERS COMPANY
Philadelphia · London · Toronto 1973

Authorized English Edition.
All rights reserved.
Original German Edition:
HAUT UND AUGE
© 1969 by Georg Thieme Verlag, Stuttgart.
English Edition published by
W.B. SAUNDERS COMPANY
Philadelphia, London, Toronto, 1973.
Library of Congress Catalog card number 76—108—370.
ISBN 0—7216—5492—4.
Printed in Germany.

Preface

The principal theme of the present monograph is not new: drawing together knowledge of dermatology and ophthalmology, and especially organizing ocular findings and impressions in a useful way. This approach has been used in numerous works on the skin and eye, both old (Groenouw, Igersheimer, Lutz, and others) and recent (Schreck, Pillat, Sprafke). Among writers of monographs, Schönfeld from the dermatologic side, Schrader from the ophthalmologic, and Casanovas and Vilanova combining both sciences have attempted to present overall views of the many special associations between the skin and the eye.

Diseases of the skin can affect various parts of the eye, and furthermore, many dermatologic diagnoses can be confirmed by examination of the eyes, or tentative assessments of disorders seen in their incomplete or atypical forms can be made certain. The reciprocal effects that appear on clinical grounds require a synthesis of both disciplines, and the author has been encouraged in this effort by the publisher, Dr. G. Hauff.

The purpose of this book, then, is not to serve as a new dermatologic or ophthalmologic text, but merely to present the most outstanding diagnostic points from both sciences in a manner that is useful to the practicing dermatologist. Schreck has previously accomplished much of this for the ophthalmologist.

The material is presented from the dermatologist's point of view, and therefore its sequence follows that of several existing monographs. Photographs of dermatologic diseases are arranged according to such a nosologic grouping as to make the oculocutaneous associations, and they are of interest to both skin and eye physicians not only on the basis of their topographic connections but also as the basis for syndromic classifications. It is therefore necessary at times to turn to the literature in order to determine whether a particular disorder is more particularly dermatologic or ophthalmologic; references are to be found following each principal subdivision. The photographs come in the main from the University Skin Clinic, Mainz, where the clinic photographer, Mr. Faber, deserves special thanks.

Several ophthalmologic photographs have also been included in order to familiarize dermatologists with such appearances as "angioid streaks." For these I am most grateful to Professor Jaeger of Heidelberg, Professor Sautter of Hamburg, and especially to my own colleague Professor Nover, to whom I also owe warm thanks for many hours spent in consultation during preparation of the work. The manuscript was prepared in its final form by Mrs. Moritz. Especially hearty thanks go to the publisher, Dr. Hauff, without whose sympathy for the theme of this monograph, it would not have been undertaken.   G. W. KORTING

Contents

ACUTE ERYTHEMAS AND EXAN-
THEMS 1

Morbilli (Measles) 1
Erythema multiforme 1
Erythema nodosum 2
Toxic-allergic exanthems 2

PAPULOSQUAMOUS DERMATOSES . 4

Psoriasis vulgaris 4
Parapsoriasis 6
Pityriasis rubra pilaris 6

THE ECZEMAS 7

Seborrheic eczema 8
Endogenous eczema 8
Lichen-simplex chronicus, Vidal 14

URTICARIA, STROPHULUS AND
PRURIGO 16

LICHEN RUBER PLANUS 18

ACUTE INFECTIOUS DISEASES OF
THE SKIN 20

Pyodermas 20
Erysipelas 22
Anthrax of the skin 22
Diphtheria of the skin 22
Toxoplasmosis of the skin 23
Tularemia 23

CHRONIC INFECTIOUS DISEASES OF
THE SKIN 25

Tuberculosis cutis 25
Sarcoidosis 28
Melkersson-Rosenthal syndrome 32
Ascher syndrome and other forms of
blepharochalasis 32
Leprosy 34
Leishmaniasis cutis (Oriental or Aleppo boil) 36

VIRUS DISEASES OF THE SKIN . . . 39

Smallpox (Variola vera) 40
Side effects of protective vaccination . . . 40

Molluscum contagiosum 42
Herpes simplex (Herpes febrilis, Fever
blisters, Cold sores) 44
Herpes zoster (Zona, shingles) 44
Varicella (Chicken pox) 45
Warts 46
Cat-scratch disease (Maladie de griffes de chat,
nonbacterial regional lymphadenitis) . . 46

MYCOTIC INFECTIONS OF THE SKIN 50

Favus 50
Microsporon infections 50
Trichophytosis 51
Candidiasis 51
Blastomycosis 52
Sporotrichosis 52
Actinomycosis 52
Dermatoses due to molds 54

ZOONOSES 55

Scabies 55
Cutaneous reactions due to Ixodidae . . . 55
Pediculosis 55
Lepidopteriasis 56
Ophthalmomyiasis 58
Vermiasis 58

CONGENITAL ABNORMALITIES OF
THE SKIN 60

Ectodermal dysplasias of the skin (Anhidro-
sis hypotrichotica and others) 60
Aplasia cutis 61
Congenital dysplasias of the skin 61
The syndrome of Thomson and congenital
dystrophy of Rothmund 62
Progeria 62
Cutis laxa (Meekeren-Ehlers-Danlos syn-
drome) 64
Malformations with particular relationship
to the periorbital region 65
Familial dermochondrocorneal dystrophy . 65
Skin and eye changes due to malformations
of the skeleton 66
Pterygium syndrome (Bonnevie-Ullrich-Tur-
ner syndrome) 69

KERATOSES AND DYSTROPHIES . . 72

Diffuse keratoses (Ichthyoses) 72
Palmar and plantar keratoses 73

Keratosis palmoplantaris diffusa 73
Keratosis palmoplantaris insuliformis seu striata. 73
Keratosis palmoplantaris papulosa . . . 73
Follicular keratoses 74
Keratosis follicularis lichenoides seu lichen pilaris. 74
Keratosis follicularis spinulosa 74
Acneiform follicular keratoses 74
Darier's disease (Keratosis follicularis) . . 74
Dyskeratosis intraepithelialis benigna hereditaria. 75
Elastosis perforans serpiginosa 75
Pseudoxanthoma elasticum 76
Acanthosis nigricans 78

BULLOUS DERMATOSES 81

Pemphigus 81
Bullous pemphigoid 82
"Benign" mucous membrane pemphigoid (Lever) 84
Familial pemphigus (Hailey and Hailey) . . 84
Dermatitis herpetiformis (Duhring, 1884) . 84
Subcorneal pustular dermatosis (Sneddon and Wilkinson, 1956) 85
Hereditary epidermolyses 85
Epidermolysis bullosa hereditaria simplex 85
Epidermolysis bullosa hereditaria dystrophica dominans 85
Epidermolysis bullosa hereditaria dystrophica recessive 86

PHYSICAL AND CHEMICAL INJURIES TO THE SKIN 87

DISEASES DUE TO LIGHT (Including Radiation Reactions) 90

Xeroderma pigmentosum 90
Light urticaria 91
Eczema solare (Unna, Veiel, Wolters) and Chronic polymorphic light eruption (Haxthausen) 91
Springtime pernio 91
Photodynamic reactions 92
Radiation reactions (roentgen, laser, and others) 92

DISTURBANCES OF PIGMENTATION 95

Melanodermas 95
Nevus fuscocaeruleus ophthalmo-maxillaris of Ota 96
Incontinentia pigmenti (Bloch-Sulzberger) 96

Leukopathias 98
Vogt-Koyanagi-Harada syndrome . . . 98
Nevus of Sutton 100
Vitiligo 100

CIRCULATORY AND VASCULAR DISORDERS 102

Endangiitis obliterans 102
Periarteritis nodosa 103
Wegener's granulomatosis 103
Arteritis temporalis (Hortan, Magath and Brown) 104
Arteriolitis "allergica" cutis 104

HEMORRHAGIC DIATHESES. 106

Angiopathies. 106
Osler's disease 106
Schönlein's purpura 108
Progressive pigmentary purpura 108
Thrombocytopathies 108
Morbus maculosus Werlhof 108
Coagulopathies 109

METABOLIC AND STORAGE DISEASES 110

Porphyrias 110
Hepatic porphyria 112
Porphyria hepatica acuta intermittens . . 112
Porphyria cutanea tarda (Waldenström) . 112
South African form of porphyria (porphyria variegata) 112
Erythropoietic porphyria 112
Porphyria congenita (Günther) 112
Familial protoporphyrinemic photodermatosis 112
Congenital erythropoietic coproporphyria 112
Lipoidoses 113
Hypercholesterinemic xanthomatosis . . 113
Hyperlipemic xanthomatosis 114
Necrobiosis lipoidica 114
Nevoxanthoendothelioma 114
Amyloidoses 116
Hyalinosis cutis et mucosae (Wiethe, 1924), Lipoid proteinosis (Urbach) 116
Mucinosis 118
Sphingolipidoses 120
Cutaneous gout 120
Calcinosis cutis 121
Angiokeratoma corporis diffusum (Fabry) 121
Ochronosis 122
Phenylketonuria 122
Carotinosis 123
Argyria 123

COLLAGENOSIS GROUP (Generalized
  Connective Tissue Diseases) 126

Scleroderma 126
  Scleredema adultorum 127
Lupus erythematosus 128
Dermatomyositis 130

ATROPHIES OF THE SKIN 131

Senile-degenerative atrophy 131
Facial hemiatrophy 131
Atrophoderma vermiculatum 132
Macular atrophy 132
Acrodermatitis chronica atrophicans 132

AVITAMINOSES 133

Vitamin A 133
Vitamin B₂ (Riboflavin) 134
Nicotinic acid (Niacin) 135
Vitamin B₆ (Pyridoxine), Pantothenic acid . 135
Vitamin C (Scurvy) 136

BENIGN TUMORS OF THE SKIN . . . 139

Nevi and nevoid syndromes including pha-
  komatoses 139
  Von Recklinghausen's disease, neuro-
    fibromatosis 139
  Bourneville-Pringle phakomatosis . . . 140
  Pigmented spots and intestinal polyposis
    (Peutz-Klostermann) 142
  Melanophakomatoses 142
  Angiomatous phakomatoses 142
  Epitheliomatous phakomatoses 142
Hard nevi 144
Organic nevi 144
Fibromas 146
Keloids 148
Lipomas 148
Leiomyomas 148
Cysts 149
Angiomas and angiophakomatoses 149
Pyogenic granuloma (Granuloma telangiec-
  taticum benignum) 150
Angiomatous phakomatoses 150
Angiomatosis retinae (v. Hippel-Lindau) . 150
Angiomatosis trigemino-cerebralis or ence-
  phalo-oculo-cutanea (Sturge-Weber syn-
  drome) 150
Klippel-Trénaunay-Weber phakomatosis . . 152
Ataxia telangiectatica (Louis-Bar syndrome) 152

MALIGNANT AND POSSIBLY MALIG-
  NANT TUMORS OF THE SKIN, PIG-
  MENTED TUMORS 155
Basal cell and squamous cell epitheliomas,
  metastatic cutaneous cancers 155
Precancers 160
Pseudocancers 164
  Seborrheic wart 166
Sarcomas 166
Pigmented tumors 167
  Nevus cell nevi 167
  Blue nevus (nevus caeruleus) 167
  So-called juvenile melanoma 168
  Melanosis circumscripta preblastomatosa 168
  (Malignant) melanoma 170

LEUKOSES, RETICULOSES, AND RE-
  TICULOENDOTHELIOSES OF THE
  SKIN 177

Leukoses 177
Reticuloses in their narrow sense 177
  Lipoid storage reticuloses of the skin . . 178
  Urticaria pigmentosa 178
  Cutaneous and ocular manifestations in
    plasmocytoma and macroglobulinemia
    (Waldenström's disease) 179
Reticulogranulomatoses 180
  Systemic reticulogranulomatoses 180
  Circumscribed lympho- and plasmocytic
    hyperplasias of the skin 182

DISEASES OF THE SEBACEOUS
  GLANDS 189

Acne vulgaris 189
Rosacea 189

DISEASES OF THE HAIR, ESPECIALLY
  ALOPECIA AREATA 190

BEHÇET'S DISEASE 194

REITER'S DISEASE 196

SJÖGREN'S SYNDROME 197

VENEREAL DISEASES 198

Syphilis (lues) 199
Gonorrhea 203
Chancroid (ulcus molle) 204
Lymphogranuloma inguinale venereum
  (lymphopathia venereum) 204

OCULAR CHANGES AS SIDE EFFECT
  OF SEVERAL DRUGS USED IN DER-
  MATOLOGICAL PRACTICE 206

INDEX 211

Acute Erythemas and Exanthems

Morbilli (Measles)

In this highly contagious children's disease, in which the infection with the viral agent occurs by way of the *conjunctiva*, an *acute conjunctivitis*, among other symptoms, develops after an incubation period of 9 to 11 days and is associated with the beginning of the cutaneous exanthem. This does not occur in scarlet fever or other infectious exanthems. Since there are many other exanthems that can show a morbilliform eruption, the *Koplik spots*, which occur only in measles, are an important aid in diagnosis. These superficial epithelial necroses appear similar to those seen in moniliasis of the mucosa or to small erosions. Their primary location is around the opening of the parotid duct, but similar spots can also be seen on the conjunctivae and also on the caruncles (BLANK and RAKE, SCHRADER). Beyond this, in addition to the typical keratoconjunctivitis, there can also be found in measles a scrofulalike inflammation of the lacrimal ducts. In recent years, BLATZ and others have reported the occasional destruction of the cornea, sometimes followed by blindness (BLATZ, MOLL).

Erythema multiforme
(Erythema exudativum multiforme)

For this disease a whole series of synonyms exists among which some ophthalmologists prefer the concept of a "mucocutaneous-epithelial erythematous syndrome", which dates back to Ernst Fuchs (1851–1930).

The primary lesion of this erythematous disease consists of a bright red round spot which continues to change in many ways, and which, because of repeated concentric exudation, rather rapidly acquires a characteristic iris or circle form. Frequently the lesions become confluent to form garlandlike erythemas (Fig. 1). The primary locations are the extensor surfaces of the upper extremities, the face, somewhat less frequently, and only rarely the eyelids. In cases where the localization is inverted (in other words in cases where the primary involvement occurs on the flexor surfaces of the arms and hands), there may develop a particular variety of this disease called *"ectodermosis erosiva pluriorificialis"* or *dermatostomatitis*.

A clinical condition in children, consisting of erythema multiforme and a course complicated by pneumonia or nephritis associated with severe constitutional symptoms and purulent conjunctivitis, has been described under the name of the *Stevens-Johnson syndrome*.

As far as the region of the eye is concerned, the conjunctiva and frequently the cornea are the sites of major involvement. The findings usually consist of a short-lasting vesicular eruption followed by a grayish yellow, gelatinous, pseudomembranous conjunctivitis (Fig. 2). Less frequently an ulcerating keratitis develops, or occasionally there may be involvement of the deeper tissues of the eye by a purulent uveitis (FRANÇOIS). This is in contrast to *Behçet's disease*, which primarily involves the uvea. While the changes noted on the eyelid and on the conjunctivae and cornea tend to heal, rather severe sequelae can develop because of involvement of the adnexal areas of the eye (ectropion, entropion, dacryocystadenitis, symblepharon, madarosis, trichiasis, for example, case SHEDDUN). Retention cysts of the lachrymal glands have also been reported. Additional serious complications that have been noted are occasional cases of cataracts (THIES) and secondary glaucoma (MÜLLER and HARTENSTEIN).

Unless a herpes virus is considered because of small groups of vesicular lesions, it is generally assumed that this clinical condition represents a cutaneous reaction which is primarily due either to a drug or to an infection allergy. It may be a secondary cutaneous reaction of leukemia or occasionally may even occur in epidemic form associated with atypical pneumonia with increase in cold agglutinins, or possibly related to infection with mycoplasma pneumoniae (KORTING, KALKOFF). In other cases, a preceding attack of tonsillitis suggests that the etiology may be streptococcal, while one sometimes finds recurrent cases caused by physical influences such as wind or marked exposure to sunshine, or recurrent spring and fall attacks.

Erythema nodosum

Erythema nodosum affects primarily the deeper cutaneous blood vessels, and thus the deep corium and the septae of the subcutaneous fatty tissue, usually on the lower extremities. Clinically, one finds either superficial infiltrates that look like contusions, or deeper, platelike or nodular, sometimes movable areas which are sensitive to pressure. Occasionally, particularly in children, the extensor surfaces of the upper arms and even the face may be involved.

From the ophthalmologic point of view, GROENOUW has reported on the formation of nodules of the conjunctivae and of circular nodules in the ciliary body; SCHRECK has reported conjunctivitis sometimes extending to the development of an episcleritis, and a purulent iridocyclitis which can continue to hypopyon and chorioretinitis. The simultaneous appearance and disappearance of the cutaneous and ophthalmic changes can be quite striking (V. RÖTTH).

From the point of view of etiology, erythema nodosum in childhood is probably most commonly of tuberculous origin. In adults it is a syndrome of multiple etiology like erythema exudativum multiforme. It can occur with any of the venereal diseases; it

may be secondary to fungal infection, particularly in young women with small nodular follicular lesions, and it can be one of the initial cutaneous signs of sarcoidosis. It has also been found associated with rheumatic fever, with Hodgkin's disease, and with other reticuloses and leukemias. Thus, this disease can be considered a *toxic exanthem* of nodose appearance.

Toxic-allergic exanthems

Toxic-allergic exanthems, which at present are among the most common dermatologic disorders and are frequently difficult to diagnose, manifest themselves in the ocular area in a relatively anonymous and monotonous form as blepharadenitis, conjunctivitis, and occasionally bullous lesions similar to ocular pemphigus, a manifestation that was particularly due in the past to sulfonamides (Fig. 3). Even when periorbital involvement is present, these manifestations of drug eruptions only rarely show anything that specifically suggests the agent that causes the disorder. In some cases there are useful criteria, such as for instance the recurrence of the exanthem in the same area each time, followed by a darkly pigmented eruption. Such *fixed drug eruptions* are commonly due to phenolphthalein, derivatives of pyrazoalon, and barbiturates. Acneiform appearance suggests one of the corticosteroids, Halogens, INH, or vitamin D; a vegetating granular appearance suggests iododerma or bromoderma, while lesions resembling lupus erythematosus have been known to be produced by griseofulvin or hydralazine.

More serious drug reactions of the skin are frequently associated with systemic symptoms such as "drug fever," depression of the white blood cell count, hemolysis, or joint pains. This multisystemic involvement has been described by SCHRECK as a "*cutaneous muco-oculo-epithelial allergic syndrome*" (also see OPPEL).

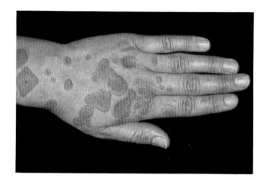

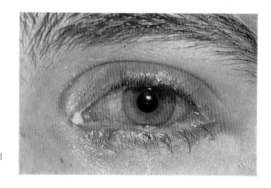

1 Erythema exudativum multiforme. Target erythema of the back of the hand and the forearm.

2 Erythema exudativum multiforme. Typical picture in the eye area.

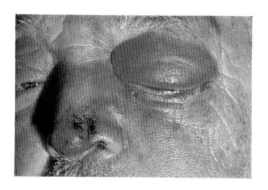

3 Toxic-allergic drug eruption.

References

BLANK, H., and RAKE, G.: Viral and rickettsial diseases of the skin, eye and mucous membranes. Little, Brown & Co., Boston, 1955.

BLATZ, G.: Hornhauteinschmelzung nach Masern. Klin. Mbl. Augenheilk. 129 (1956) 762–772.

FRANÇOIS, J.: Des ectodermoses erosives pluriorificielles. Acta ophthal. (Kbh.) 32 (1954) 5–36.

FUCHS, E.: Herpes Iris conjunctivae. Klin. Mbl. Augenheilk. 14 (1876) 333–351.

GROENOUW, A.: Beziehungen des Auges zu den Hautkrankheiten. In: Handbuch der Haut- und Geschlechtskrankheiten, Bd. XIV/1, hsg. von J. Jadassohn. Springer, Berlin, 1930.

KALKOFF, K. W.: Zur Nosologie und Ätiologie des syndroma muco-cutaneo-oculare acutum Fuchs. In: Haut und innere Krankheiten, hsg. von Schwabe, Basel, 1968.

KORTING, G. W., and TADŽER, J.: Zur Ätiologie des idiopathischen, polymorphen Erythems (Kälteagglutinationsbefunde, atypische Pneumonien). Makedon. Med. Pregl. (1949) 36.

MOLL, H.: Erblindung nach Masern. Arch. Kinderheilk. 155 (1957) 186–191.

MÜLLER, W. A., and HARTENSTEIN H.: Die Beteiligung innerer Organe beim Erythema exsudativum multiforme maius. Dtsch. med. Wschr. 85 (1960) 879–882.

OPPEL, O.: Augenbeteiligung bei Toxicodermie, Bericht über die 65. Zusammenkunft der Dtsch. Ophthalm. Ges. in Heidelberg, 1963, 52.

v. RÖTH, A.: Über Erkrankung der Bindehaut bei Erythema nodosum. Z. Augenheilk. 66 (1928) 323–327.

SCHRADER, K. H.: Auge und Allgemeinleiden. Schattauer, Stuttgart, 1966.

SCHRECK, E.: Über einander zugeordnete Erkrankungen der Haut, der Schleimhäute und der Deckschicht des Auges (cutaneo-muco-oculoepitheliale Syndrome). Arch. Derm. Syph. (Berlin) 198 (1954) 221–257.

STEVENS, A. M., and JOHNSON, F. C.: A new eruptive fever associated with stomatitis and ophthalmia. Amer. J. Dis. Child. 24 (1822) 526.

THIES, O.: Das Auge bei Erythema exsudativum multiforme. Klin. Mbl. Augenheilk. 116 (1950) 44–53.

Papulosquamous Dermatoses

Psoriasis vulgaris

Psoriasis attacks the entire skin, with characteristic individual lesions which may be localized or widespread. Involvement is frequently symmetrical, with predilection for the knees and elbows, and in general the face is spared unless light or other external irritants produce an isomorphic reaction. However, the eyelids are frequently the location of erythematous lesions, which cover them with fine scales of a "seborrheic" appearance and are frequently missed. If such scaling lesions are located in the vicinity of the eyelashes, an inflammatory ectropion, madarosis, or trichiasis may develop (Figs. 4 and 5). Very rarely granulating areas may develop on the conjunctiva (FONTANA, KALDECK). SCHRECK has also noted a nodular episcleritis in his studies on ocular psoriasis, and ulcerations of the cornea have been reported in severe cases (GROENOUW). The involvement of the cornea in the manifestations of psoriasis vulgaris has in recent years been described by a number of authors, among whom PILLAT suggested that the corneal involvement in psoriasis should be known as "*keratitis psoriatica.*" Finally, LUTZ has described totally transparent, subcapsular, punctate, or disclike opacities of the lens located in the deeper layers of the outer cortical area. There is at least one report of exacerbation of psoriasis following removal of a cataract in a patient in her mid-forties.

In psoriasis one is dealing with an erythematous, scaling skin disorder, in which after removal of the lamellar scale, a deep erythema topped with dewlike drops of blood is noted (Auspitz's phenomenon). In spite of the fact that the individual lesions are quite similar to each other, the general appearance of psoriasis has an extremely variable expression, so that in addition to the major variants such as erythrodermic, arthropathic, and pustular psoriasis, one can quite reasonably speak of psoriasis vulgaris, punctata, guttata, nummularis, geographica, and so forth. From the standpoint of etiology, psoriasis is considered an irregularly dominant, probably multifactorial hereditary disease. This genetic predisposition is markedly affected by external as well as internal factors.

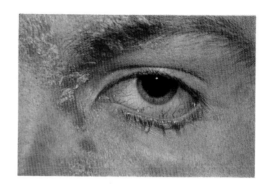

4 Psoriasis vulgaris.

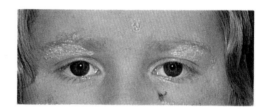

5 Psoriasis vulgaris.

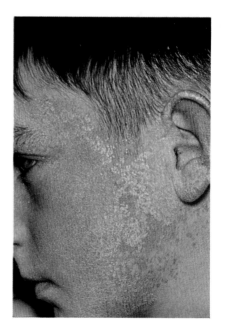

6 Pityriasis rubra pilaris.

Parapsoriasis

Under this title are grouped three rather morphologically distinct disorders which really are not comparable to psoriasis. The three major types — *pityriasis lichenoides chronica* (parapsoriasis guttata), *parapsoriasis en plaque*, and *parakeratosis variegata* (parapsoriasis lichenoides) — usually affect the trunk and only rarely the face. FRIEDE has reported involvement of the eyelids in two cases, characterized by fleshy-red injection and mucous ulceration of the conjunctiva occurring simultaneously with the cutaneous manifestations. The typical cutaneous findings in these patients were gray-to-brownish red, originally more coppery red, lichen planus-like nodules. Fully developed, they were covered with a wafer or colloidinlike scale, which could be removed as a translucent flake. Typical are the chronicity and the resistance to therapy, the absence of any subjective manifestations, and the occasional healing with hypopigmentation.

Pityriasis rubra pilaris

This disease usually affects children and young persons, and occasionally individuals in their fifties. The skin lesions consist of densely crowded acuminate nodules (particularly on the fingers), psoriasiform areas mainly located on the elbows, and confluent grayish and doughy scaling of the scalp (Fig. 6). In addition to these, there are frequently flat hyperkeratoses on the palms and soles and, rarely, a generalized erythroderma.

As far as the eye area is concerned, the face may be generally covered with a fine, flourlike scale which, however, is more marked on the eyebrow area and around the nares. In the beginning, typical skin lesions may also be found on the lid margins. In cases with particularly intensive scaling of the face, fissures may form in the periocular region and may lead to ectropion of the lower eyelids. The possibility of associated conjunctivitis and iritis has been reported by SPRAFKE.

References

FONTANA, C.: Alterazioni vascolari del limbus sclero corneale nella psoriasi. G. Ital. Oftal. (1953) 316–321.

FRIEDE, R.: Über einen Fall von Pityriasis lichenoides chronica der Lider und der Conjunctivae. Zschr. Augenheilk. 44 (1920) 253.

GROENOUW, A.: Beziehungen des Auges zu den Hautkrankheiten. In: Handbuch der Haut- und Geschlechtskrankheiten, Bd. XIV/1, hsg. von J. Jadassohn, Springer, Berlin, 1930.

KALDECK, R.: Ocular psoriasis. Arch. Derm. Syph. (Chic.) 68 (1953) 44–49.

LUTZ, W.: In: Lehrbuch der Augenheilkunde, hsg. von M. Amsler, A. Bruckner, A. Franceschetti et al. Karger, Basel, 1948.

PILLAT, A: Zur Frage der Miterkrankung der Hornhaut bei Psoriasis (Keratitis psoriatica), Klin. Mbl. Augenheilk. 93 (1934) 751–765.

SCHRECK, E: Veränderungen des Sehorgans bei Haut- und Geschlechtskrankheiten, in: Dermatologie und Venerologie, Bd. IV, hsg. von H. A. Gottron u. W. Schönfeld. Thieme, Stuttgart, 1960.

SPRAFKE, H.: In: Der Augenarzt, Bd. VI, hsg. von K. Velhagen. Dermatosen und ihre Beziehungen zum Auge. Georg Thieme, Leipzig, 1964.

The Eczemas

The group of clinical manifestations characterized as *eczemas* have been classified by GOTTRON, as (1) plain eczema, (2) seborrheic eczema, and (3) endogenous eczema.

The ordinary form of eczema is primarily due to external causes, while the seborrheic and endogenous eczemas are based primarily on internal causes (KORTING). *Contact dermatitis* consists of an exogenous skin reaction, primarily erythematous in nature, which in its primary relationship as well as its distribution is clearly related to the agent eliciting it. In contrast, the ordinary eczemas are characterized by various lesions, of which the papulo-vesicle is the most common. The ordinary eczema is also a nodule-forming dermatosis.

The term "eczema" is of ancient origin. The modern concept of eczema dates back to the Englishmen Willan and Bateman. This concept was further extended by the Viennese school which was able to produce eczema experimentally, for instance with croton oil, and finally it has been influenced by the concept of *diathesis* of the French, and at the turn of the century by the concept of an allergic reaction proposed by JADASSOHN and B. BLOCH. Based on the studies of KREIBICH, who felt that eczema was a vasomotor reflex neurosis, GOTTRON and HALTER suggested that the spongy accumulation of serum in the epidermis that was called so aptly *spongiosis* by BESNIER, is primarily dependent on the state of the nervous system innervating the blood vessels (Fig. 7). In contrast to this, LETTERER considered eczemas to be based on a dysregulative allergy, as has been described for example by KORTING as the basic cause of endogenous eczema. The present state of knowledge of the pathologic cause of the process of *eczema* is not completely clear, but it certainly does not only depend on an allergic reaction. At the present time the demonstration of fluorescent antibodies in eczema has not been entirely successful. However, the importance of lymphocytes and of the lymph system in the development of eczema can be considered assured, as has been shown by the experimental studies of HAXT-HAUSEN, FREY and WENK in animals. Finally, in addition to changes in the lymphatic system, the nerve endings of the terminal blood vessels and even central autonomic nervous influences and individual predisposition on the part of the patient such as age, status of the endocrine system, psyche, and so forth play a distinct role.

Dermatitis, ordinary iatrogenic eczema, and contact dermatitis due to medication in the area of the periorbital region are quite common. In particular, the skin around the eyes, like the skin of the scrotum, tends to react with massive edema, so that in many cases a clinical similarity to erysipelas is marked (Fig. 8). A unilateral distribution in the periocular region suggests a contact dermatitis to some medications, while bilateral involvement of both orbital areas is more suggestive of airborne contacts. Allergic reactions to cosmetics affect primarily the area of the eyebrows and the upper lids because of the way in which they are applied, while involvement of the retroauricular and nasolabial areas suggests a reaction to eyeglass frames. The most common allergens affecting the periorbital area are chloramphenicol, penicillin, neomycin, procaine, paraoxybenzoic acid compounds, and also such commonly used folk medications as chamomile tea compresses (Fig. 9). As far as cosmetics are concerned, mascara, eyebrow pencils, and various face creams, among others, may act as allergens, and among the ingredients, the pigments of the paragroups such as parahydroxybenzoic acid esters, etheric oils, balsam of Peru, synthetic resins or bleaching materials such as mercury and hydroquine are the most common allergens. It is generally known that materials which are first applied to the fingers may, by appropriate touching, cause a reaction in the vicinity of the eyes. Most commonly this is due to nail polish or the phosphoric acid sesquisulfide used in matchboxes.

Of the chemical agents which occur in the frames of eyeglasses, BANDMANN and DOHN mention primarily nickel, cobalt, formaldehyde, and camphor, and again dye materials of the para groups. Of the airborne materials, turpentine is the most common sensitizer. There are in addition plants which can cause contact dermatitis in the facial area; in particular dermatitis due to primrose, poison ivy, and pollen is known to occur, although pollens are more likely to cause conjunctivitis than periorbital dermatitis.

The best way to determine the particular allergen is, of course, by means of the patch test. The reaction to a positive patch test is in essence a miniature eczema. Patch tests should be performed by means of the appropriate patch test material on normal, unaffected skin. In contact dermatitis, in addition to the lymphocytes, monocytes also infiltrate the area, while among the toxic cutaneous reactions mostly polymorphonuclear leukocytes accumulate. Of greatest importance is the knowledge of concentrations of agents appropriate for patch testing, so that primary irritant contact dermatitis will not occur. Concentrations for patch testing are published in appropriate reference books.

Seborrheic eczema. In seborrheic eczema (UNNA, 1887) or dysseborrheic dermatitis (GANS, BELISARIO) the primary lesions are yellowish-brown, erythematous plaques with peripheral accentuation, occurring symmetrically and primarily affecting areas of the skin with dense distribution of sweat and sebaceous glands, and the so-called intertriginous areas of the skin. One of the hallmarks of seborrheic eczema is the tendency to develop small, generalized patches (dissemination, microbid) which commonly occur

in addition to the gradual extension of the initially figured individual primary lesions (pityriasis marginata). Furthermore, there are minimal variants which can be considered preseborrheic and occur primarily in the face. These are characterized by the presence of some hypertrophy of the tonsils and a faint discoid, frequently only pale or hypopigmented, discoloration of the skin of the face that is frequently activated by spring sunshine. (They have been called pityriasis simplex or streptodermia superficialis.) Beyond the above described cutaneous manifestations, one notices, particularly in infants, an eczematous blepharitis of the eyelids which has a rather typical character, consisting of fine scaling in between the eyelashes and a tendency to painful fissuring.

Endogenous eczema

The endogenous eczema (neurodermatitis disseminata, atopic dermatitis, among others) is a disease entity which either at the same time or consecutively comprises symptoms of asthma, hay fever, and eczema.

The term *endogenous* used to characterize this form of eczema is used like endogenous asthma in internal medicine, or endogenous psychosis in psychiatry to characterize a personal predisposition. This personal predisposition then conditions the normal function and determines the type of manifestation that is presented by this disease.

The cutaneous manifestations of endogenous eczema usually begin as exudative eczematous lesions often called "milk rash." We consider milk rash the spongy, eczematous first reaction of endogenous eczema and not a seborrheic dermatitis.

As endogeneous eczema continues, its appearance changes in quality. Thus, one finds as the next phase of endogenous eczema, an entirely different localization and reaction — namely, the so-called flexural eczema (Fig.

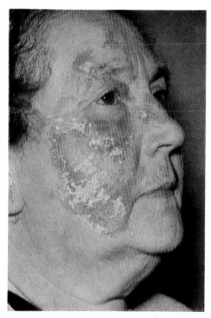

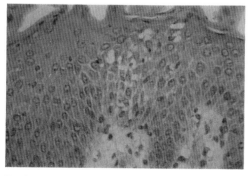

7 Spongiosis, the histologic manifestation of eczema.

9 Subacute contact dermatitis secondary to eye drops.

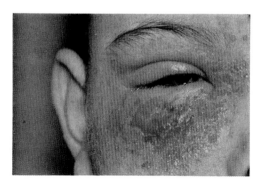

8 Acute contact dermatitis.

10 Seborrheic exfoliative dermatitis in the aged.

12). Here HEBRA and KAPOSI report: "When eczemas occur in the antecubital spaces, they are very frequently associated with similar reactions on the popliteal space." This very common association of lesions in popliteal spaces occurs in our experience in about 40 per cent of those patients who have flexural eczema of the antecubital spaces, while the localization to the wrists and the back of the hands occurs in about 25 to 30 per cent of all patients. Localization of this disease to the flexural aspects, particularly common during school age, results in the typical clinical manifestation of endogenous eczema which is called lichenification, and which develops either secondarily, that is, following an acute eczematous reaction, or primarily by the coalescence of round, flat, pinhead-sized nodules (lichen simplex). In this particular tendency toward lichenification there is mirrored already in morphologic fashion the functional basis of the sufferer from endogenous eczema, namely, his marked hypo-regulation. Considering the chronic fashion in which the patient's skin reacts, it is not surprising that patients with such stigmata show a tendency to subacute furuncles or deep infections of fungi, or that it is most difficult to produce blisters or exudations in these patients experimentally. An additional reaction factor present in lichenification is the obligatory itching and scratching, originally described by HEBRA. The clinical appearance is characterized by accentuation of the parallel and crossing lines of the skin that now develops a dirty gray color. Later on, the affected areas start to scale. At first this consists of a fine lamellar sticky scale, but later the scales become coarser and are more adherent. In some areas and at some times the basic character of isolated nodules is demonstrated by the appearance of individual lesions which, instead of having a flat shagreen leatherlike appearance consist of a closely packed plaque of dully gleaming small squares.

In the differential diagnosis from neuro-dermatitis circumscripta (lichen simplex), it is important to point out that the typical triad of pigmentation, lichen simplex nodules, and flat lichenification is not present in endogenous eczema but that rather lichen simplex-like and eczematous reactions appear side by side in irregular fashion. This reaction phase has been described as of disseminated, eczematous lichen simplex-like appearance. Finally, we see later, in addition to deep pigmentation and hyperpigmentation, areas similar to papular urticaria and monomorphic prurigo. Later on, one form of reaction in this disease is the development of nodules the size of rice grains, more palpable than visible, which frequently are topped by a small, not very visible, vesicle which subsequently forms a central crust. Such prurigo nodules occur as the primary lesion in endogenous eczema, particularly on the extensor surfaces of the extremities; they are frequently associated with moderate to marked swelling of the lymph nodes, and affect in general adults rather than infants. Occasionally, such nodules will occur after the fourth decade as the only persistent prurigo nodularislike reaction of this disease. From this description it is apparent that the cutaneous reactions of the patient with endogenous eczema change (as it were) in the course of his life. Furthermore, it is not at all uncommon to find that prurigo is the only expression of endogenous eczema that may be useful in the differential diagnosis from diseases of the hematopoietic system, Duhring's disease, and hepatic prurigo. It may be of interest that SABOURAUD suggested the name *asthma prurigo* for endogenous eczema some time ago.

In addition to general dryness of the skin, the patient with endogenous eczema shows a rather specific alteration in sweating limited sharply to the areas involved or about to be involved by manifestations of the disease; this can be demonstrated by stimulating central regulation of sweating and can be reduced by narcotic medications which depress the brain stem. This alteration of areas of decreased or increased sweating (termed oligobradyhidrosis) differentiates patients with endogenous eczema not only from patients with seborrheic dermatitis (who have diffusely increased secretion, particularly of the middle body line), but also from patients with vegetative dystonia who also have anhidrosis and hyperhidrosis, but without any specific areas being involved. The importance of this oligobradhidrosis for the pathogenesis of endogenous eczema rests primarily on a faulty distribution of sebum that is functionally coupled to decrease in sweating, so that in addition to decreased sweating there is also decreased sebum delivery. Finally, acute and subacute manifestations of endogenous eczema may occur primarily in the area of the face. In this area, endogenous eczema may have more or less similarity to seborrheic dermatitis, but the constriction of cutaneous blood vessels and the thinning of the lateral third of the eyebrows are helpful in the diagnosis of endogenous eczema.

There are frequently striking changes in hair growth that are useful for rapid diagnosis in patients with endogenous eczema. In these patients there is often a dense hair growth on the scalp, particularly on the lateral aspects of the forehead, in contrast to the situation in seborrheic dermatitis. Furthermore, the difference in the amount of hair on the nasal and lateral part of the eyebrows is worth noting. However, these stigmata, which are well known to dermatologists (they occur in alopecia due to thallium, in secondary syphilis, and so on), are not merely of external origin, that is, due to abrasion in patients with endogenous eczema. When the face is markedly involved with eczema, these phenomena are more common than in the average population and can be produced by rubbing. However, it is often noted that the lateral parts of the eyebrows remain sparse even in intervals when the patient does not have eczema and this suggests that this manifestation is based on some endocrine or vegetative alteration. It is historically important to point out that HERTOGHE in his

original description in the year 1900 pointed to some inflammatory components in the loss of eyebrows ascribing it to hypothyroidism. Thus, today we can call some of these manifestations pseudo-Hertoghe signs (Fig. 13). We also know that an idioapathic, apparently genetically determined HERTOGHE phenomenon exists with a genotypic development of eyebrows and lashes more marked in the nasal region (BÖRNER). Since in the male there is frequently a decrease in the development of the lateral part of the eyebrows, the Hertoghe phenomenon can be useful diagnostically in older persons with endogenous eczema only in the female (Fig. 14). As far as concerns the changes in the lateral eyebrow areas in endogenous eczema, we must be reminded that the temporal part of the eyebrows is phylogenetically younger and that, while the medial eyebrow area is innervated by the trigeminal nerve, the lateral part is innervated vegetatively. POLEMANN and PELTZER relate the rarefaction of the lateral eyebrow area to the distribution of the sensory trigeminal nerves of the face. According to CLARA, the temporal part of the eyebrows belongs to the third segment of the face while the second segment of the face is probably responsible for the butterfly localization which can be seen in some skin diseases such as scarlet fever and rosacea, among others. The second and third segmental areas correspond to the middle nuclei of the nucleus tractus spinalis of the fifth nerve in the medulla oblongata. It is also possible to show the particular nervous irritability of the third face area in patients with endogenous eczema (KORTING), for instance in an appropriate thermoregulatory sweating experiment by which means the previously discussed interpretation of the Hertoghe sign can be considered to be specially related to endogenous eczema.

Alterations of the eye itself can be seen in endogenous eczema primarily in more or less transient irritation of the conjunctiva.

They are related in these patients both to season and later to the rhinitis and asthmalike symptoms that many of the patients develop. Both eyes are always affected, and the manifestations appear simultaneously. Only after these changes have been present for some time does one find a cobblestonelike change in the tarsal conjunctiva of the upper lid or a marked proliferation of the connective tissue on the conjunctival limbus. In acute cases one can also find eosinophilia in the secretions of the conjunctiva, even though it may be minimal.

Participation of the cornea occurs more commonly in the ordinary type of eczema than in endogenous eczema where there occur rarely atopic keratitis and other anomalies, such as for instance keratoconus (KAREL et al., BRUNSTING et al.). The clinical appearance of corneal eczema originally described by PILLAT in 1937 was also observed by HEYDENREICH. In this disease these changes can be noted: first, a massive, nonbacterial serous conjunctivitis as it also occurs in eczema erythematosum; second, a keratitis epithelialis vesiculosa disseminata punctata superficialis, interstitial edema of the cornea and wrinkling of Descemet's membrane similar to that which occurs in eczema vesiculosum madidans, and third, infiltration of the cornea with ulcerations as occurs in eczema pustulosum.

The most important disease of the eyes in the patients with endogenous eczema is opacity of the lens. The major reviews on cataracts in patients with endogenous eczema were published by BRUNSTING, LÖHLEIN, JANZEN, SCHÖNFELD, REED, and BAIR. The most recent reports on this subject are summarized by KORTING.

Ophthalmologists do not classify every opacity of the lens that occurs in the patient with endogenous eczema as a dermatogenous cataract. It is possible to differentiate the atopic caratact both in its course as well as in its appearance from other juvenile and pre-senile cloudiness of the lens and from other clouding of the lens that occurs with a variety of skin diseases. For instance, the atopic cataract primarily affects males below the age of 30. Usually, it is at first unilateral but frequently soon involves the other eye. The cataract develops gradually, accelerated by acute exacerbations, and is in many cases not noted subjectively and usually discovered only by routine examination with a slit lamp. The cutaneous expression of the disease, that is, the eczema, is apparently a prerequisite for the development of this type of cataract which almost always follows the development of the skin disease. Rarely, monosymptomatic cataracts related to eczema have been reported. Morphologically, one sees in the early stages a subcapsular disclike clouding of the anterior pole of the lens with a concave border that later also develops in the posterior pole, to be followed by homogeneous total clouding. According to SAUTTER, cataract formation in endogenous eczema consists of "the abnormal development of an axial cataract of the capsular epithelium that stands out because of its size, its intense white color, and its shieldlike appearance."

In addition to the atopic cataract which, in our own experience, occurs in about 0.4 per cent of all patients, other kinds of cataracts of the type which have been more often described in the presence of endocrine disturbances develop at least as frequently. These in the past have usually been called x-ray caratacts, since many of these patients frequently received ionizing radiation of the face because of their chronic eczema.

BEETHAM separates two types of cloudiness of the lens associated with endogenous eczema — type A, which is nonspecific, and type B with the characteristic shieldlike opacity at the anterior pole of the lens that later extends to subcapsular and cortical cloudiness. From the point of view of classification, the typical dermatogenous cataract is not likely to be based on endocrine disturbances, but caused by a deep continuous disturbance which parallels the skin disease and may even be on an immunologic basis.

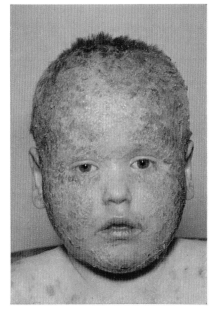

11 Endogenous eczema: "Milk rash." Hertoghe sign.

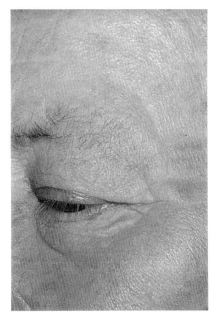

14 Hertoghe sign.

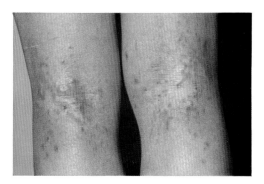

12 Endogenous eczema: "Flexural eczema."

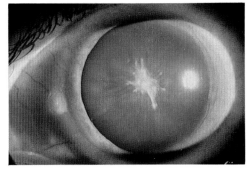

15 Atopic cataract. (From the colletion of Prof. Sautter, Heidelberg.)

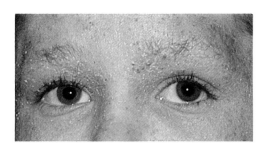

13 "Pseudo-Hertoghe" sign.

From the practical point of view, the primary function of the dermatologist is to bring the patient to a state in which skin manifestations are minimal and there is no itching, so that operative intervention may be carried out under the best circumstances.

A number of authors have pointed out the occurrence of retinal detachment (and we have also noted it twice), which certainly should not be considered as only a secondary complication of cataract extraction, since it occurs just as frequently before such an operation has been carried out. Circumstances contributing to the detachment of the retina in these patients may be an *equatorial* degeneration of the retina or liquefaction and destruction of the vitreous (COLES and LAVAL, INGRAM).

As far as the color of the iris is concerned, CARNEY recently reported that two-thirds of all patients with endogenous eczema over five years old had brown eyes, while this occurred in only two-fifths of three separate control groups. In contrast to this, GERTLER and HARNACK were unable to come to the same conclusion based on a study of a very large group of patients and controls. In their patient material, blue eyes occurred 5 per cent less frequently in patients with endogenous eczema, but 16.8 per cent had green eyes while only 8.3 per cent of the controls had this iris color. It thus appears that this observation of Carney may only apply to his own patient material.

quently is an underlying gastrointestinal dysfunction in lichen simplex chronicus, in the presence of which minor trauma localizes the cutaneous manifestations of the disease. In particular, the lack of association with bronchial asthma or the development of cataracts separates lichen simplex chronicus from endogenous eczema. As far as the clinical appearance is concerned, in lichen simplex one usually notes areas consisting of round or flat, pinhead-size papules (rarely slightly larger), which have become confluent in the center of the lesion to form the primary area of lichenification and which are surrounded by a grayish-brown pigmented area. Single lesions of this kind can be found in round or linear form, particularly on the lateral surfaces of the neck and back of the neck, or on the medial surfaces of the thighs and calves, and are usually associated with intractable pruritus. When it occurs on the face (which is the only reason why lichen simplex chronicus is described here) there may develop diffuse masklike lichenification which sometimes even gives rise to a leonine face as described by BROCQ, PAUTRIER, and HOFFMANN. Occasionally, the conjunctiva is also involved. Studying these reports in detail, one comes to the conclusion that because of the frequent association with involvement of the flexor surfaces and loss of the lateral eyebrows, these patients most likely represent isolated forms of endogenous eczema as described previously.

Lichen-simplex chronicus, VIDAL (Neurodermatitis circumscripta of BROCQ and JACQUET)

Lichen simplex chronicus is a disease which resembles endogenous eczema, but can generally be separated from it. For instance, lichen simplex chronicus is not of familial origin, there is no preceding infantile eczema, and it occurs primarily after age 40 or 50. In contrast to endogenous eczema, there fre-

References

BANDMANN, H. J., and DOHN, W.: Die Epicutantestung, Periorbitalekzem. Bergmann, München, 1967.

BEETHAM, W. P.: Atopic cataracts. Arch. Ophthal. 24 (1940) 24.

BLOCH J.: Der älteste Gebrauch des Wortes "Ekzem." Mhf. Prakt. Dermat. 53 (1911) 690.

BRUNSTING, L. A.: Atopic dermatitis (disseminated neurodermatitis) of young adults: analysis of precipitating factors in 101 cases and report of 10 cases with associated juvenile cataract. Arch. Derm. Syph. (Chic.) 34 (1936) 935–957.

BRUNSTING, L. A., REED, W. B., and BAIR H. L.: Occurrence of cataracts and keratoconus with atopic dermatitis. Arch. Derm. Syph. (Chic.) 72 (1955) 237–241.

CARNEY, R. G.: Eye color in atopic dermatitis. Arch. Derm. Syph. (Chic.) 85 (1962) 17–21.

COLES, R. S., and LAVAL J.: Retinal detachments occurring in cataract associated with neurodermatitis. Arch. Ophthal. 48 (1952) 30–39.

GERTLER, W., and HARNACK K.: Irisfarbe und endogenes Ekzem. Derm. Wschr. 153 (1967) 1260.

HEYDENREICH, A.: Die Beteiligung der Augen beim Eczema vulgare. Hautarzt 6 (1955) 460–464.

INGRAM, R. M.: Retinal detachment associated with atopic dermatitis and cataract. Brit. J. Ophthal. 49 (1965) 96.

JANZEN, L.: Inaug.-Diss., Breslau, 1940.

KAREL I., MYŠKA, V., and KVIČALOVÁ, E.: Ophthalmological changes in atopic dermatitis. Acta derm.-venereol. (Stock) 45 (1965) 381–386.

KORTING, G. W.: Zur Pathogense des endogenen Ekzems. Thieme, Stuttgart, 1954.

KUGELBERG, I.: Juvenile Katarakt bei Dermatosen (Cataracta syndermatotica). Klin. Mbl. Augenheilk. 92 (1934) 384–508.

LÖHLEIN, W.: Klinischer Beitrag zu den mit Dermatosen begegnenden Starformen. Derm. Wschr. 109 (1939) 859–865.

PILLAT, A.: Zur Mitbeteiligung der Hornhaut beim Ekzem. Wien. klin. Wschr. 1937/I, 768–770.

POLEMANN, G., and PELTZER, L.: Das Augenbrauenzeichen von Hertoghe als endokrin-vegetativ gesteuertes Symptom. Medizinische (1952) 856–860.

ROSEN, E.: Atopic Cataract. Thomas, Springfield, Ill., 1959.

SAUTTER, H.: Die Trübungsformen der menschlichen Linse. Thieme, Stuttgart, 1951.

Urticaria, Strophulus and Prurigo

Strophulus (lichen urticatus) consists of an acute papulo-vesicular eruption which primarily affects the palmar and plantar and gluteal areas but usually spares the face, as do the classical papules of prurigo (with the exception of summer prurigo which is really a photodermatosis). In contrast to this, urticaria, (the ordinary hives) irregularly affect the face also (Fig. 16). In cases of generalized urticaria, the loose connective tissue around the eyes, as well as around the external genitalia, is so markedly affected that not infrequently the patients are unable to open their lids. When urticaria becomes very extensive, it is referred to as angioneurotic edema (Quincke's edema), which is sometimes familial and which can cause asphyxia because of edema of the larynx and glottis (Fig. 17). When this affects the eye itself, exophthalmos produced by periorbital edema or edema of the papilla of the optic nerve has been described in the past. In the more recent literature, chemosis of the conjunctiva and of the cornea is more commonly reported (BOTTERI, LUNDBERG).

Except for the special cases quoted above, ordinary urticaria has many varying symptoms. For instance, today urticaria associated with fever is usually due to an allergic reaction to penicillin, particularly when the last dose was given eight to ten days previously. Acute urticaria is most commonly due to foodstuffs, and in general may be produced by fruit — apples, apricots, tomatoes, strawberries, and so on. Frequently, the disease can be found to be due to breakdown products and to allergens which have been inhaled, and the possibility that it may be produced by worms (ascaris, trichuriasis) should not be excluded. In the differential diagnosis, the periorbital angioneurotic edema is relatively easily differentiated from edema of the eyelids secondary to myxedema, diseases of renal and cardiac origin, as well as trauma and contact allergies (Fig. 18). Edema of the eyelids associated with chemosis occurs in trichinosis and "pseudotrichinosis" (dermatomyositis). In the presence of emphysema of the skin, palpation of the edematous area of the eyelid leads to crackling. Unilateral edema of the eyelids may also be due to osteomyelitis of the maxilla. In rare cases one must consider unilateral lid edema of Chagas' disease (schistosomiasis) and following filariasis. Unusual swellings around the eyes occasionally also occur in the so-called Melkersson-Rosenthal syndrome (the latter consists of a symptom complex primarily characterized by episodes of paralysis of the facial nerve, lingua plicata, cheilitis, and parotitis) (GAHLEN and GILLMAN).

Chronic persistent swelling of the eyelids may be a form of elephantiasis, secondary to hypertrophy of the connective tissue of the tarsus and overdevelopment of the meibomian glands in a disease called pachydermoperiostosis (TOURAINE, SOLENTE, GOLE).

References

BOTTERI, A.: Urticaria und Auge. 6.–12. Versammlg. Oto-Neuro-Ophthalm. Ges. Zagreb. Klin. Mbl. Augenheilk. 82 (1929) 295.

GAHLEN, W., and GILLMANN, H.: Melkersson-Rosenthal-Syndrom bei Filiariasis. Münch. med. Wschr. 96 (1954) 189–191.

LUNDBERG, A.: Einige Fälle von allergischer Conjunctivitis bei Patienten mit Urticaria. Hygiea (Stockh.) 98 (1936) 562; Ref. Zbl. Haut- u. Geschl.-Kr. 56 (1937) 55.

Lichen ruber planus

The term lichen is used to describe disseminated small papules, affecting a circumscribed area of the skin, that in general show no tendency to develop any further. In lichen planus the diagnosis is based especially on the observation of sharply marginated, polygonal, pinhead-sized, flat nodules with coloration ranging from yellowish pink to purple pink to slate purple (Fig. 19). Furthermore, the lesions of lichen planus frequently are somewhat depressed in the middle. They show a waxy dull or shiny surface and frequently are found in groups. Histologically, these lesions consist of an epidermal-subepidermal (that is, mixed) papule, in which the increase in keratohyalin results in whitish fernlike figures on the mucosae. They are less easily seen on the papules on the skin (in general only when these papules are either moistened or oiled). On the top of the skin papules they resemble a spider web (Wickham's striae). The eruption has a predilection for the flexors of the forearms, the flanks, the sacrum, the genitals, and particularly the extensor surfaces of the legs where even hypertrophic or verrucous lesions may occur.

The skin of the face and thus the general area of the eyes, eyebrows, and eyelids as well as the conjunctiva is only rarely affected by the lesions of lichen planus, although there are recent reports concerning this type of location (MICHELSON and LAYMAN). An unusual case was reported by LUHR in a 75-year-old patient with generalized lichen planus which involved the palpebral and bulbar conjunctiva and in which tapioca nodules were found on the cornea, and pinhead-sized, bright red spots on the eyelids. When lichen planus affects the face, which is rather rare, the typical lesions are not noted but rather, as has been pointed out by Fellner, "uncharacteristic annular lesions or skin manifestations which resemble seborrheic dermatitis or flat warts," even though other parts of the body are affected by typical classical lichen planus lesions. As far as is known, lichen planus of the face generally lasts only a short time. As far as etiology is concerned, these itching, purplish red nodules are a cutaneous reaction, that is, a form of expression in which the same clinical picture can be produced by a variety of causes. Emotional influences frequently play a role in this markedly itching skin disease, and infectious, drug-induced and auto-toxic causes must be sought for and ruled out.

Lichen nitidus is still by and large only a variant of lichen planus, although the histologic appearance is tuberculoid. The small primary lesions are rounded and affect different areas, while itching and the predisposition for development of a epidermolysis bullosae is lacking. The skin of the face is even more rarely involved with lichen nitidus than with lichen planus.

References

FELLNER, M.: Zur Morphologie des Lichen ruber planus an der Gesichtshaut. Derm. Wschr. 92 (1931) 1—6.

LUHR, A. F.: Lichen planus of the conjunctiva. Amer. J. Ophthal. 7 (1924) 456; Ref. Zbl. Haut- u. Geschl.-Kr. 16 (1925) 414.

MICHELSON, H. E., and LAYMON, C. W.: Lichen planus of the eyelids. Arch. Derm. Syph. (Chic.) 37 (1938) 27—29.

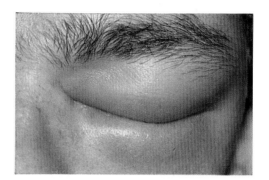

17 Angioneurotic edema.

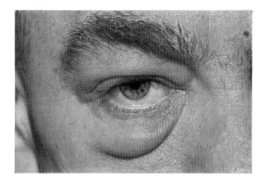

18 Subacute elephantiasis of the eyelids in allergy to cobalt.

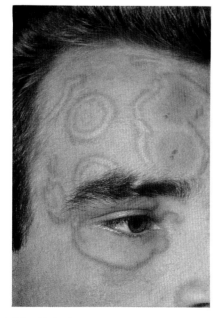

16 Urticaria.

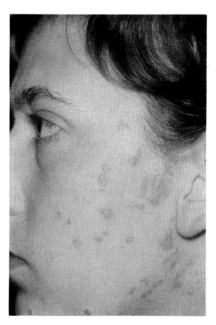

19 Lichen ruber planus.

Acute Infectious Diseases of the Skin

Pyodermas

Pyodermas are infectious disorders of the skin, primarily produced by Streptococcus pyogenes or Staphylococcus aureus. They affect different layers of the skin, and are either perifollicular, or constitute independent surface changes (Figs. 20 and 21). When one considers the distribution of the pyodermas, it is very striking that even infection of the periorbital area is generally associated with remarkably little conjunctival involvement, so that one really sees a purulent conjunctivitis associated with a Pyoderma only in cases of staphylococcal pemphigoid of the newborn or in Ritter's disease (dermatitis exfoliativa), which in reality is the most extensive form of staphylococcal impetigo contagiosa. Equally rare are reports concerning infection of the eyes with Pseudomonas aeruginosa, although the conjunctiva in general is considered to be one of the sites of predilection for bacillus pyocyaneus (CALLOMON).

In Ritter's disease extensive bullous elevations of the epidermis develop, leaving behind characteristic eroded oozing surfaces or on occasion even total epidermolysis with a positive Nikolsky sign — the epidermis can be sloughed off the underlying tissue as one can peel the skin off a ripe peach by rubbing, or as occurs in exfoliative erythrodermas of other kinds. More recently, it has been considered that in this form of superficial bullous staphylodermia the clinical appearance stimulates what in the last century was described in adults as pemphigus febrilis, and what has been described in the last few years as the Lyell syndrome or epidermolysis acuta toxica. In the adult, most cases seem to be the maximal variants of a toxic allergic bullous epidermolytic exanthem, while in infants and small children the staphylococcal relationship is all-important. Symblepharon has been described following this form of the disease by AUBERTIN and co-authors.

It must again be emphasized that even in the presence of extensive involvement of the face, which is often affected by acute streptococcal or staphylococcal pyodermas, conjunctivitis is usually lacking, even when the lesions are located in the vicinity of the eyelids or on the eyelids themselves. Chronic vegetative pyodermas occur much more commonly on the backs of the hands but an exception is the chancrelike pyoderma described by HOFFMANN, which, when fully developed, results in the more or less round, erosive ulcerated erythemalike pyoderma localized in the vicinity of the eyelids or the nasolabial folds near the lip area (Fig. 22).

In the differential diagnosis of the chancrelike pyoderma, which at the beginning produces a significant amount of pus and then for several weeks results in either a blistering or crusted surface, one must consider primary syphilis in the vicinity of the eyes and also the oculoglandular form of tularemia.

Of the follicular pyodermas, *Hordeolum* (sty) is particularly common in the area of the eyelids. This is usually due to a staphylococcal or streptococcal infection of the sebaceous glands or sweat glands of the external surface or of the meibomian glands in the internal surface of the eyelids, resulting in the picture of a purulent folliculitis or perifolliculitis which resembles a furuncle. In association with this, there may be marked reactions consisting of a diffuse swelling and redness of the connective tissue of the affected area. The prognosis of furuncles of the eyelids is without question significantly better than that of the ominous upper lip or nasal furuncle, which in the not too distant past not infrequently resulted in death. According to TACHAU, furuncles may also occur at the lacrimal carunde, in whose vicinity follicles also exist.

Sycosis vulgaris is the name given to multiple recurrent follicular infections which are

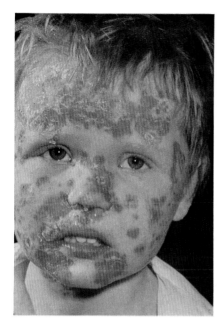

20 The crusting form of impetigo contagiosa due to staphylococci.

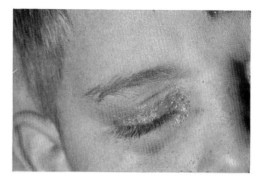

21 Eczematous pyoderma in the area of the eyelids.

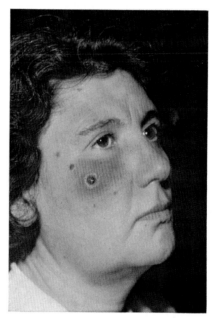

22 Pyoderma simulating chancre.

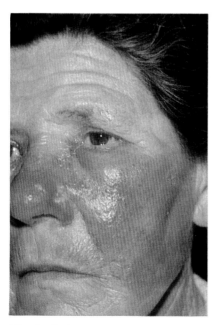

23 Erysipelas.

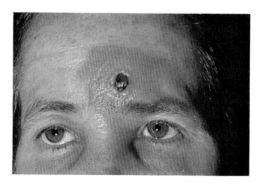

24 Anthrax of the skin.

usually perforated in the center by a hair. In some cases these pustules are preceded by follicular nodules which eventually become topped by a pustule. Although the cheeks, chin, and upper lip are the areas of predilection for sycosis vulgaris, it may not infrequently affect the eyebrows. The formation of such localized nodules and pustules may be followed later by inflammatory infiltration of the border of the eyelids, resulting eventually in madarosis and trichiasis. In this fashion there may develop on the lower lid marked scarring ectropions. The upper lid is usually not so often affected.

Multiple abscesses of the *eccrine sweat glands* affect primarily those areas of infants exposed to pressure, such as the back of the head, the back, the feet, and the buttocks, while the apocrine sweat gland abscesses of the adults affect primarily the axilla or other areas where apocrine glands are found. These diseases are mentioned since recently BERGERON and STONE have again pointed out the development of an interstitial keratitis similar to that occurring in congenital lues in connection with the presence of hydroadenitis profunda. Similarly, JESS also described an interstitial keratitis occurring in the presence of ecthyma gangrenosum and pyocyaneus infection of the skin.

Erysipelas

Erysipelas when it first begins is accompanied by constitutional signs of disease such as sudden high fever, shaking chills, headache, vomiting (Fig. 23). When it affects the face, the initially flaming red, sharply demarcated, edematous erythema becomes diffuse after several days and leaves grotesque edema, particularly in the periocular area, so that the eyes cannot be opened at all. If the disease progresses to gangrene, gangrene of the eyelids may occur. A late result on the face, as in other areas, may be a persistent

lymphostatic edema such as *elephantiasis palpebrarum*. The conjunctiva and the eye itself are frequently unaffected even though erysipelas affects the eyelids and although DELBANCO and CALLOMON have described bullous or purulent inflammations of the cornea and dacryocystitis.

Anthrax of the skin

This systemic disease produced by Bacillus anthracis, primarily a disease of animals, occasionally affects men, particularly those in certain occupations such as veterinarians, workers in bristle and brush factories, farmers, and butchers. Two or three days after infection, a blood blister the size of a cherry stone forms and rapidly turns coal black and dries to form a blue pustule (Fig. 24). Infiltration of the surrounding tissue results in an appearance simulating a furuncle. Occasionally, edema of the tissue occurs without a preceding blister or pustule. Although involvement of the eyelids with primary anthrax is extremely striking, the frequency with which this occurs seems to vary greatly, depending on the origin of the reports.

Diphtheria of the skin

Cutaneous infections with the gram positive Corynebacterium diphtheriae generally begin with uncharacteristic eczematous, impetiginous, or ecthymatous lesions, which gradually progress into local areas of gangrene or ulceration covered with a yellowish white pseudomembrane. In addition to the primary infection of the skin, id reactions can occur secondary to diphtheria of the throat. The organism itself can be found with some frequency on wound surfaces, particularly burns. The regional lymph nodes may be enlarged. Cutaneous manifestations of diphtheria in children are most commonly intertriginous, such as behind the ear, near the axillae or in the folds of the neck of small children, or between the fingers and in the genital area. However, attention must be called to the possibility of the development of a pseudomembranous conjunctivitis with marked edema of the eyelids. In these cases the problem arises of differentiating the diphtheria bacilli from saprophytic organisms which frequently are found in the conjunctival sac. Purely diphtheritic blepharitis and keratoconjunctivitis, resulting in ulceration and perforation of the cornea, are rare. It must also be remembered that even in the presence of purely cutaneous diphtheria, toxin production of the bacteria may cause early injury of the myocardium, neuritis and particularly paralysis of the muscles of accommodation. This seems to be due to toxic injury in the brain at the site of the ocular motor nuclei. In diphtheria, only the ciliary muscle is usually affected, while in botulism both eye muscles innervated by parasympathetic nerves are involved (m. ciliaris and m. sphincter pupillae).

Toxoplasmosis of the skin

Toxoplasmosis, an anthropozoonosis caused by the protozoon Toxoplasma gondii, is usually acquired either from an animal host, particularly the dog, or via the placenta. The early lesions are noncharacteristic, acute, febrile, miliary nodules or gummalike skin reactions. This is associated with a subacute or mildly febrile lymphadenopathy of the nodes of the neck. In such cases "granulomatous" uveitis has been described.

Congenital toxoplasmosis, which is usually not apparent, may give rise to an inflammatory hydrocephalus and later to calcification of the brain (plexus choriordeus) and chorioretinitis with a salt and pepper appearance simulating that found in congenital syphilis, with associated visual disturbances which may flare up at a later date.

Skin manifestations may appear anywhere on the body and in cases where this disease is suspected and the organisms cannot be directly demonstrated, complement fixation tests and antitoxoplasma antibody tests are useful.

Tularemia

The oculoglandular form of tularemia (rabbit fever) was first described by the French ophthalmologists PARINAUD and GALEZOWSKI in 1889 as a particular variety of conjunctivitis in association with lymphadenopathy. Similar phenomena may be seen in cat scratch disease. Depending upon the portal of entry of the organism, several different clinical pictures have been described — namely, an ulcerated glandular, a glandular, and a typhoidlike form of tularemia. Primary infection of the skin begins with a small, frequently rather painful and crusted, ulcerating lesion which is associated with marked regional lymphadenopathy. In the oculoglandular form, one finds unilateral swelling of the preauricular submaxillary and cervical lymph nodes associated with unilateral edema of the eyelids and massive conjunctivitis. Furthermore, one notes on the eyelids and folds yellowish nodules and verrucous protuberances. This leads to an ulcerogranulomatous conjunctivitis which generally heals in months without scarring. The diagnosis of tularemia is based on the history — that is, on contact with wild rodents which harbor ectoparasites, usually ticks or lice.

The diagnosis may be made by laboratory tests consisting primarily of agglutination, complement fixation, thermo-precipitation tests, and so on. In addition to primary tularemia, an exanthematous form (tularemid) is known. The exanthem begins in the vicinity of the primary complex and generally occurs in the first three weeks of the disease. The primary lesions of tularemia last only four days.

References

AUBERTIN, E., AUBERTIN, J., VÉRIN, Ph., and APARICO, M.: Syndrome de Lyell. Bull. Soc. Franc. Derm. Syph. 73 (1966) 61.

BERGERON, J. R., and STONE, O. J.: Interstitial keratitis associated with hidradenitis suppurativa. Arch. Derm. Syph. (Chic.) 95 (1967) 473—475.

CALLOMON, F. T.: Die Pyocyaneuserkrankungen der Haut. In: Handbuch der Haut- und Geschlechtskrankheiten, hsg.

von J. Jadassohn, IV/1A. Springer, Berlin, 1965.

DELBANCO, E., CALLOMON, F. T.: Erysipel. In: Handbuch der Haut- und Geschlechtskrankheiten, hsg. von J. Jadassohn, IX/1. Springer, Berlin, 1929.

JESS, A.: Augenheilk. 105 (1931) 39.

TACHAU, P.: Pemphigoid der Neugeborenen und Kinder. In: Handbuch der Haut- und Geschlechtskrankheiten, hsg. von J. Jadassohn, IX/2. Springer, Berlin, 1934.

Chronic Infectious Diseases of the Skin

Tuberculosis cutis

The various cutaneous disease forms produced by the gram positive Mycobacterium tuberculosis consist of primary tuberculosis, miliary tuberculosis, lupus vulgaris, scrofuloderm (tuberculosis colliquativa cutis), tuberculosis verrucosa cutis, and tuberculosis miliaris ulcerosa cutis et mucosae, as well as lupus miliaris disseminatus faciei. The tubercular skin diseases develop either by direct exogenous infection or continuously by means of the lymphatics or blood stream. *Tuberculids* are transient specific hematogenous disseminations tending to spontaneous involution. They appear in late generalized stages of tuberculosis at a time when there apparently exists an intense immunobiologic resistance on the part of the patient, as determined by the very small content of bacteria of such lesions. Using these criteria, one must particularly consider as tuberculids the lichen (eczema) scrofulosorum, erythema induratum, and the papulonecrotic tuberculid. However, today we are more and more aware of the fact that these lesions are related to vasculitides, particularly because of their relative lack of a tubercular histologic structure. The eyelids are occasionally (but rarely) involved with the primary infections of tuberculosis such as lupus, scrofuloderma, or tuberculous ulcerations, and usually this infection occurs by injury to the skin of the eyelids. Similarly, the tuberculids, particularly lichen scrofulosorum or papulonecrotic tuberculid, may occur rarely on the eyelids. Primary tuberculosis of the conjunctiva is extremely rare. When it does occur in infants or small children, a livid papule or pustule appears, which may turn into a small yellowish nodule and which is associated with lack of an allergic reaction. In association with this, softening and later calcification of the postauricular lymph nodes occur. More commonly the conjunctiva is infected secondary to involvement of the adjacent skin, the tear gland, or the tear sac. Soft nodules may be noted which break through to the outside. Schreck reported alterations of the cornea in association with tuberculosis of the skin, and even a tuberculous alteration of the fundus oculi. Other alterations have been interpreted as tuberculids of the iris. Of great practical importance is the keratoconjunctivitis scrofulosa which may occur after measles or whooping cough. According to RIEHM, phlyctenules are a particularly important allergic reaction of tuberculosis, and are due to an abortive deposition of tubercle bacilli.

As far as the specific types of primary tuberculosis are concerned, lupus vulgaris, which has been known since the middle ages and is particularly well described by WILLAN and BATEMAN, begins in 90 per cent of the cases in the middle portion of the face as a pinhead-sized, firm, brown-red macule covered by thinned skin. This macule is firm, and on pressure with a glass plate appears in a typical case the color of apple jelly (Fig. 25).

Patients with lupus vulgaris are ten times more likely to have pulmonary tuberculosis than those without skin lesions, and patients with pulmonary tuberculosis are four times more likely to have lupus vulgaris than normals. Patients with lupus vulgaris have seven times the chance of death from pulmonary tuberculosis than controls. For this reason it is important to investigate every patient with tuberculosis of the skin, particularly those with lupus vulgaris, for the presence of non-cutaneous infection with tubercle bacilli (lungs, bones, urinary tract, sputum, G. I. tract, and so on), and to study his blood picture, his subfebrile temperature and sensitivity to tuberculin.

For tuberculin testing we prefer the old tuberculin of KOCH and believe that the intracutaneous testing with 0.1 ml. of a

1:10,000 dilution of tuberculin is most effective. Testing with purified tuberculin may, of course, also be carried out. From the diagnostic point of view, one may generally assume that some patients with tuberculids such as lichen scrofulosorum or conjunctivitis phlyctaenulosa generally show a very high-grade sensitivity to tuberculin, while other tuberculids such as the papulonecrotic type may vary between high sensitivity and almost no sensitivity to tuberculin. Lupus miliaris disseminatus faciei is also characterized by varying sensitivity, while primary lesions of tuberculosis, miliary tuberculosis, and tuberculous ulcers of the mucosa usually show little or no resistance.

Lupus vulgaris, which is comparable to the tuberoserpiginous syphilid of Lewis, may occur primarily on the eyelids, although this is not as common as its appearance on the nose or ear. More commonly the lower eyelid and conjunctiva are secondarily involved by extension from lesions in the area of the mucosa of the mouth and nose, by way of the tear ducts or by the lymphatic route. In addition to the tear duct, which may be contiguously involved by extension from the infected nasal mucosa, the tear sac may also be involved. In general, the picture of lupus vulgaris of the conjunctiva is very difficult to differentiate from other kinds of tuberculous manifestations, particularly since phlyctenular reactions are not uncommon in people with lupus vulgaris. Furthermore, in patients with lupus vulgaris there occur tuberculous granulomata and ulcers of the cornea and less commonly iritis and iridocyclitis. Tuberculosis verrucosa cutis, which is due to exogenous re-infection of a person previously exposed to tuberculosis, is usually an occupational form of tuberculosis occurring in butchers, farmers milking cows, and workers in anatomy laboratories, and is primarily restricted to the hands and fingers. Its appearance on the face would be an extreme rarity and would have to be differentiated from lupus vulgaris verrucosus.

Tuberculosis fungosa serpiginosa is an auto-inoculation type of tuberculosis occurring in advanced age.

Scrofuloderma (tuberculosis cutis colliquativa) is the gummalike manifestation of cutaneous tuberculosis and is most commonly seen in the lymph nodes between the lower jaw and the sternocleidomastoid muscle. It may occur purely cutaneously or subcutaneously, in advanced age on the chest and back, and it has been reported occurring on the eyelids.

Lupus miliaris disseminatus faciei occurs as an expression of late generalization of tuberculosis more commonly in young males than females and in the presence of varying degrees of cutaneous sensitivity to tuberculin. The clinical appearance simulates a follicular accentuated rosacea or rosacea lupoid, consisting of pale red or brownish red, pinhead to 1 or 2 mm sized, soft, individual nodules which never become confluent (Fig. 26). On compression with a glass plate they look like lupus and are easily ruptured with a probe. In contrast to acne vulgaris is the total lack of comedones. Frequently, the lesions tend to be grouped in the center of the face with a particular predilection for the upper lip and the upper eyelids. These nodules remain unchanged for months and years and histologically show the rather typical appearance of tuberculosis – namely, epithelioid cell tubercles with giant cells and a surrounding zone of lymphocytes and central caseation, an appearance that can be found in all tuberculous or tuberculid forms (Fig. 27).

The lichen scrofulosorum, first described by HEBRA in 1860 as a lichenoid form of tuberculid, may occur anywhere on the skin surface although the lateral aspects of the thorax generally are the most common location. Clinically, the eruption consists of an exanthem of tiny, skin-colored or yellowish red, sometimes acuminate papulosquamous or follicular nodules, which frequently can be seen only by close inspection. HEBRA called attention to the tuberculous appearance of

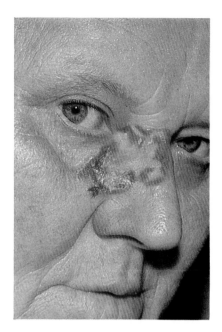

25 Lupus vulgaris.

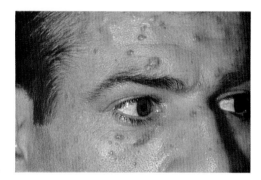

26 Lupus miliaris disseminatus faciei.

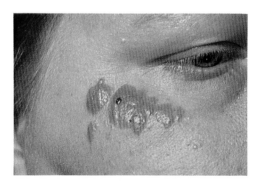

28 Sarcoidlike foreign body granuloma.

the patients and in addition, to the preponderance of old tuberculosis and lymph node swelling. The eye may be affected in the form of a *phlyctenular conjunctivitis* or eczematous blepharoconjunctivitis, iritis, iridocyclitis and choroiditis. Even a parenchymatous keratitis has occasionally been described in lichen scrofulosorum (KREY). Further details about diseases of the eye in the presence of lichenoid tuberculosis are reported by VOLK.

primarily necrotic. Because of the depth of involvement, healing occurs with the formation of scars surrounded by dirty brown residual pigmentation. Involvement of the face is rare and in such cases it is difficult to differentiate between papulonecrotic tuberculid and lupus miliaris disseminatus faciei.

In erythema induratum, parenchymatous keratitis has occasionally been reported as a complication (MEISSNER, NAKAZAWA).

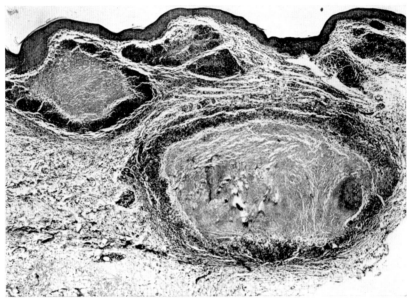

27   Lupus miliaris disseminatus faciei. Histologic appearance.

Eye disorders similar to those found in lichen scrofulosorum can also be found in the presence of papulonecrotic tuberculid, which extends deeper into the cutis and subcutis than the previously described group of tuberculids and which develops in a peculiar kind of distribution affecting primarily the trunk, the buttocks, and the lower extremities. The individual lesions begin as pale, later blue-red or brown-red, rice-sized nodules which change into acneiform pustules, rarely ulcerate, but rather dry with crusts or they may be

Sarcoidosis

The diagnosis of true sarcoidosis can be made in the presence of a generalized disease with manifestations in more than one organ (lymph nodes, liver, skeleton, cardiac muscle, and so on). Anatomically, this disease manifests itself in the form of an epithelioid cell granuloma which differs from the typical tubercle by the lack of central caseation, a sparse inflammation with lymphocytes and the presence of few Langhans-type giant cells

(Fig. 28). A further criterion for such epithelioid cell granulomas is, among others, the appearance of asteroid bodies or Schaumann bodies in the cytoplasm of the giant cells (Fig. 29). The above cited definition of sarcoidosis is necessary, since epithelioid metamorphosis of histiocytes resembling sarcoidosis may without doubt occur at the site of parenteral absorption of a variety of basically different materials such as quartz, beryllium and the pollen of pine trees, among others. Boeck's disease, in contrast, is a frequently familial, systemic reaction pattern which may be due to a virus or may be a peculiarly atypical form of tuberculosis.

For instance, it has recently been suggested by MANKIEWICZ that in patients with sarcoidosis non-neutralized mycobacterial phages exist which alter the tubercle bacilli so that they cannot be demonstrated by the usual methods.

In spite of its systemic character, the disease process in sarcoidosis begins either with fever and joint pain or unobtrusively without symptoms. Clinically, it primarily affects the bases of the lungs, more commonly on the right side.

In association with this first involvement of the hilar or paratracheal lymph nodes, erythema nodosum may appear on the skin, a combination which has been called the Loefgren syndrome. A common expression of the disease is the frequent association of bone involvement in the form of an osteitis cystica multiplex. Kidney changes are characterized by epithelioid cells, granulomatous foci, hyalinization of the glomeruli, and particularly, calcinosis of the kidney, while the associated involvement of the central nervous system is clinically suggested by the development of diabetes insipidus.

The cutaneous lesions, which occur in about 40 per cent of patients with sarcoidosis, are characterized by deep brown-red to blue-red nodules which are commonly traversed by telangiectatic vessels and which as a rule neither scale nor break down. Under pressure

with a glass plate they do not resemble a spot of lupus vulgaris, but rather appear dusty. They are resistant to pressure by a probe. The disappearance of such specific cutaneous lesions occurs with only minimal loss of pigment and sometimes with very little scarring. On the other hand, even BOECK called attention to the frequency of involvement of pre-existing scars with lesions of sarcoidosis, with the exception of the scars of vaccination or of burns. The lesions themselves may be either diffusely infiltrated and nodular or lichenoid and annular, and very occasionally consist of generalized erythroderma (Fig. 30). Of the types that infiltrate profusely, one must make particular mention of *lupus pernio* occurring in the nasal region (Fig 31) and the *angiolupoid*, which consists of large nodules located at the inner canthi. They both occur more commonly in women of middle age. In the periorbital regions of the skin the author's own personal experience suggests that the lesions are primarily annular or, as shown in Fig. 30, consist of small nodules.

An important component of the symptomatology of sarcoidosis is the eye changes already reported by BOECK, SCHUMAKER, and BERING. Of these changes, uveitis is probably the earliest evidence of generalization and, judging from the large series of patients reported, eye changes are frequently the initial signs of sarcoidosis. Changes of the uvea, papillitis, and periphlebitis retina, constitute the most common signs. Glassy nodules of sarcoidosis may also be noted on the margins of the lids and the conjunctivae bulbi. When the eye alone is affected, it can be extremely difficult to differentiate between tuberculosis and sarcoidosis. The frequency of iridocyclitis in sarcoidosis varies between 5 and 20 per cent of all reported cases. Other manifestations are exophthalmos due to infiltration of the retrobulbar space by sarcoid lesions, infiltration of the optic nerve, calcification of the cornea and conjunctiva, and bleeding into the vitreous. Most diagnostic for ocular sarcoidosis is the subchronic, febrile uveopa-

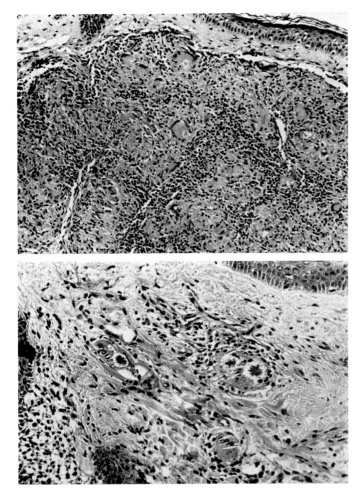

29 Histologic appearance of a sarcoidlike foreign body granuloma (see illustration 28). In the lower picture are typical "asteroid bodies."

rotitis of the type called Heerfordt syndrome, described in detail in an excellent paper by WEBER. The primary symptoms of this form of sarcoidosis are protracted low-grade fever, chronic enlargement and hardness of the parotid gland, nodular iridocyclitis, and similar changes in the other salivary glands or the breast as well as pareses of the spinal nerves sometimes associated with meningitis and signs of brain stem disorder. Among the cranial nerve symptoms, a unilateral paralysis of the facial nerve dominates, and the eye changes on the same side as the paralysis of the facial nerve are characteristic.

From the ophthalmologist's point of view, one finds in addition to the already noted iridocyclitis which is associated with precipitates on the back of the cornea, also retinal inflammation and alteration of the optic nerves. The Heerfordt syndrome as a component part of sarcoidosis has only been known since 1935, and it is interesting that this particular syndrome is rather uncommonly associated with cutaneous involvement.

Mikulicz syndrome is a morphologic entity consisting of indolent swellings of the lacrimal and salivary glands, and occasionally also other glands, sometimes associated with an occasional iridocyclitis. A part of the observations described as Mikulicz syndrome undoubtedly are related to the Heerfordt syndrome and therefore etiologically related to sarcoidosis. However, some late variants of the symptom complex described by Mikulicz seem to be more related to Sjögren's syndrome, since they are characterized by a progressive atrophy of the salivary glands and dryness of the mucosa, dry conjunctivitis, superficial keratitis punctata, xeroderma, nail changes, and many manifestations such as polyarthritis, arthritis, serum protein abnormalities, iron deficiency anemia, and the like.

As far as laboratory studies are concerned, the most important are biopsy of the skin, skin testing for tuberculin (which is negative or decreased in about 70 per cent of all patients) and the skin test according to KVEIM-NICKERSON. As in the *Frei* or *Mitsuda* tests, in this test suspensions of tissue are used as a test antigen (prepared for instance from lymph nodes of patients with sarcoidosis) and injected intracutaneously in about 0.1 ml quantities. Four to six weeks later in the positive case a nodule develops at the site of the test, about 1/2 cm in diameter, that histologically has to be examined for the presence of a typical epithelioid cell granuloma. STEIGLEDER, SILVA, and NELSON found in 99 well-studied patients with sarcoidosis 16.2 per cent with negative Kveim tests. In general, the demonstration of tubercle bacilli usually rules against the diagnosis of sarcoidosis. The blood picture is not characteristic, except for a general, sometimes excessive, increase in gamma globulin. Less frequent is an increase in total protein, and sometimes the serum calcium is elevated and perhaps more frequently there is an increase in urinary calcium.

Melkersson-Rosenthal syndrome

The *"recurrent edematous granulomatosis"* of MELKERSSON-ROSENTHAL is probably closely related to sarcoidosis (Fig. 32). The syndrome consists of a complex of signs beginning with first intermittent and later permanent swelling of the lips, resulting finally in a full-blown cheilitis granulomatosa, episodic facial paralysis of the facial nerves, and finally the degenerative phenomena of lingua plicata or scrotalis. As the disease progresses, there may be periodic edema followed by induration of the tissue of the buccal mucosa (pareiitis granulomatosa), of the hard palate (uranitis granulomatosa), and particularly of the periorbicular tissue. These phenomena may be associated with a variety of neurologic and neurovegetative symptoms such as acroparesthesia, attacks of migraine, scotomas and hyperacusis, among others. The protruding thickening of the lips resembles local elephantiasis such as occurs after erysipelas, recurrent herpes simplex, or contact dermatitis. The ophthalmologic abnormalities occurring in Melkersson-Rosenthal syndrome have been described as retrobulbar neuritis, unilateral ptosis, increased or decreased flow of tears, a variety of keratitic abnormalities, such as those due to prolonged lagophthalmos, or an alteration in the sensitivity of the first branch of the trigeminus nerve, anisocoria, unilateral connatal amblyopia or coloboma. In general, the majority of the ophthalmologic aberrations seem to be due to segmental vegetative alterations in addition to other changes of the nervous system.

Ascher syndrome and other forms of blepharochalasis

The following symptom complexes must be briefly described since the periorbicular swelling of the lids occurring in Melkersson-Rosenthal syndrome and the associated blepharitis granulomatosa may result in a

blepharochalasis, a localized dysplasia of the skin (Fig. 33).

KLEINE-NATROP stated that blepharochalasis may be the expression of a primary chalazodermia (Fig. 34), as already described by ALIBERT under the term of dermatolysie palpébrale. Furthermore, sagging of the lid was already described by the ophthalmologist SICHEL in 1844 as ptosis atonica-lipomatosa, and the ophthalmologist FUCHS has called attention to real, non-elastic increase of the surface of both upper lids of which the skin was of cigarette paper consistency and allowed dilated veins to shine through the reddened skin. This blepharochalasis of FUCHS is a morphologically well defined entity described by OPPENHEIM as blepharitis atrophicans progressiva cum chalasi which seems realistic, since the blepharochalasis of FUCHS seems to be the edematous early stage which eventually results in atrophy with crinkling of the eyelids and the shining through of vessels.

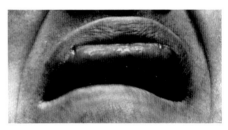

35 "Double lip" in the Ascher syndrome.

Blepharochalasis is also a clinically constant symptom of a combination of manifestations described in 1920 by ASCHER on the basis of eight cases. The *Ascher syndrome* includes, in addition to blepharochalasis, a variable alteration of the lips, either in the form of a double lip by duplication of the mucosa because of constant edema of hyperplasia of the glands of the lips, or in form of recurrent and later persistent edema of the lips. In addition to this, there exists an increase in

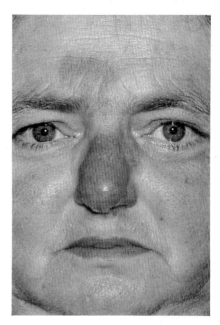

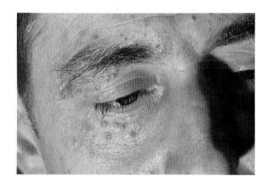

30 The lichenoid form of sarcoidosis.

31 Lupus pernio.

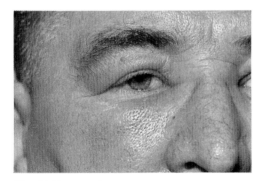

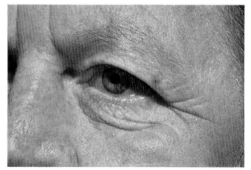

32 Melkersson-Rosenthal syndrome: "Blepharitis granulomatosa."

33 Blepharochalasis.

the size of the thyroid glands not associated with hyperthyroidism and also, somewhat more rarely, endocrine abnormalities such as acromegalic appearance, dysmenorrhea, and so on. In contrast to cheilitis granulomatosa of Miescher and the Melkersson-Rosenthal syndrome, in the syndrome of Ascher the upper lip is almost exclusively affected (Fig. 35). Occasionally the Ascher syndrome even occurs in families.

Leprosy

Leprosy, a disease which affects between 10 and 15 million patients the world over and which in Germany is the most carefully controlled infectious disease, is characterized in spite of its marked clinical diversity by a close association of alterations of the skin and the nerves and also by frequent changes in the eyes (Fig. 36). In contrast to this, the visceral components of this disease are relatively minor.

After a long incubation period, which on an average lasts three to five years, there develop at first indeterminate areas of depigmentations, occasionally red-brown, macular lesions, more rarely extensive papular or lichenoid areas associated with diagnostic hyperplasia of some of the nerves, particularly the ulnar or the auricular nerve. In contrast to these early symptoms, the full-blown cutaneous manifestations of leprosy are relatively much more diagnostic. The major cutaneous types of characteristic leprosy are the highly contagious, serious lepromatous form (with nodular lepromas), which when it affects the eyebrows and nose may progress to the classical picture of the "leonine face" and occasionally with involvement of the testes and manumary gland and tuberculoid leprosy, previously called maculoanesthetic.

The dimorphic type is a mixed form of tuberculoid and lepromatous lesions which may or may not behave as a contagious disease.

The causative organism, the mycobacterium leprae discovered by HANSEN in 1872, is an acid-fast rod 4 to 6 microns in length which,

in contrast to the tubercle bacillus, is not pathogenic for guinea pigs and presumably enters by means of the upper respiratory tract or the oral mucosa and tonsils (Fig. 37). In order to find the causative organism, scrapings with a scalpel from suspicious skin lesions and particularly, as described by JAUSION, from the anterior part of the nasal septum are perhaps done under local anesthesia. Bleeding should be prevented if possible. The organisms are demonstrated by staining the scrapings with the Ziehl-Neelsen technique as acid-fast rods which occur sometimes in masses or in cigarette bundle groups in the nasal secretion. In relatively benign tuberculoid leprosy, specific changes in the nose are uncommon and for this reason negative findings from the nose should not be the deciding point for or against the diagnosis. Immunologic studies are of importance in the diagnosis of leprosy. In the tuberculoid form of leprosy, the Mitsuda reaction is frequently positive. It is elicited by the intracutaneous injection of a standardized extract of lepromatous organs and may be positive at the end of 48 hours or at the end of 2 or 4 weeks. The positive skin test consists of a papule appearing at the site of injection. Unfortunately, false-positive tests occur in tuberculin positive individuals. Less useful tests are those testing the secretion of sweat or the contraction of arrector muscles of the hairs at the sites of suspected skin lesions.

Eye involvement has been reported by many authors. All agree on the frequency of ocular manifestations, the incidence of which reaches almost 90 per cent in lepromatous types and which result in blindness in almost one third of the patients. GROENOUW reports the frequency of involvement of the various portions of the eye as follows: eyebrows, eyelids, sclera, cornea, and iris. KLINGMÜLLER in his monumental summary reports that involvement of the eye is less common in lepromatous leprosy. In contrast to this, the muscles of the eyelids and of the iris are particularly affected in tuberculoid leprosy. In any case, the eye-

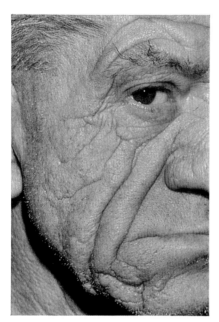

34 Chalazodermia.

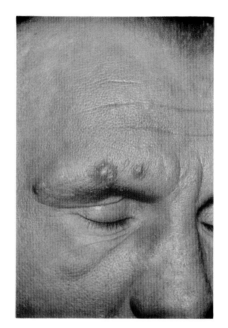

36 Lepromatous leprosy.

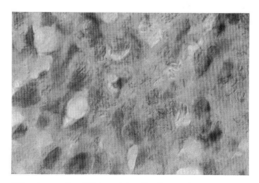

37 Lepra bacilli in a histologic section.

brow region is an area of predilection for leprosy in contrast to the other portions of the hairy scalp. Alopecia and wrinkling of the eyebrows have been considered stigmata of leprosy since antiquity. Lepromatous alterations of the eyelids are similar in general to those anywhere else on the skin, although the smaller lepromas at the free edge of the eyelids may at first glance look like wheat grains. The conjunctiva is involved primarily by hyperemic catarrhs and later by shrinking, while lepromatous alterations of the conjunctiva of the eyelids are supposed to be rare. Because of the common local nerve paralysis, ptosis develops and is followed by lagophthalmic ulcerations of the cornea. Ectropion and symblepharon are seen quite frequently. Furthermore, as in congenital syphilis or in tuberculosis, interstitial keratitis may be noted. Nodules may develop on the cornea, generally starting from the limbus. There is much evidence that the eye may be infected from the inside by leprosy, yet the bulbus oculi is primarily affected in its anterior parts. Involvement of the iris may be acute and inflammatory or it may be slow and nodular with the nodules being single or multiple (iridocyclitis granulomatosa). Chorioretinitis occurs rather frequently in some series of cases but involvement of the optic nerve appears to be rare. Specific involvement of the tear ducts and tear glands is uncommon. As previously noted, lepromatous eye involvement frequently leads to amaurosis which will depend in each individual case primarily on the maintenance of the transparent media (formation of pannus). Apparently the degree of blindness also varies geographically, perhaps according to different climatic conditions.

Leishmaniasis cutis
(Oriental or Aleppo boil)

Cutaneous leishmaniasis occurs primarily in the Mediterranean areas (Fig. 38). How-

ever, we must call attention to the American leishmaniasis which in its severe cutaneous and mucosal form is called *Espundia*. This infectious disease produced by Leishmania tropica and transmitted by the sand fly *Phlebotomus* occurs primarily in the summer or in the fall and has no fixed time for incubation since this seems to depend primarily on the number of organisms transmitted. About 60 to 70 per cent of all the primary lesions of leishmaniasis occur on the face, although this seems to depend on ethnic and geographic variations. At the location of the insect bite there develops at first a dirty brown, then blue-red infiltrate which at first resembles a chancrelike pyoderma and frequently ulcerates in the center.

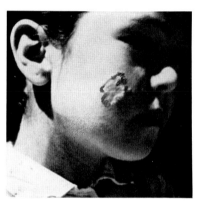

38   Leishmaniasis.

The disease extends by the lymphogenous route and forms satellite lesions which may be small nodules, large nodules, verrucous papulomatous lesions or, in the presence of lowered resistance, even ulcerated and serpiginous areas which finally resemble the disclike shapes of lupus vulgaris or lupus erythematosus. Such individual lesions commonly regress in about one year and leave behind a scar. In prolonged primary lesions, leishmanids may appear in the form of small nodules distributed as they are in an exanthem.

The causative organisms may be found in scrapings or in sections with Giemsa staining. Dogs and cats are common intermediate hosts. As far as the ocular lesions themselves are concerned, one can occasionally see primary lesions in this area which, however, basically have the same attributes as the primary lesions found anywhere else on the body. In any case, involvement of the eyelids was noted more commonly in the early reports and has not been reported very frequently recently. Plaquelike specific lesions have been described on the conjunctiva by CHU and on the tarsus by SILVA.

References

ASCHER, K. W.: Blepharochalasis mit Struma und Doppel-lippe. Klin. Mbl. Augenheilk. 65 (1920) 86.

CHU, T. S.: Leishmaniose mit ungewöhnlichen Haut- und Schleimhauterscheinungen. China Med. J. 71 (1953) 354.

FINDLAY, G. H.: Idiopathic enlargements of the lip; cheilitis granulomatosa, Ascher's syndrome and double lip. Brit. J. Derm. 66 (1954) 129–138.

FUCHS, H.: Über Dermatochalasis. Wien. klin. Wschr. 39 (1926) 1331.

HEERFORDT, C. F.: "Über eine Febris uveo-parotidea sub-chronica." v. Graefes Arch. Ophthalm. 70 (1909) 254.

KEEY, O.: Tuberculöse Keratitis parenchymatosa und Lichen scrofulosorum. Klin. Mbl. Augenheilk. 76 (1926) 520–524.

KLEINE-NATROP, H. E.: Über Lidsäcke. Derm. Wschr. 135 (1957) 521.

KUDLICH, H.: Zur primären Tuberkuloseinfection der Binde-haut des Augenlids. Z. Tuberk. 43 (1925) 66.

MEISSNER, I.: Einseitige Keratitis parenchymatosa bei Ery-thema induratum scrophulosorum Bazin. Z. Augenheilk. 31 (1914) 547–548.

MORGAN G.: Case of cutaneous leishmaniasis of the lid. Brit. J. Ophthal. 49 (1965) 542–546.

NAKAZAWA, T.: Ein Fall von Erythema induratum Bazin mit Keratitis parenchymatosa tuberculosa. Jap. J. Derm. 30 (1930) 44. Ref. Zbl. Haut- u. Geschl.-Kr. 35 (1931) 525.

REIS, W., and ROTHFELD, J.: Tuberkulide des Sehnerven als Komplikation von Hautsarcoiden . . . v. Graefes Arch. Ophthalm. 126 (1931) 357.

RICHTER, R.: Die Lepra. In: Dermatologie und Venerologie, Bd. V/1, hsg. von H. A. Gottron u. W. Schönfeld. Thieme, Stuttgart, 1963.

SCHIMPF, A.: Das Ascher-Syndrom. −1086. Derm. Wschr. 132 (1955) 1077.

SCHRECK, E.: Veränderungen des Sehorgans bei Haut- und Geschlechtskrankheiten. In: Dermatologie und Venerolo-gie, Bd. IV, hsg. von H. A. Gottron u. W. Schönfeld. Thieme, Stuttgart, 1960.

SILVA, Fl.: Leishmaniose tegumentar. Ann. Fac. Med. (Bahia) (1942–1943).

STEIGLEDER, G. K., SILVA, JR., A., and NELSON, C. T.: Histopathology of the Kveim test. Arch. Derm. Syph. (Chic.) 84 (1961) 828–834.

VOLK, R.: Tuberkulose der Haut: In: Hdb. der Haut- und Geschlechtskrankheiten, Bd. X/1, hsg. von J. Jadassohn. Springer, Berlin, 1931.

WEBER, G.: Zur Teilsymptomatologie der Boeckschen Sar-koidose: das Heerfordtsche Syndrom. Derm. Wschr. 144 (1961) 925–933.

Virus Diseases of the Skin

In a general way it is particularly important to refer to the observations of the ophthalmologist GREGG and the work done by subsequent authors, which pointed out for the first time in 1941 that nongenetic factors, particularly infections in the first third of pregnancy, can be of importance for the development of congenital malformations. Of the various new concepts based on this observation, the term *embryopathy* is probably the most important. This concept, as already noted, dates back to GREGG, who noted a relationship between German measles infection of the mother and congenital malformations (cataracts, lack of formation of the brain in infants with idiocy, defects in the septum of the heart), an observation that was subsequently confirmed by many observers. Of particular importance for the understanding of intrauterine injury to the fetus is the observation that the *time* element at which some injury occurs is of greater importance than the type of injury. From the type of malformations which have been reported, it is possible to draw conclusions about the gestational period at which these abnormalities occurred. This allows one to form a sort of timetable of abnormalities, which for instance fixes the time for the development of eye changes in the vicinity of the fifth week, of heart changes in the vicinity of the third to the seventh week and of deafness in the seventh or twelfth week of pregnancy. In addition to this more or less specific time at which certain kinds of injuries may occur, there is also apparently an organ affinity, since time and time again the lens, the organ of Corti, or the septum of the heart is the primary site of malformations secondary to infectious agents.

As to the cause of such embryopathies, the reader must be reminded that embryonal tissue in principle is extremely sensitive to injurious effects of any kind, and that particularly injury to the central nervous system at a time of as yet limited fetal specialization

Table 1 Ocular Manifestations of Viral Disease of the Skin (modified from Blank and Rake, 1955)

Viral Disease	Ocular Localization				
	Lids (skin)	Conjunctiva	Cornea	Uveal tract	Retina
Lymphogranuloma venereum	++	+++	+++	++	
Cat-scratch fever	++	+++			
Smallpox	++++	+++	++	++	
Alastrim-variola minor	++				
Vaccinia	++	++	++	++	
Molluscum contagiosum	++++	+++	++	++	
Herpes simplex	++	++	++++	++	
Herpes zoster	+++	+++	++	+	
Chickenpox	+	++	++	+	
Measles	+++	+++	++		++
German measles (rubella)	++	++	+		
Infectious mononucleosis	++	++		++	+
Foot and mouth disease	+	++	++		
Dengue fever	++	+++		+	+
Sandfly fever		+++			+
Verrucae and condylomata acuminata	++++	+++	++		

of tissues is of primary pathogenetic importance for subordinate alterations of development.

As far as the frequency of eye changes in a variety of virus diseases of the skin is concerned, Table 1 shows the latest available information (from NASEMANN, T.: The Infections Due to Herpes Simplex Virus, Fischer, Jena, 1965).

Smallpox (Variola vera)

The regularly cyclic course of human smallpox begins after an incubation period of 8 to 14 days (quarantine period 18 days). The virus is transmitted by droplets, dust, and sometimes even flies. The disease usually begins acutely with pain in the sacrum, in the extremities, or in the testicles (Fig. 39). This is frequently associated with an uncharacteristic pre-exanthem, which occurs on the inner side of the upper thighs and is associated with catarrhal manifestations of the mucosa such as pharyngitis, conjunctivitis, and so on. Fixation of the virus to organs is manifested three to four days later by a characteristic drop in temperature. The actual smallpox exanthem then begins on the forehead, the nose, the upper lip, the scalp, the trunk, and the extremities, usually sparing the axillae, the groin area, and the genital areas. The initialy tiny blisters may also occur on the conjunctiva and may cause edema of the eyelids and photophobia. On the seventh to the eighth day of the disease the blisters begin to dell in the middle and suppurate. At this point the severity of the general symptoms again rises. Following the healing of the smallpox exanthem, the typical smallpox scars develop which may cause ectropion if they are located in the vicinity of the eyes. Otherwise, in contrast to conjunctivitis, the cornea is only secondarily involved and primary uveitis is rare. In the presence of smallpox purpura, there may be bleeding into the orbit. In India even today 20 per cent of all blindness is due to smallpox.

In order to make a definite diagnosis, which today should certainly not be made only on clinical grounds, epidemiologic and laboratory studies are necessary. The epidemiologic studies should consider whether the patient has stayed in areas where the disease is known to be endemic and delve into the possibility of previous vaccination.

Laboratory studies consist of spreading a thin layer of the blister content on a flat, presterile microscope slide, which is to be covered by a second microscope slide and sent to an appropriate laboratory. Very careful wrapping of the entire material in a metal or wooden box, and appropriate contacts with the laboratory so that they know that the material is coming are obviously required. In addition to this, 5 to 10 ml of venous blood with citrate should be sent.

Side effects of protective vaccination

As vaccinial diseases, or deviations from the usual reaction to vaccination, one must include first of all, Area migrans, Area bullosa, erysipeloid erythemas following vaccination, and the so-called vaccinial ulceration which may be associated with adjacent lymphangitis. Furthermore, self-inoculation may result in secondary vaccinia which may be serpiginous. Hematogenous dissemination may, particularly if there is a lack of antibody-forming capability, result in generalized vaccinia and a variety of exanthems. Eye complications of vaccinations develop also either as autoinoculation, or secondarily by the hematogenous route, the latter primarily in association with postvaccinial encephalitis.

As far as vaccinial lesions on the eyelids are concerned, these are usually located at the edges of the lids or at the canthi, probably because small injuries to the skin allow the virus to enter (Fig. 40). Because of their

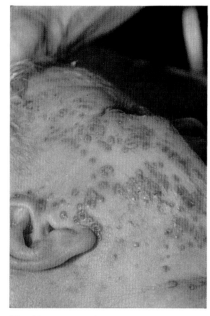

39 Smallpox. (Photograph courtesy of Dr. Stüttgen, Berlin.)

41 Eczema vaccinatum.

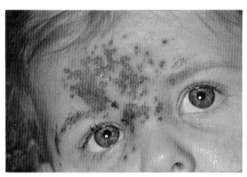

42 Eczema herpeticum.

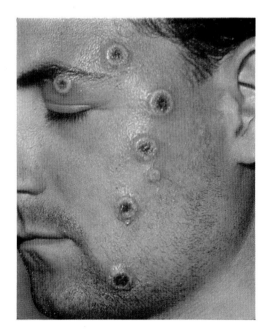

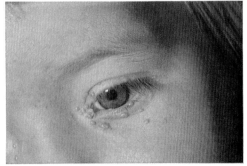

40 Vaccinia of the eyelids.

43 Molluscum contagiosum.

particular location, the pustules on the eyelid margins open very early, so that one notices in general only a large ulceration covered with exudate which is soon covered by slimy pus (BRUENS). The diagnosis is, however, made relatively simple in that almost always typical vaccinia lesions can be found on other parts of the body. Frequently, but not always, the ulcerations from the eyelid margin extend onto the conjunctiva of the lids, and this results in marked chemosis so that the clinical appearance of conjunctival vaccinia may resemble diphtherial keratoconjunctivitis or an eyelid abscess (LINNEN). The cornea may also be the location of a real vaccination, which may occasionally be accompanied by a serous iritis. In any case, the patient is markedly inconvenienced by great swelling, pain, and itching. "The eyeball may be walled in for days behind swollen eyelids (HART-LEIB)." Furthermore, a usually partial madaro-sis may follow vaccinia of the eye as a permanent change. ROSEN reports rare side effects of vaccination consisting of isolated paralysis of eye muscles, chorioretinitis, and thrombosis of the central vessels of the retina.

Of equal importance as far as seriousness of prognosis is concerned is eczema vaccina-tum, whose mortality rate may run up to 30 per cent (Fig. 41). This disease develops almost exclusively in patients with chronic eczema or atopic dermatitis, who therefore should never be vaccinated if they have the least cutaneous change or if their disease has been controlled only recently by cortisone. In eczema vaccinatum, as in true smallpox, the light-exposed areas of the skin are the location of multilocular, persistent, delled blisters with thick tops, which appear after an incubation period of about 5 to 12 days and con-tinue to develop in crops. In the presence of periocular pustules in particular, conjunc-tivitis and keratitis may occur, and also ectropion, which develops after healing (LANGANKE).

Because of its great clinical similarity, eczema herpeticum (SCHÖNFELD) (Fig. 42)

must be mentioned here. This disease has in the past also been called pustulosis varicelli-formis (KAPOSI, JULIUSBERG). After an in-cubation period of about 2 to 7 days, the eczematous or atopic dermatitic areas are affected by the virus of herpes simplex. The blisters are in general smaller than in eczema vaccinatum, and usually unilocular. Eczema herpeticum is fatal in about 10 to 20 per cent of cases, and occurs three quarters of the time in childhood. As far as differential diagnosis is concerned, particularly in the absence of general systemic symptoms, one must consider ordinary impetigo of the type often called varioliform pyoderma. As in eczema vaccinatum, eye complications are possible, and may consist of conjunctivitis and keratoconjunctivitis herpetica (literature quoted by SCHRECK).

An infection with ordinary cowpox occurs in general in farmers and milkers. It usually consists of individual varioliform lesions, which may be complicated by lymphangitis. In these occupations, the spontaneously in-voluting milker's nodules occur (paravaccinia nodules — PIRQUET). These people also have milker's calluses or milker's granulation nodules which are generally foreign body reactions to cow hair which has entered the skin (GOTTRON). Very rarely milker's nodules may be noted on the eyelids, as demonstrated histologically in a case of Ludwig's, which had the appearance of a granulating tumor compared to a lesion present on the hands.

Molluscum contagiosum

This generally harmless infectious acan-thoma, first described by Bateman in 1817, is due to a brick-shaped virus and affects primarily children. After an undetermined incubation period (several weeks or months) it involves primarily the thin skin areas (Fig. 43). The single, slightly transparent or waxy, miliary

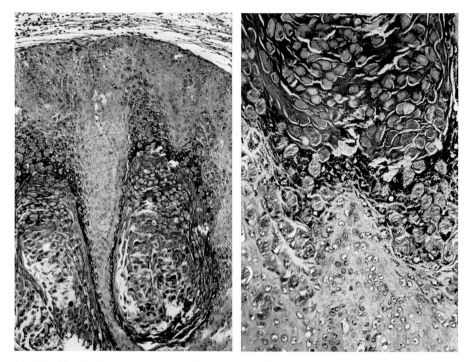

44   The histologic appearance of molluscum contagiosum.

to pea-sized hemispherical molluscum nodules are characterized, in addition to their small size, by the central delling which has resulted in the German name *"delled wart."* If one squeezes such a molluscum nodule from the sides, it will empty (with very little bleeding) whitish granular masses which may be doughy, the so-called *molluscum bodies.* As can be seen histologically, in molluscum contagiosum proliferation of cells and degeneration of cells go together, so that each individual lesion acquires a typical lobular form (Fig. 44). Furthermore, massive homogenized cell masses develop, which

finally are formed into the already mentioned molluscum bodies. By confluence these enlarge so that they can be noted macroscopically.

Quite commonly such molluscum lesions are localized on the eyelid and may result in resistant inflammation of the conjunctiva, noted as early as 1895 by STEFFAN. The literature since then has been reviewed by NOVER, who has pointed out that longlasting, recurrent keratitides may be produced by mollusca of the eyelid. Molluscum contagiosum nodules have even been seen on the cornea (SYSI).

Herpes simplex
(Herpes febrilis, Fever blisters,
Cold sores)

The herpes simplex virus may cause the development of a whole variety of clinical manifestations (Fig. 45). Primary infection with herpes virus frequently is associated with high fever and may result in a disease with poor prognosis (herpes sepsis, meningoence-phalitis herpetica, and the secondarily developing eczema herpeticum). Less severe primary manifestations of herpes are stomatitis and aphthous vulvo-vaginitis herpetica, and finally the aphthoid of Pospischill-Feyrter (secondary vegetating ulcerations following measles, whooping cough, or panmyelophthisis). The secondary manifestations of herpes infection are mentioned only briefly here, since they particularly affect 20- to 40-year-old persons who already have specific antibodies in their serum. They are generally characterized by unpleasant, but relatively harmless eruptions in the form of herpes labialis, genitalis, and the like. The typical appearance consists of grouped blisters on slightly reddened skin or mucosa, whose development is preceded by a slight local sensation such as a feeling of tightness or faint itching. Recurrent attacks suggest herpes simplex, while segmental development and neuralgio suggest herpes zoster.

Primary, but also secondary, recurrent manifestations of herpes simplex are well known to occur on the cornea. As a matter of fact, the field of experimental study in herpes virus research began with the experiment of GRÜTER on the cornea. It is well known that herpetic keratoconjunctivitis, which may occur primarily as well as secondarily, is the most common specific keratitis (for literature review see NASEMANN, 1961). Furthermore, it appears that herpes simplex involvement of the eye has been increasing in recent years, with a peak involvement in the age groups from 1 to 15 and from 35 to 50, while after the sixth decade the frequency of herpetic corneal disease is very much lower. As far as clinical manifestations are concerned, herpetic involvement of the eye can be very polymorphic and it ranges from superficial epithelial changes (keratitis punctata, vesicularis, filiformis, dendritica, and serpiginosa) to the deep interstitial and then usually chronic types, which may be secondarily infected by bacteria (keratitis disciformis). Occasionally, diffuse interstitial or ulcerating keratitis may develop, associated with hypopyon, which may develop into an iridocyclitis or be complicated by glaucoma. GROENOUW (and more recently SCHRECK) showed that in all nontraumatic inflammations of the cornea which are produced by herpes simplex virus, there exists a reduction in the touch sensitivity of the cornea, while SCHÖNFELD pointed out that corneal herpes frequently occurs simultaneously with herpes simplex of the nose or the eyelids.

Recently it was shown, first in experimental herpes simplex keratitis of the rabbit eye, and then therapeutically in the human disease, that 5-codu 2′ desoxyuridine interferes with formation of new virus particles since it substitutes for thymine which is necessary for the development of normal DNA. In contrast to the good results on the cornea, my own experiments on skin and on mucosae have not been very convincing. This may be due to differences in the ability of the compound to be absorbed (for literature review see HOLZMANN).

Herpes zoster (Zona, shingles)

Since the studies of VON BÄRENSPRUNG (1861 — for details see LEIBBRANDT) we know that herpes zoster, produced by the varicella-zoster virus, begins in a spinal ganglion and projects to the periphery. It is also the clinical manifestation of reinfection with the varicella virus in a partially immune person (DOWNIE). However, it is much more common that varicella develops after contact with the patient who has herpes zoster than the reverse. HOPE SIMPSON has therefore very appropriately compared the contagiousness with a one-way street. Furthermore, FEYRTER has pointed out that herpes zoster is a virus-dependent, hematogenous inflammatory disease which is characterized by a hyperergic capillaritis and also a hyperergic arteritis and phlebitis with or without hemorrhage or necrosis.

The clinical course is quite typical — after some prodromes stretching over three to five days there develop suddenly, usually unilaterally, bandlike groups of blisters on a pale red skin that are preceded at first only by an area of edema and later develop into the small unilocular blisters, clear as water, of herpes zoster. The age distribution is quite diffuse, with a peak in the vicinity of the seventh decade. Occasionally, the lesion may be hemorrhagic, necrotizing, or gangrenous, while a tendency for generalization (varicelliform herpes zoster) occurs primarily in the presence of leukemia, reticulogranuloma, cancer, or abnormalities of protein metabolism.

Blood studies show an increased sedimentation rate, moderate leukocytosis, and the development of an atypical mononuclear cell in the peripheral blood.

Although, as studies by HAUSER show, the nerve segments C3 and C4 (that is, the dermatomes of the phrenic nerve) are predominantly involved, nevertheless in recent years ophthalmic herpes zoster has increased without question, particularly in my material, in contrast to the lack of such increase in zoster oticus (which peculiarly never becomes generalized). According to the data of SCHRECK, the cornea is involved in 50 to 60 per cent, and the eye muscle nerves in 7 to 24 per cent of cases of zoster ophthalmicus (see PRIGGERT, BÖRNER et al.). GRAEBER reported that 85 per cent of 48 patients with zoster ophthalmicus had involvement of the eyes. In my primarily dermatologic and not always ophthalmologically controlled observations, this high percentage of involvement of thel esions

eye was not found. NASEMANN, 1961, properly points out the great variety of changes that can be found in zoster ophthalmicus: epiphora, photophobia, lid edema, conjunctivitis, keratitis, cornea ulcers, iritis (black precipitates, produced by the destruction of the pigmental layers of the iris), retinitis, neuritis, and later leukoma, staphyloma, secondary glaucoma, and finally involvement of the ciliary ganglion which can be demonstrated by alterations in the reaction of the pupils. Unilateral fixation of the pupil has been described by, among others, HAMBURGER, and by RADNÓT and BAJNOK, and paralysis of the facial nerve as a sequef to zoster has been described by MEYER-ROHN. Finally, the visual acuity may be diminished because of the already noted increase fo the internal pressure in the eye (HALLERMANN) secondary to corneal scarring. Very rarely, ophthalmic herpes zoster may occur in children (TUCKER, THOMANN: a 12-day-old infant), and we ourselves have noted two cases in the last few years (see case of CABRÉ). HOLMBERG states that when the nasociliary nerve is affected, eye involvement is particularly likely to occur.

Varicella (Chicken pox)

Two to three weeks after the infection a macule develops in this very contagious typical childhood disease following minimal prodromes and commonly following some skin eruption, on top of which nodules and then thin-walled clear blisters appear after several hours which later become cloudy (Fig. 47). Because new blisters keep appearing, one sees usually various stages of development at the same time (HEUBNER: "Chart of stars"). The patient with varicella must be considered contagious until the blisters (which have usually dried to a yellowish-brown crust) have completely disappeared and the crusts have fallen off. Primary location of the lesions is the trunk; much less often the head or the face is involved. The oral mucosa and the eyelids may be involved and show typical blisters. Involvement of the cornea or its perforation is a rare occurrence. Phlyctenular reaction in the region of the limbus has been described.

Warts

Among these contagious hyperkeratotic acanthomas (JADASSOHN, 1896) there are now differentiated in general several different clinical basic types (including the pointed condylomas). Although the same virus apparently causes each disease, they differ clinically because of the location and the particular conditions under which the tumor develops. Ordinary hard warts, verrucae vulgares, are at the beginning generally greyish-yellow; later on when fully developed and markedly keratinized they become blackish and finally show the typical lumpy clefts on the surface (Fig. 48). We will not discuss in more detail the special forms of these verrucae vulgares, the so-called mosaic wart of the hand or the very painful plantar wart, but we should point out the clinical appearance in the eye area, particularly on the eyelids (as well as on the lips and on the neck). Here a paint-brush-like (verrue en pinceau) or filiform wart (cutaneous tag) develops (Fig. 49). The spontaneous disappearance of warts located on the palpebral and bulbar surfaces has recently been described by NOOJIN. *Verrucae planae Juvenilis* (BESNIER and DOYON) which, as their name shows, appear particularly in children and young persons, never occur singly, but in general develop in groups and are eruptive (Fig. 50). These appear as round or polygonal, flat, more or less skin-colored or pink, and sometimes itching elevations, which can easily be scratched off and leave a small punctate area of bleeding. They are localized particularly at the volar areas of the forearm and the back

of the hands, and appear on the face, on the forehead and cheeks while the periorbital region is usually spared.

Cat-scratch disease (Maladie de griffes de chat, Nonbacterial regional lymphadenitis)

The spread of this disease has been attributed to the house cat, which incidentally is also responsible for the transmission of a variety of infectious diseases such as rabies, toxoplasmosis, leptospirosis, and pasteurellosis (DEBRÉ). The disease apparently is also caused by an injury from thorns (MOLLARET) (Figs. 51 and 52). There is not much question that there is a close relationship

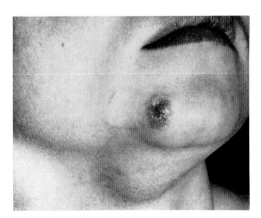

51   Cat-scratch fever.

between cat-scratch disease and lymphogranuloma inguinale, both of which are considered to be Miyagawanelloses. In both these diseases the primary lesion is circumscribed and barely noticeable; it occurs at the site of innoculation, and is followed by regional lymph node swelling. In contrast to this, in tularemia the cutaneous changes at the site of the scratch are much more marked, and the general well-being of the patient much

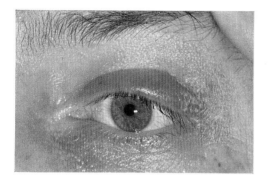

45 Herpes simplex.

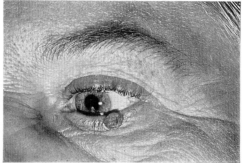

48 Verruca vulgaris of the lower eyelid.

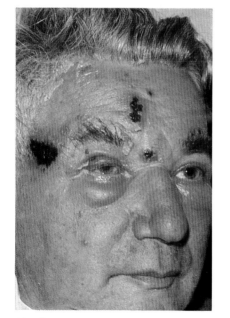

46 Ophthalmic herpes zoster.

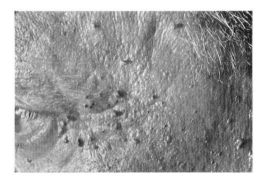

49 Filiform warts.

47 Varicella.

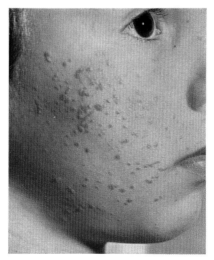

50 Juvenile plane warts.

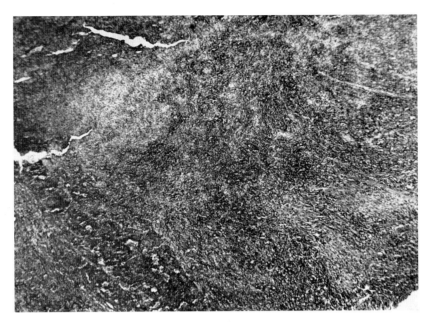

52   The histologic appearance of the lymph node in cat-scratch fever.

more impaired. In both diseases the diagnosis can be made with an antigen made from the purulent material of the lymph nodes that, when injected, produces a papular skin reaction (Tularin or Mollaret-Debré antigen, respectively), which resembles the Frei-, Mitsuda-, Kveim-reactions among others.

Histologically, we find a typical abscess forming granulomatous lymphadenitis (HEDINGER, LENNERT), while the cutaneous primary lesion presents of an ulcerating tuberculoid reaction.

In addition to the most commonly noted cutaneous granular form (see BREHM and JÜNGST, 1965), ocular, buccopharyngeal and pseudovenereal forms have been noted. The ocular-glandular form of cat-scratch fever develops as a unilateral conjunctivitis (follicularis) with indolent enlargement of the ipsilateral preauricular lymph node groups (see ALFANO and PEREZ) and thus differs primarily with regard to the portal of entry and location of the primary lesion from the usual cutaneous manifestation of cat-scratch fever. It is not surprising, in view of its clinical appearance, that the ocular-glandular form in the past was considered in part to be Parinaud's syndrome.

As far as the clinical course of ocular-glandular cat-scratch fever is concerned, the reader is referred to the description by RUGE or NASEMANN, 1961.

References

GROENOUW, A.: Beziehungen des Auges zu den Hautkrankheiten, Kap. Zoster. In: Hdb. der Haut- und Geschlechtskrankheiten, Bd. XIV/1, hsg. von J. Jadassohn, Springer, Berlin, 1930.

GRÜTER, W.: Experimentelle und klinische Untersuchungen über den sog. Herpes corneae. Ber. Dtsch. Ophthal. Ges. 42 (1921) 162.

HARTLEIB, R.: Vaccine-Erkrankungen des Auges. Med. Klin. 44 (1949) 994—996.

HOLZMANN, H.: Zur Herpes-simplex-Therapie in der Dermatologie mit 5-jod-2' desoxyuridin Haut- und Geschl.-Kr. 35 (1963) 86—88.

LANGANKE, E.: Klinischer Beitrag zum Eccema vaccinatum. Münch. med. Wschr. 97 (1955) 1467.

LEIBBRAND, W.: Der Herpes zoster als Geburtstagskind. Med. Klin. 56 (1961) 233—235.

LINNEN, H. J.: Zur Pathogenese und Klinik der Vakzinia des Auges und seiner Adnexe. Ein Beitrag zum Impfschädengesetz. Medizinische 1 (1957) 407—415.

LUDWIG, A.: Über Melkerknoten am Oberlid. Arch. Augenheilk. 109 (1935) 346.

NASEMANN, Th.: Die Viruskrankheiten der Haut. In: Hdb. der Haut- und Geschlechtskrankheiten, Bd. IV/2, hsg. von J. Jadassohn. Springer, Berlin, 1961.

ROSEN, E.: Some new concepts concerning ocular complications following vaccination. Ophthalmologica (Basel) 115 (1948) 321—332.

SCHÖNFELD, W.: Viruserkrankungen. In: Dermatologie für Augenärzte, Thieme, Stuttgart.

SCHRECK, E.: Veränderungen des Sehorgans bei Haut- und Geschlechtskrankheiten. In: Dermatologie und Venerologie, Bd. IV, hsg. von H. A. Gottron u. W. Schönfeld. Thieme, Stuttgart, 1960.

SYSI, R.: Molluscum corneae. Acta ophthalm. (Kbh.) 19 (1941) 25.

Mycotic Infections of the Skin

The general concept of the dermatomycoses was formulated by RUDOLF VIRCHOW and includes all those disorders of the skin and mucous membranes which are caused by fungi. While it is possible to divide the pathogenic fungi on a botanical basis into filamentous, budding, and ray fungi, it is considerably more difficult to determine the etiology in any one dermatologic picture on clinical aspect alone. Thus, at the present time, it is customary to use the less prejudicial term "tinea" instead of the historical botanic name "epidermophytosis" and thus speak about tinea corporis, tinea interdigitalis, tinea inguinalis, and so on. At present the Epidermophyton of Kaufmann-Wolf is generally considered a form of Trichophyton mentagrophytes, since, like Trichophyton rubrum it can also affect the hair. Of the previously very large group of epidermophytons only Epidermophyton floccosum has survived, since this, as implied in its name "epidermophyton," actually only affects the keratin of the epidermis and does not involve the hair. As mykids (WILLIAMS) one considers exanthems which, like tuberculids, are produced by toxins derived from fungi or by a hematogenous dissemination of fungal elements.

Fungous infections of the eye (mycoses in the area of the eyes) have only recently been described in detail in a monograph by HOFFMANN to which interested readers are referred for details. Parts of this problem have also been considered in the dermatologic-mycologic literature, particularly by POLEMANN. Although the question of infection by hyphomycetes was already considered on an experimental basis quite some time ago (for instance by W. JADASSOHN and REHSTEINER), increasing clinical experience has shown that fungal organisms, whose pathogenicity for other areas of the body is not particularly clear, may be important for the eye. For instance, species of cephalosporon may at times cause serious injury to the eye. The frequent use of corticosteroids and broad-spectrum antibiotics probably has increased the frequency of mycotic infections of the eye. It has been reported that dermatophytes may cause disease of the cornea. However, it is not certain whether fungi, like certain bacteria, normally reside on the edges of the eyelids, the conjunctiva, and the tear passages.

Favus

This reportable scarring disease of the hair and scalp (which characteristically is supposed to smell like mouse urine) is caused by ACHORION or Trichophyton schönleini. It slowly produces sulfur-yellow, curved scales which are called scutula (small shields) and are followed after healing by a typical scarring alopecia. Favus has been noted frequently on the eyelids, eyebrows, and eyelashes in association with favus of the scalp. Primary and solitary appearance of favus of the eyelid region has been reported on a number of occasions.

Microsporon infections

Microsporosis is a disease affecting children and feared because of its high degree of contagiousness, particularly in school children. Microsporon audouini causes sharply delineated areas of alopecia on the hairy scalp which heal spontaneously when the patient undergoes puberty, probably because of a change in the milieu of the scalp (change in the composition of fatty acids?). The disease begins with round, coin-shaped grayish white areas which appear as if they were dusted with ashes or flour. In the affected areas, the lusterless hairs are broken off to short stumps having a white sheath. Involvement of other areas can usually be found by examination with Wood's light, under which the hairs infected with microsporon fluoresce blue-green (MARGAROT and DÉVÈZE, 1925).

Microsporon infection of the eye area is very rare. GÖTZ as well as HOFFMANN were able to find in the more recent literature reports on a series of cases of infection of the eyebrows and eyelids (resembling blepharitis), sometimes with rather deep involvement.

Trichophytosis

In this dermatomycosis which has been known since the middle of the nineteenth century (GRUBY and MALMSTEIN, 1844 and 1845), there may develop either a superficial or a deep inflammatory clinical picture (Fig. 53). The deep type is particularly common in the adult where it involves the bearded area (Sycosis parasitaria) and in the child where it produces the honeycomb-like kerion celsi on the hairy scalp. Not uncommonly, single or multiple nodular perifollicular trichophytoses occur on the calves of young girls or women, who have minor vascular diseases of the legs. The superficial form of trichophytosis produces circular or maplike patches with peripheral exudation and central scaling. On the eyelid one finds the stumps of broken eyelashes or open or closed desquamating and crusting areas (GAILLETON, 1889). Such a mycotic blepharitis (trichophytosis ciliaris, sycosis parasitaria ciliaris) is rarely primary or solitary. It occurs almost always unilaterally, although it occasionally can be present bilaterally with or without involvement of the conjunctiva. Allergic conjunctivitides are possible in the presence of trichophyton lesions of other areas, apparently analog conjunctional mykyds. This type of lesion is rare, and MONTGOMERY and WALZER noted only one case involving the eyelid among 560 cases of deep trichophytosis. In 1968 Lejman and Bogdaszlivska-Czabonowska published 7 of their own observations.

Candidiasis

While dermatophytosis, tinea versicolor, erythrasma, and trichomycosis generally are not found in the area of the eye, diseases due to candida (of which at least 30 subspecies exist) are well known. The first observation of ophthalmic candidiasis is supposed to have been made by BILLARD in 1830 (quoted by HOFFMANN), which is truly remarkable if one considers that the organism causing thrush was discovered as late as 1839 by LANGENBECK and confirmed by BERG in 1841.

Organisms of the species Candida, particularly Candida albicans, cause a great variety of clinical manifestations. In addition to candidiasis of the lungs, yeast septicemia, specific meningitis and endocarditis, the skin and mucous membrane manifestations are at times erythematous, pustular, erosive, intertriginous; they may involve nails and rarely are granulomatous. Other manifestations of Candidiasis are perlèche, paronychia, colpitis, or balanitis (Figs. 54 and 55). On the mucosa the appearance of Candida infection is characterized by whitish or yellowish, easily removable, superficial plaques while on the skin the appearance is characterized by sharply demarcated, deeply red erythemas with a scale surrounding the edges or the appearance of satellite lesions.

In the area of the eye, the tear duct may be involved. The eyelids are usually secondarily involved with edema, redness, pustules, and eczematous or ulcerative lesions (VOZZA and BAGOLINI). A pseudomembranous candida conjunctivitis tarsi et bulbi is well known. Since 1950, concomitant with the marked increase in yeast infections, more reports have been published concerning candidiasis of the cornea. It is either superficially infiltrative and erosive or may manifest itself as a vascular keratitis, or it may appear in the form of an ulcus serpens with hypopyon. Metastaic inflammations intraocularly and toxic allergic reactions of the uvea have been thought to be caused by yeasts.

Blastomycosis

Originally, all diseases caused by budding yeast or yeastlike organisms were called blastomycosis. At the present time only the European blastomycosis of BUSSE-BUSCHKE (cryptococcosis, torulosis) and North and South American blastomycosis as well as the keloidal blastomycosis of JORGE-LOBO are included (for details and classification see KÄRCHER). In addition to involvement of internal organs such as the lungs and the central nervous system, the clinical appearance of cryptococcosis expresses itself by affecting the skin and mucosa in the form of papules and pustules, which are frequently centrally depressed and may resemble psoriasis or impetigo with marked leukocytic infiltration. Cryptoccocci have been found in ulcers of the eyelids and the tear ducts, and in chronic conjunctivitis characterized by stringy or foamy secretion. The organisms have also been found in ulcers of the cornea and occasionally have been isolated from intraocular infections. For details of purely ophthalmologic interest and clinical descriptions, the interested reader is referred to HOFFMANN.

Sporotrichosis

The knowledge of this subacute, chronic, progressive, deep mycosis is based primarily on the work of DE BEURMANN and that of GOUGEROT and SCHENK. The disease is inherently very variable and in its early stages reminiscent of the cutaneous picture of tuberculosis or syphilis. As it progresses, the worm – or glanderslike, lymphangitic involvement with individual destructive lesions becomes very characteristic. Gummalike forms with involvement of adjacent lymph nodes may occur on the eyelids and this heterogeneous picture has been given the name of "Parinaud's syndrome." Sporotrichosis of the conjunctiva

has occasionally been reported as a laboratory infection. In such cases the conjunctiva develops yellowish-white nodules, from which one may culture the organism. The clinical appearance may be dominated by the development of ulcers. Concomitant with involvement of the eyelids it is possible that sporotrichotic dacryocystitis, ostitis, and periostitis, as occur in actinomycosis, develops. In conjunction with systemic sporotrichosis, intraocular inflammatory changes such as iritis and retinitis may occur.

Actinomycosis

Actinomycosis was originally described by VON LANGENBECK (1839), and later by ISRAEL and BOLLINGER and PONFICK. It occurs most commonly in a cervicofacial form which results from involvement of the jaw or face with a hard infiltrate, which simulates scrofuloderma and which after some time characteristically develops multiple fistulae (lumpy jaw). The grainy consistency of the pus on account of the drusen is rather typical. The laboratory diagnosis of actinomycosis can be made by culturing the pus under microaerophilic conditions on Fortner medium in the presence of contaminants, which use up the remaining oxygen. Furthermore, the organism is usually associated with a comet-taillike mixture of bacterial organisms in which Actinobacterium actinomycetem (described by KLINGER in 1912) is dominant.

In the area of the eye the tear duct apparently offers the best conditions for the anaerobic development of this organism. Usually, actinomycosis of the eyelids is secondary to actinomycosis of the jaw. According to GROENOUW, SCHÖNFELD, and others, actinomycosis of the conjunctiva is characterized by a small blister or small yellowish nodules. Actinomycosis of the cornea may be superficial (dull, granular yellowish areas with shallow ulcers at the edge, keratitis super-

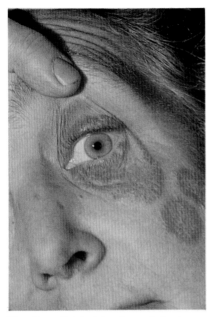

53 Superficial trichophyton infection.

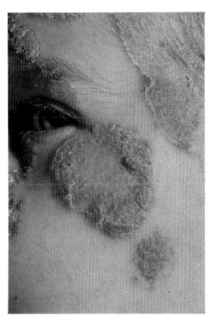

54 Granulomatous candidiasis.

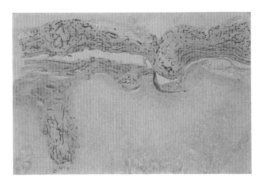

55 Demonstration of organisms in granulomatous candidiasis (McManus reaction).

ficialis punctata) or deep (yellowish disclike infiltrates). Involvement of the orbit has as its prevalent symptoms an inflammatory displacement of the bulb associated with infiltrates and the formation of fistulas and alterations of nerve function.

In contrast to the worldwide distribution of actinomycosis of the type israeii, one must particularly consider nocardioses in tropical and subtropical areas. The causative organism is gram positive, usually acid fast and aerobic. The clinical picture of nocardiosis tends to hematogenous dissemination, frequently produces fistulous tumors (CARTER: mycetoma) and generally resembles actinomycosis.

The commonly used ophthalmologic term "Streptotrix" (COHN, 1875) is identical with the term actinomycosis, in contrast to the use of the name "leptotrix," which denotes nonpathogenic bacteria of the family chlamydobacteriaceae (HOFFMANN). This must be noted here, since in the earliest ophthalmologic-mycologic literature on fungus diseases of the lower tear duct both these terms were used, particularly by VON GRAEFE (1854, 1855, 1869).

Dermatoses due to molds

As has already been stated, molds are rarely considered disease-producing organisms, particularly by dermatologists, and are usually considered to be facultative pathogens. More often such organisms are considered important allergy-producing agents. The cutaneous manifestations that have been described as due to mold infections are generally eczematous or verrucous granulomatous. They are thought to be produced by molds if organisms such as aspergillus can be cultured. When the lesions are gummatous or ulcerative they are thought to be related to cephalosporon (for review see KADEN). In contrast to the rarity of skin lesions produced by molds, ophthalmologists have reported convincing findings relating molds to mucormycosis of the cornea (superficial infiltrates, deep ulcerations, orbital syndromes secondary to thrombosis of blood vessels) or eye infections secondary to cephalosporium (daeryocystitis, or stone formation in the tear duct; for details see HOFFMANN and THÖNE).

References

FRIEDRICH, E.: Die Sproeßpilze des Menschen. Beiträge zur Hygiene und Epidemiologie, Heft 16. Barth, Leipzig, 1962.

GAILLETON zit. nach R. SABOURAUD: Les Teignes. Masson, Paris, 1910.

GÖTZ, H.: Pilzkrankheiten durch Dermatophyten. In: Hdb. der Haut- und Geschlechtskrankheiten, Bd. IV/3, hsg. von J. Jadassohn. Springer, Berlin, 1962.

HOFFMANN, D. H.: Mykosen im Bereich der Augen. Haut- u. Geschl.-Kr. 33 (1962) 434–435.

HOFFMANN, D. H.: Pilzinfektionen des Auges. Fortschr. Augenheilk. 16 (1965).

HOFFMANN, D. H.: Strahlenpilzerkrankungen am Auge. In: Krankheiten durch Aktinomyzeten und verwandte Erreger. Wechselwirkung zwischen pathogenen Pilzen und Wirtsorganismus. Vorträge d. 4. Wiss. Tagg. Deutschspr. Mykol. Ges. Freiburg/Brsg. 30. u. 31. 10. 1964, hsg. von H.-J. Heite. Springer, Berlin, 1967.

JADASSOHN, W., and REHSTEINER, K.: Experimentelle Untersuchungen über die Infektion des Auges mit Achorion Qinckeanum. Klin. Mbl. Augenheilk. 85 (1930) 280.

JADASSOHN, W., and REHSTEINER, K.: Experimentelle Hypho-

myceteninfektion am Auge. (Ein Beitrag zum Problem der Organotropie) Klin. Wschr. 10 (1931) 308–310.

KADEN, R.: Die Schimmelpilzdermatosen. In: Hdb. der Haut- und Geschlechtskrankheiten, Bd. IV/4, hsg. von J. Jadassohn. Springer, Berlin, 1963.

KÄRCHER, K. H.: Zur Begriffsbestimmung der Blastomykosen. In: Hdb. der Haut- und Geschlechtskrankheiten, Bd. IV/4, hsg. von J. Jadassohn. Springer, Berlin, 1963.

LEJMAN, K. J. and BOGDASZEWSKA-CZABONOWSKA: Trichophytia blepharociliaris et peripalbebralis. Hautarzt 19 (1968) 264–269.

MONTGOMERY, R. M., and WALZER, E. A.: Tinea capitis with infection of the eyelashes. Arch. Derm. Syph. (Chic.) 46 (1942) 40–43.

POLEMANN, G.: Klinik und Therapie der Pilzkrankheiten. In: Mykosen des Auges. Thieme, Stuttgart, 1961.

THÖNE, A. W.: Mycotic infection of the eye. In: Handbook of Tropical Dermatology, Vol. II, D. G. Ph. Simons, ed. Elsevier, Amsterdam, 1953.

VOZZA, R., and BAGOLINI, B.: Su un caso di grave ulcerazione bilaterale delle palpebre da Candida albicans. Boll. Oculist. 43 (1964) 433–439.

Zoonoses

Scabies

The extremely pruritic disease produced by Acarus scabiei (Sarcoptes hominis) has again become somewhat more common in our area after a decade-long pause following the Second World War (NÜRNBERGER). In these recent cases the classic appearance of the disease has been completely confirmed by new epidemiologic studies: the cul-de-sac-like, linear or angular darkish lines produced by the tracts of the mites appear in men on the glans and scrotum, and particularly on the anterior axillary folds, in women on the nipples, and in children particularly on the soles of the feet, the palms, and the buttocks. Involvement of the face and neck occurs for practical purposes only in the nursing infant who has contact with a scabietic mother. Mite tracts on the eyelids are extremely rare. Nevertheless, mites have been found in a corneal infiltrate (SAEMISCH, quoted by GROENOUW). Generally the tracts of the mites are scratched off and therefore usually covered with papules and vesicles or are secondarily infected. A mite of Norwegian scabies in the area of the eyelid is shown in Fig. 57.

A scabieslike appearance may be produced by animal mites (Dermanyssus avium), by vegetable mites such as Trombiculus autumnalis, or pyemotids (Pediculoides ventricosus, the wheat mite). These are usually characterized by erythematous, urticarial, or finely papular and vesicular exanthems, which involve the trunk, particularly on areas of contact with clothing.

Cutaneous reactions due to Ixodidae

The bite of Ixodes ricinus, a genus of ticks which can transmit spirochetes, rickettsiae, and viruses, is considered the eliciting cause by dermatologists of erythema migrans (Fig. 56), of lymphocytoma, and of acrodermatitis chronica atrophicans. This tick, which is usually swept from branches while walking in the woods, attaches itself to folded areas of the skin, such as the area behind the ear or the scrotum. This may explain the predilection of lymphocytoma for such local areas beyond the particularities of these tissues. In any case, the author noted a more frequent occurrence of erythema migrans on the face including the periorbital area in a tick-infested part of the country. HAUSER considers that 7 per cent of erythema migrans occurs on the face.

Pediculosis

The dark gray head louse, which is still very common and which is smaller by one posterior body segment than the body louse, most often infests the dense hair of women and children where it attaches its eggs (the nits) to the lower parts of the hair shaft. With increasing infestation by lice, pruritus and, secondarily, infected scratch marks develop, which reach their peak in the form of dense matting of the hair, the "plica polonica." Together with the accompanying lymphadenitis nuchalis there frequently exist an eczematous blepharoconjunctivitis and formation of phlyctenulae so that the disease may resemble tuberculosis. The parasite and eggs are visible with the slit lamp biomicroscope.

The body louse, responsible for pediculosis vestimentorum (which also is far from extinction), can be diagnosed easily by the "vagabond skin" it causes. This appearance may, however, also occur in patients with lymphogranulomatosis, pruritus senilis, or dermatitis herpetiformis. Because of this, it is necessary to demonstrate the parasite. It must be looked for particularly in the seams of clothing, since

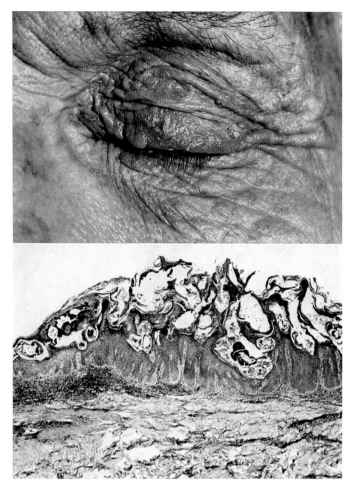

57 a) Norwegian scabies of the upper eyelid.

b) Histologic picture of the lesion. Groups of parasites in and under the horny layer.

the lice themselves are found on the skin only when they are feeding. Only pediculosis pubis (morpionosis, phthiriasis) is of ophthalmologic interest, since these brownish lice lie flat and without motion on the skin, are intertwined with hair and skin and occasionally may be found on the eyelashes and eyebrows. This probably is due to the fact that these parasites like areas which have apocrine glands (such as the meibomian glands of the eye area). Phthiriasis palpebrarum was already known in the Middle Ages (for details see KORTING), and has been reported up until the most recent time (literature in CASANOVAS) (Figs. 58 and 59).

Lepidopteriasis

As far as other epizoonoses (pulicosis, culicosis, cimicosis, and so on) are concerned, it will be mentioned only that with appropri-

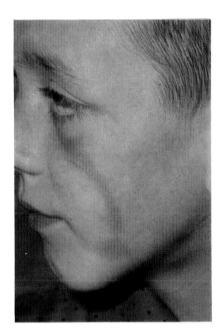

56   Erythema migrans.

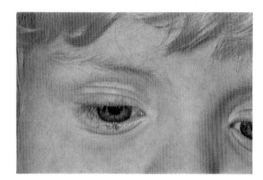

58   Lice of the eyelids.

59   Louse with nits.

ate attack by bedbugs (that is, the entrance of the cosmopolitan and light-phobic bedbugs into the conjunctiva) irritation and even ulceration of the cornea may occur (GROENOUW). More important are irritations secondary to caterpillars whose chitinous hairs may, after some latent period, produce very pruritic papulovesicular reactions which may be associated with general symptoms such as albuminuria, and in the area of the eyes may even produce ophthalmia nodosa; in other words, a disease that involves the entire bulb. As SCHRECK points out, the hairs which are frequently equipped with barbed hooks may penetrate into the depths and thus cause uveitis chronica granulomatosa with nodule formation (WEVE, DREYER).

Ophthalmomyiasis

Larva migrans, a disease well known to the dermatologist and known as "creeping eruption" because of its characteristic zigzag figures, is caused by the larva of various flies (particularly Gastrophilus equi) and occurs primarily on the feet of children who are playing barefoot in the sand. It has its ophthalmologic counterpart in ophthalmomyiasis. Here the eyelids, the conjunctiva, the tear ducts, and even the inner part of the eye may be reached by the larva of such flies, as has been noted particularly in children in the North Sea area. A marked catarrh of the conjunctiva, tumors of the eyelid and serous iritis follow. However, ophthalmomyiasis is most commonly found in the tropical areas of the world.

Vermiasis

Among the patients of the dermatologist, oxyuriasis causes primarily perianal itching, eczema, and vulvovaginitis, while usually ascaris or trichiuris causes urticaria or prurigo associated with sometimes very marked eosinophilia, and, somewhat more rarely, asthma or eosinophilic lung infiltrates. The ophthalmologist, in contrast, is accustomed to see mydriasis as the primary symptom of ascariasis, although blepharitides are known to occur together with alterations of accommodation, and inflammation of the tear ducts, in infestation with oxyuris. In trichinosis, the first symptom coming to the attention of the ophthalmologist is the flat swelling of the periorbicular region which reminds the dermatologist of dermatomyositis (HEPP: pseudotrichinosis). Rarer symptoms of trichinosis are conjunctivitis, proptosis, and paresis of the eye muscles.

In particular it is worth pointing out the ophthalmic involvement of filariasis, since these nematodes during their meanderings through the skin not only involve the skin of the eyelid or enter under the conjunctiva, but beyond this may cause exudative iridocyclitis, retinal bleeding, and emboli of the blood vessels of the eye. Such eye changes become more common the closer to the eye typical skin lesions are found. Particularly feared are Onchocerca volvulus and Onchocerca caecutiens. Onchocerca volvulus is transmitted in Africa by flies of the genus Simulium breeding in water. The skin lesions begin with itching, scleradenitis, and dryness of the skin. In the American type, lichenification of the face and extremities may appear, while in both the development of the nodular onchocercomas on the side of the thorax or above the hip is typical. Puncture with a thick needle usually demonstrates the filarias. Characteristic also is the marked eosinophilia, which may reach sixty percent of all white blood cells. In the area of the upper lid such nodules may appear similar to a sebaceous gland (PEARS and FASAL). Details about the ocular changes which often result in very high morbidity and blindness can be found in a review by SCHRECK.

References

CASANOVAS, J.: Dermatosis causadas por parasitos, hongos, bacerias y virus. In: Dermatooftalmologia, hsg. von J. Casanovas u. X. Vilonava. Alhacen, Barcelona, 1967.

DREYER, V.: Ophthalmia nodosa. Acta ophthal. (Kobh.) 31 (1953) 421—436.

GROENOUW A.: Beziehungen des Auges zu den Hautkrankheiten. In: Hdb. der Haut- und Geschlechtskrankheiten, Bd. XIV/1, hsg. von J. Jadassohn. Springer, Berlin, 1930.

HAUSER, W.: Wahrscheinliche Infektionskrankheiten der Haut. In: Hdb. der Haut- und Geschlechtskrankheiten, Bd. IV/1A, hsg. von J. Jadassohn, Springer, Berlin, 1965.

KORTING, G. W.: Phthiriasis palpebrarum — und ihre ersten historischen Erwähnungen. Hautarzt 18 (1967) 73—74.

NÜRNBERGER, F.: Skabies. Med. Welt (Stuttg.) 19 (1968) 575—576.

PEARS, F., and FASAL, P.: Onchocerciasis. In: Handbook of tropical dermatology, Vol. II, ed. D. G. Ph. Simons, Elsevier, Amsterdam, 1953.

SCHRECK, E.: Veränderungen des Sehorgans bei Haut- und Geschlechtskrankheiten. In: Dermatologie und Venerologie, Bd. IV, hsg. von H. A. Gottron u. W. Schönfeld. Thieme, Stuttgart, 1960.

STARGARDT K.: Phthiriasis der Lider und Follikularkatarrh. Z. Augenheilk. 38 (1918) 288.

WEVE, H.: Über eine durch die Nesselhaare der Goldafterraupe (Euproctis chrysorrhoea) erzeugte Augenentzündung. Arch. Augenheilk. 104 (1931) 192—221.

Congenital Abnormalities of the Skin

From the morphologic point of view, malformations are significant, and anomalies are insignificant deviations of the individual from the normal variation of his kind. They occur, if compatible with life, in about one-half percent of the general population. Recent major reviews on this subject have been prepared for the ophthalmologists by KLEIN and FRANCESCHETTI, and for the dermatologists by KORTING.

Ectodermal dysplasias of the skin (Anhidrosis hypotrichotica and others)

The major combinations of abnormalities which make up the concept of ectodermal dysplasia (WEECH) or ectodermal polydysplasia (TOURAINE) are hypohidrosis, hypotrichosis, and hypodontia (Fig. 60). In the minor types of this complex of anomalies, there is a combination of hypoplasia of hair and nails and there are no changes in the sweat glands. In such cases one speaks of "hidrotic" ectodermal dysplasia. At first glance the key anomalies of anhidrosis hypotrichotica are reminiscent of patients with congenital syphilis (olympic forehead, prominent glabella, satyr ears). Functionally these patients have a markedly reduced thermoregulatory capacity, which is of practical importance since it manifests itself by increased body temperature following more extensive physical exercise. The breasts are often malformed, and such patients have papular elevations on the cheeks or temples which represent rudimentary follicular openings which are hyperacanthotic (Fig. 61).

As far as the eyes are concerned, a mongoloid eyelid axis (PACHE), microophthalmia and nystagmus (VILLA and et al.), coloboma (FREEMAN, FLECK), strabismus (FREEMAN, COLE, GIFFEN and STROUD) and cataracts (VILLA et al., KIRMAN, COLE et al.) have been

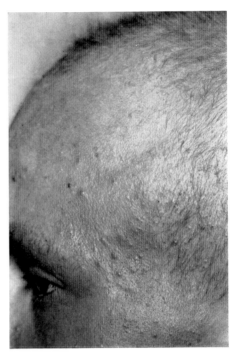

61   Ectodermal dysplasia. Rudimentary follicles appearing like papules on the temple.

noted in anhidrosis hypotrichotica. Many of the cases of cataracts are suspect of being a Rothmund's poikiloderma. HOFFMANN and SCHIRREN noted a thinning of the upper edge of the cornea with slight clouding of the superficial layers, a GROENOUW type of corneal dystrophy (KLINE, SIDBURY and RICHTER). PRETO found vacolated dystrophy of the epithelium of the cornea and reduced secretion of tears; FRIEDERICH and SEITZ also noted a marked decrease in secretion of tears. The punctae of the tear duct in their case were found to be flat, shallow dells which could not be probed, so that most probably a malformation of the tear duct existed. Furthermore in these patients a dystrophic epithelium of the cornea was found with a particular predilection for the development of lesions. Further details about the dystrophies of the cornea were described by FRANCESCHETTI and THIER as well as JUNG and VOGEL, who noted in their case a central macula of the cornea which extended deep into the parenchyma. Histologically, JUNG and VOGEL described in their case a circumscribed central narrowing of the epithelium and absence of Bowman's membrane as well as a circumscribed thickening and lack of nuclei of the corneal fibers. FUCHS reported on dysplasia of the iris.

In the previously described minor type, the hidrotic ectodermal dysplasia, there exist several variants in which the prime findings from the ophthalmologic point of view are hypertelorismus or amblyopia such as is found for instance in the Curtius syndrome (Fig. 62). Rarefaction of the lateral parts of the eyebrows and decrease in the number of eyelashes also are found more commonly in such cases.

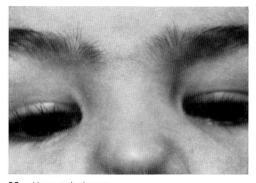

62   Hypertelorismus.

Among single case reports, the following are examples: KORTING and RUTHER described the combination of ichthyosis vulgaris and acrofacial syndactylic dysostosis which showed in addition to hypertelorismus a tapetoretinal degeneration. Of other combinations of abnormalities, in which the dermatologic primary symptom is hypotrichosis, the so-called dyscephaly with congenital cataract and hypotrichosis of ULLRICH and FREMERY-DOHNA is noteworthy. It includes deforma-

tion of the calvarium with separation of the sutures, hypoplasia of the lower jaw, bilateral cataract, and localized hypotrichosis. It was classified by ULLRICH as the fourth case of the bio-type of mandibulofacial dysostosis described by LUDWIG and KORTING which is similar to the Vogt-Koyanagi syndrome. The fifth complete case of this syndrome (which incidentally so far has not been described as hereditary) was reported by WEYERS in 1953. A multiple ectomesodermal degeneration characterized by sparse hair, eyelids, and eyebrows, localized areas of orange peel skin, microophthalmia, and cataract was described by TOLENTINO and BUCALOSSI.

Aplasia cutis

In aplasia cutis congenita circumscripta the lesion can usually be noted in the newborn and is in later life characterized only by a hairless scarred round defect located by predilection in the area of the posterior fontanella or to either side of the midline on the posterior skull (Fig. 63).

From the ophthalmologic point of view, the occasional association with coloboma is of importance. DUGOIS et al. report frequent coloboma formation in pseudobullous skin hypoplasia consisting of sacculated herniation of the tissue, pigment disturbances, telangiectasia, periorificial papulomatosis, hypotrichosis, and nail dystrophies. VOGEL and KIESSLING described white rings on the cornea in association with congenital hypoplasia of the skin of the scalp.

Congenital dysplasias of the skin

Pachyonychia congenita (JADASSOHN and LEWANDOWSKI, 1906). This consists of primarily congenital polykeratosis which is characterized by signs of pachyonychia (very hard compact nail plate), acneform follicular keratosis (over the elbows, knees, buttocks, or shoulder blades), keratoses on the palms and soles and finally leukokeratoses of the oral mucosa and hoarseness because of leukoplakia of the vocal cords. Associated signs in the area of the eye are epiphora and alopecia of the eyelashes (FONTAINE and WELLENS). Related variants were later reported as Siemens-Schäfer or Riehl syndrome. In the case of Schäfer congenital formation of cataracts was found; in other cases clouding of the cornea has been described. Furthermore, dystrophies of the cornea were present in the cases of BRÜNAUER and FUHS.

Dyskeratosis congenita (ZINSSER, 1906). In this type one finds poikilodermatous changes on the face, neck and chest, and the extremities, leukoplakia of the mucosa, which may include the rectal and urethral mucosa, and dystrophies of the nails. From the ophthalmologic point of view, leukokeratoses of the conjunctiva with obliteration of the tear ducts (CALMETTES, DEODATI and DARAUX, PASTINSZKY et al.), loss of eyelashes (GARB, COSTELLO) and hyperpigmentation with melanin of the fundus (APLAS) and finally blister formation of the conjunctiva (COSTELLO, COLE) have been reported.

The syndrome of Thomson and congenital dystrophy of Rothmund

The syndrome of congenital poikiloderma described by THOMSON in 1923 and 1936 is similar or even identical to the picture described by the ophthalmologist AUGUST ROTHMUND, JR. in 1868. Thomson described the syndrome that begins in early childhood, is prevalent in the female, and is characterized by poikiloderma and the absence of leukoplakia and cataracts. ROTHMUND reported a series of cases from the Walser valley under the title "On Cataracts in Combination With a Peculiar Skin Degeneration," in which the condition is closely related or identical to Thomson's syndrome, except that it shows

clouding of the lens and does not show bullae or crusting.

In Thomson's syndrome the eyebrows can occasionally be missing and trichiasis (GRUPPER and ZELLER) or hypertelorismus (KORTING) may be present. As previously noted, in contrast to the congenital poikiloderma of THOMSON, there develops in the Rothmund dystrophy in about half the cases (mostly simultaneously and bilaterally) a rapidly progressing cloudiness of the lens. It begins in children three to six years of age. Nothing certain is known about the early stages of this cataract formation. KLEIN and FRANCESCHETTI suspect a subcapsular cataract of the endocrine type. Individual reports describe the appearance of corneal degeneration (MAEDER) or atrophy of the iris (RUSSO).

Progeria

This very striking combination of dwarfism, small extremities, and atrophy of the skin and subcutaneous tissue, as it occurs in the aged, is relatively rare. Certain eye anomalies have been noted: exophthalmus (CLEMENT), microphthalmus (SCHONDEL) and coloboma and strabismus (CURTH).

Werner's syndrome. In spite of the existing sclerodactyly and the birdlike face which simulates scleroderma, Werner's syndrome probably belongs in the realm of progeria (progeria adultorum) (Fig. 64). The syndrome was described in 1904 by Werner under the title "On Cataract in Combination with Scleroderma" and since then over 120 cases have been reported. It is characterized by sclerosing atrophy of the skin, is familial, and occurs more often in males, almost exclusively beginning in the second decade of life. Important parts of the syndrome are: stunted growth, the birdlike face, presenile appearance, premature graying, torpid trophic ulcerations, and hypogonadism. (For details see KORTING and HOLZMANN, 1967.) As

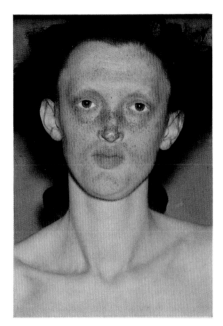

60 Ectodermal dysplasia.

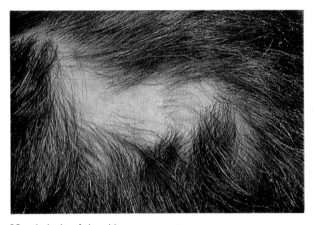

63 Aplasia of the skin.

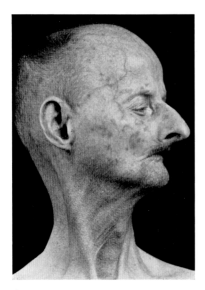

64 Werner's syndrome

Cutis laxa
(Meekeren-Ehlers-Danlos syndrome)

The clinical appearance, which in its complete form is quite rare, consists of grotesque stretchability of the skin, hyperextensibility of the joints, and abnormal response to trauma of the surface layers that is manifest after minor injury by subcutaneous hematomas or elliptically gaping wounds (Fig. 65). As far as pathogenesis is concerned, it is suspected that in most cases hereditary factors decrease the regular development of the collagen, so that inferior connective tissue bundles develop. As far as eye symptoms are concerned, the ease with which the upper lids can be turned over (BOSSU and LAMBRECHTS) is easily understandable as a special symptom of universal cutis laxa. The appearance of keratoconus or spontaneous dislocation of the lens (THOMAS, NEIMANN, CORDIER and ALGAN), and the observation of epicanthus (FREUND, ROSSI and ANGST, KINGLEWIS and POLUNIN, MIESCHER and STORCK, REYNAERS, BOSSU and LAMBRECHTS) denotes general degenerative phenomena and indicates the similarity of cutis laxa to Marfan's syndrome. The nosologic relationship of some cases of cutis laxa to osteogenesis imperfecta is demonstrated by the occasional observation of blue scleras (DURHAM, SUMMER, THOMAS, BOSSU and LAMBRECHTS, for details see GRIMALT and KORTING). Furthermore, COTTINI and SÉZARY and HOROWITZ noted strabismus, SCHACHTER noted nystagmus, FREUND noted anomalies of the blood vessels and COTTINI noted streaked changes of the fundus suggesting similarity to pseudoxanthoma elasticum. Furthermore, GELDMACHER noted hyperemia of eyegrounds and BOSSU and LAMBRECHTS noted an extensive retinitis proliferans with detachment, which probably developed because of repeated bleeding without tearing of the retina. Finally, BROBERGER, ERIKSSON, and WEDIN noted intraocular hematomas.

apparent from its original description, juvenile cataracts are a prime sign in Werner's syndrome, are almost always manifest by the third decade, and become bilateral within a few years of the onset. The skin lesions otherwise closely resemble a syndrome described by ARNDT and JAFFE as scleropoikiloderma. KNOTH, WOLFRAM, VALERA and GELLEI all have reported clouding of the posterior wall of the lens associated from time to time with punctate cloudiness and vacuoles in the anterior wall of the lens or subcapsular streaklike clouding. The histologic examination shows no decisive difference compared with the lenticular changes found in cataracts of other origin (WOLFRAM).

More rarely, keratopathies (PETROHELOS), changes in the conjunctiva (GERTLER; WOLFRAM; STOJANOV and BAZDEKOW), changes in the sclera (GERTLER), clouding of the vitreous (THIERS) or retinitis pigmentosa (VALERO and GELLEI) have been reported.

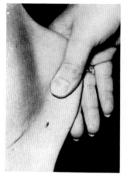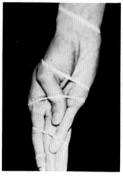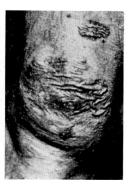

65 Ehlers-Danlos syndrome (Cutis laxa).

Malformations with particular relationship to the periorbital region

The more recent reviews of FRANCESCHETTI and KLEIN, and KLEIN and FRANCESCHETTI as well as that of RIEGER are instructive about the congenital anomalies of the human eye. SALOMON separately reviewed the hereditary diseases of the eyebrows and compiled the literature on hyperplasia superciliorum, synophrys, duplicatio superciliorum and hypoplasia of the eyebrows. It must be pointed out that congenital entropion is extremely rare. In 1935 ARKIN reported on such a bilateral anomaly associated with microphthalmus, epiblepharon, epicanthus, occlusion of the tear ducts, hypertelorismus, brachysyndactyly and other malformations in the presence of normal tarsi. In a case described by REDSLOB, the tarsi appeared thickened because of an increase in meibomian glands. DEVOE and HORWICH reported on a family with congenital anomalies which, in addition to congenital entropion of the upper lids, consisted also of tetrastichiasis, trichiasis, palpebral hyperpigmentation, and imbecility. CIVEDDA noted a piriform appendix under the right eye.

The combination of purpura simplex and ptosis in four generations was documented by FISHER, ZUCKERMAN, and DOUGLAS. SCHACHTER described congenital atresia of the lower lids with the appearance of a clown face and multiple malformation of the skeleton in a 7 year old mentally deficient child. A hereditary stigma of the outer eye, described by Ehrhardt, is the so-called "coverings crease" (Deckfalte) of the human eye that, depending on the amount of fat in the sulcus orbitopalpebralis, more or less hangs down over the upper eyelid. It changes in size with aging, a characteristic which is more marked on the right eye than the left, and more striking in men than in women. A peculiar triangular variant of the shape of the eyelid fissure that may occur with or without coloboma occurs in dysostosis mandibulofacialis (see p. 67), while obliteration of the tear duct punctum has already been described as characteristic of ectodermal dysplasia. As far as other inborn anomalies of the tear pathways are concerned, the reader is referred to the papers by BACSKULIN.

Familial dermochondrocorneal dystrophy

This complex malformation of the cornea was originally described by the ophthalmologist FRANÇOIS and later again observed by WIEDEMANN and by REMKY and ENGELBRECHT. As far as the skin is concerned, there are symmetrical xanthomalike nodules localized on the ears, elbows, fingers, and nose; there are alterations of ossification of the bones, dislocations and retractions of tendons, and in the eye there are superficial, sub-epithelial central opacities.

Keratoconus. In some cases of endogenous eczema, and occasionally in mongoloid conditions, keratoconus may occur as an associated anomaly of the cornea.

Hypertelorismus. Hypertelorismus requires special mention. This anomaly, originally described by FAGGE in 1870 in a case of ichthyosis vulgaris, so named by GREIG in 1924 and renamed by GÜNTHER in 1933 as euryopia, consists of marked broadening of the intraocular distance in which the eyelid fissure frequently resembles that found in mongolism. Hypertelorismus in its abortive form certainly is not rare and is frequently a partial symptom in a variety of complex malformation syndromes. For instance, in Crouzon's or in renofacial dysplasia (BRAUN and GROSS), it is produced by a marked tendency to growth of the anterior part of the original cranium (for details see STRACKER and GROSS). The importance of hypertelorismus for the dermatologist consists primarily in the retarded impression the patient makes, a characteristic which has to be considered in the differential diagnosis of syphilitic changes of the bones of the head. In syphilitic disease, the middle parts of the face are primarily affected (KRÜCKMANN) and generally produce a "flat face" due to insufficiently developed maxillary sinuses.

Skin and eye changes due to malformations of the skeleton

With particular reference to the periorbital region, one must consider first of all osteogenesis imperfecta, that is, the complete syndrome of blue sclerae or the Eddowe-symptom complex. In this syndrome the skin changes consist of anetodermas (BLEGVAD and HAXTHAUSEN; GRIMALT and KORTING) and the ophthalmologic changes of keratoconus, zonular cataract, hypoplasia of the iris, and glaucoma. In the syndrome described in 1896 by MARFAN as dolicho-stenomely and by ACHARD in 1902 as Arachnodactylia, the congenital malformation consists of changes in the heart and blood vessels (media necrosis of the aorta), changes in the function of the kidneys, a variety of facultative changes of the skin (hyper- and hypotrichosis, striae, verrucous nevi, and so on) and eye abnormalities of which ectopia lentis is the most important since it occurs in 70 to 80 per cent of all cases. The following eye changes have also been mentioned by KLEIN and FRANCESCHETTI: the small spherical lens, coloboma of the lens and cloudiness, miosis, hypoplasia of the iris, remnants of the pupil, and aniridia, myopia, paralysis of accommodation, ptosis, tapeto-retinal degeneration, syntropy with a syndrome of blue scleras, among others. Another systemic disease of the mesenchymal tissue is a hyperplastic variant of the Marchesani syndrome (spherophakia with brachydactyly) in which up to the present time special skin changes have apparently not been noted.

In Pfaundler's disease (dysostosis multiplex), the cause appears to be a mucopolysaccharide storage disorder associated with the granulation anomaly of Alder (see the case of KORTING and KORINTHENBERG). The characteristic manifestations of Pfaundler-Hurler's disease, besides innumerable changes in the skeleton, are disproportionate growth with kyphosis and misshapen, pawlike hands, a large plump skull with depressed roots of the nose, and large lips and tongue so that the general impression to the observer trained in internal medicine is acromegaloid or cretinoid or even suggests atypical chondrody-strophy. Because of their resemblance to Gothic waterspouts, DE RUDDER coined the term "gargoylism". The dermatologist when observing opacities of the cornea, will think less of changes secondary to chloroquin, the François syndrome, or a syphilitic parenchymatous keratitis, but will rather take into consideration the shape of the head and particularly the "electric plug" nose and the increased width between the eyes, and thus he is more likely to think of the olympic forehead of congenital syphilis or the appearance of anhidrotic dysplasia.

According to KLEIN and FRANCESCHETTI, about 75 per cent of people affected with this disease have clouding of the cornea. The cornea is affected alsoindepth by many, frequently closely spaced, grayish white spots so that visual acuity is reduced. Histologically, there is partial destruction of Bowman's membrane and deposition of spindle-shaped or spherical storage cells, while the stroma of the cornea shows basophilic granulations (WAGNER).

In the syndrome of diencephaloretinal degeneration (obesity, debility, retinitis pigmentosa, dysgenitalismus) described by LAURENCE, MOON, BIEDEL and BARDET, the dermatologic manifestations are a low hairline, decrease of hair in the genital area, and ichthyotic skin changes as well as telangiectasia.

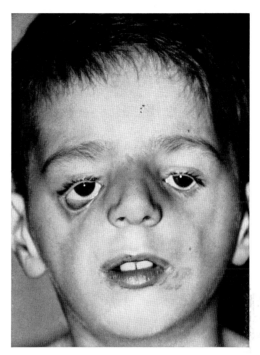

66 Mandibulofacial dysostosis.

Among the syndromes characterized by prognathy, the one called dysostosis mandibulofacialis deserves special mention (Fig. 66). This was originally known in the United States as Treacher-Collins syndrome and was described in detail by FRANCESCHETTI in 1949. It is characterized by a birdlike facies or a fish profile, antimongoloid position of the eyelid axis, macrostomia, and hypo- or aplasia of the zygoma. From the dermatologist's point of view, the frequent combination of the following is important: malformation of the external ear, sometimes associated with atresia of the external ear canal, pseudo-ear formation in the area of the cheek, formation of fistulas, as well as the development of lipodermoids, lipomas, irregular hair border and a skin texture resembling cutis laxa with hyperextensibility of the joints (KORTING, 1951).

In 1950 LUDWIG and KORTING noted a variation of this disorder in Macedonia. It was characterized by abnormalities of the skeleton, primarily by dysostoses of the mandible

and face, scalp changes consisting of multiple poliosis circumscripta and patchy scarring alopecia with sparse eyebrows and eyelashes (compare case of WEYERS, 1954). These variants were later named by ULLRICH and FREMERY-DOHNA "dyscephaly with congenital cataract and hypotrichosis." Further signs of this extremely rare syndrome were lentil- to coin-sized bluish-red spots located from the buttocks to the lower portion of the shoulder-blades; a case of HALLERMAN had a folded loose consistency similar to chalazodermia.

Since dysostosis mandibulofacialis of FRANCESCHETTI and ZWAHLEN may be combined with deformities of the upper extremities, it is noteworthy that NAGER and DE REYNIER proposed for this combination of syndromes the name "dysostosis acrofacialis." However, this term has been used by KORTING and REUTHER to describe a combination of hypertelorismus, poly- and syndactyly together with tapetoretinal degeneration and ichthyosis vulgaris, and it has been used by WEYERS (1953) for a combination of hexadactyly, separation of the lower jaw, and oligodontia. Finally, HANHART reported on the combination of micrognathia and congenital defects of the extremities (peromely) corresponding to the lethal "acroteriasis congenita" of cattle.

Other classifiable malformations of the first visceral arch were described by PETERS and HÖVELS as dysostosis maxillofacialis, which is characterized by bilateral hypoplasia of the zygoma, antimongoloid position of the lid axis, extreme hypoplasia of the maxilla, open bite, and shortening of the anterior part of the base of the skull. This abnormality probably develops prior to the fifth week of development in utero and appears to be familial.

BRAUN and GROSS describe two cases of bilateral or unilateral hypoplasia of the kidney associated with hypospadias or bicornuate uterus and extensive malformations of the internal organs which also showed alterations of the skull, hypertelorismus, microphthalmia, and congenital opacity of the cornea. They named this syndrome dysplasia renofacialis. Multiple malformations of the skeleton (among others shortening of the lower arm, deformed hands, absence of the thumb bones, deformity of the knees) in combination with hypospadias and dystopic shrunken kidney were noted by LEHMANN and LÖHR. The skin in these patients (who were not fully grown) showed decrease in the development of the subcutaneous fat tissue, minimal axillary and pubic hair formation, absence of eyebrows and eyelashes as well as brownish-gray, spotty pigmentation. Finally, the Hallermann-Streiff syndrome (also known as dysmorphia mandibulo-oculo-facialis) consists of a combination of dyscephaly, anomalies of teeth, hypotrichosis, nanismus, and primarily bilateral microphthalmus and congenital cataract.

In mongolism, three subtypes are recognized today, based on chromosome typing (the classic trisomy 21, translocation mongolism, and mosaic mongolism). The signs consist of epicanthus, hypertelorismus, tetradactyly crease, mongoloid eyelid axis, hypothelia, and enlarged big toes. KORTING and HOLZMANN found in serial examinations of mongoloid children a variety of skin changes. These consisted of ichthyotic skin alterations, acrocyanosis, or cutis marmorata, cutis laxa with hyperextensibility of the joints, occasionally synechiae and pterygia, particularly of the lateral neck area. Frequently, vasomotor persistent erythemas in a butterfly form develop, which give such children a doll-like appearance. Often noted are follicular keratoses of acneiform character, and the tendency to superficial infection or eczematization of the chafing type is striking. Very characteristic are grouped, garlandlike, pale bluish telangiectasias of the shoulder girdle area. The changes in appendages are dominated by the low neck hairline, and small, alopecia areatalike areas of thinning scalp hair. Scrotal tongue is often seen. Parallel to this is the development of a striking transverse fissure of the protruding upper lip in 10 per cent of patients with mucosal changes. Not all of these skin changes are always represented.

From the ophthalmologic point of view, half the patients have cloudiness of the lens, which may be congenital or develop after birth, without the presence of a specific mongoloid cataract (KLEIN and FRANCESCHETTI). SKELLER and OSTER found in 6 per cent of mongoloid patients keratoconus that may even occur acutely (HOFMANN).

Finally, attention should be called to the dermatologic and ophthalmologic signs of the so-called "familial dysautonomia" (Riley-Day syndrome), which was extensively reviewed in the German literature by TYNDEL and OSTER. According to available observations, this disease occurs almost exclusively in Jews and is characterized, as is acrodynia, by excessive sweating, excessive flow of saliva, abnormal reflexes and arterial hypertension. In addition to this, there are periodic attacks of vomiting, apnea, pollakisuria, neuroparalytic ulcers of the cornea and general readiness to infection. In particular there is an absence or decrease of tear secretion, which makes it necessary for this disease to be differentiated from Sjögren's syndrome. As far as skin changes are concerned, the sudden appearance of localized, sharply circumscribed, coin-sized erythemas on the face, neck, arms, and beltline following excitement or intake of favorite foods has been described by MINTZER and RUBIN.

Pterygium syndrome (Bonnevie-Ullrich-Turner syndrome)

The anomaly originally described in 1883 by KOBYLINSKI as pterygium or patagium is the primary sign of the Bonnevie-Ullrich-Turner syndrome. Other findings are: lymphangiectatic edema of the hands or feet or both, hypoplasia of the muscles and a valgus position of the elbows. As far as the eyes are concerned in this dysgenesis of the gonads, MULLER and BRUNSTING reported five cases with cataract, SCHRECK found cases in the literature with pupillary coloboma, increased appearance of cilioretinal arteries and strabismus, epicanthus and alterations of accommodation.

References

ACHARD, M. C.: Arachnodaktylie. Bull. Soc. méd. Hôp. (Paris) 19 (1902) 834.

AGATSON, S. A., and GARTNER, S.: Precocious cataracts and scleroderma (Rothmund's syndrome, Werner's syndrome) Arch. Ophthal. 21 (1939) 492.

APLAS, V.: Zur Kenntnis der Poikilodermie, Parapsoriasis und Atrophia cutis reticularis cum pigmentatione, dystrophia unguium et leukoplakia oris Zinsser-"Dyskeratosis congenita." Arch. klin. exp. Derm. 202 (1956) 224—237.

ARKIN, W.: Seltenes Zusammentreffen von Mißbildungen: Angeborenes Entropion der Oberlider; Quadrato-cephalosyndaktylie, Pes equino varus, Vitium corids. Klin. oczna 13 (1935) 331.

BACSKULIN, J.: Punctum lacrimale duplex, triplex und quadruplex. Klin. Mbl. Augenheilk. 144 (1964) 418—428.

BACSKULIN, J., and BACSKULIN E.: Klinik und Therapie der congenitalen Tränensackfisteln. Ophthal. Jap. 17 (1967) 1026—1038.

BLEGVAD, O., and HAXTHAUSEN, H.: Blaue Scleren und Tendenz zu Knochenbruch mit fleckförmiger Hautatrophie und zonulärem Katarakt. Hospitalstidende 64 (1921) 609. Ref. Zbl. Haut- u. Geschl.-Kr. 5 (1922) 150.

BOSSU, A., and GROSS, H.: Manifestations oculaires du syndrome de Ehlers-Danlos. Ann. Oculist. (Paris) 187 (1954) 227.

BRAUN, O., and GROSS, H.: Zur Kenntnis der eigenartigen mit Nierenfehlbildungen kombinierten Gesichtsveränderungen. Virchows Arch. path. Anat. 329 (1956) 433.

BROBERGER, O., ERIKSSON, G., and WEDIN I.: Contribution to the knowledge of the Ehlers-Danlos syndrome. Acta derm.-venereol. (Stockh.) 39 (1959) 196—206.

BRÜNAUER, S. R.: Zur Symptomatologie und Histologie der congenitalen Dyskeratosen. Derm. Z. 42 (1924) 6.

BRÜNAUER, S. R.: Dyskeratosis congenita palmaris et plantaris, Onychogryphosis cutanea, Leucokeratosis mucosae oris, Schmelzdefekte der Zähne, Dystrophia congenita corneae utr. Derm. Wschr. 80 (1925) 134.

CALMETTES, L., DÉODATI, F., and DARAUX H.: Angeborene Dyskeratosen und Atresie der Tränenpünktchen. Arch. Ophthal. (Paris) 3 (1957) 259.

CLÉMENT, R.: Sénilité précoce et nanisme — progeria de Gilford. Presse méd. 63 (1955) 155—157.

COLE, H. N., RAUSCHKOLB, J. E., and TOOMEY, J. A.: Dyskeratosis congenita with pigmentation, dystrophia unguis and leukokeratosis oris. Arch. Derm. Syph. (Chic.) 21 (1930) 71—95.

COLE, H. N., and TOOMEY, J. A.: Dyskeratosis congenita with pigmentation (dystrophia unguis and leukokeratosis oris. Arch. Derm. Syph. (Chic.) 25 (1932) 1159—1160.

COLE, H. N., DRIVER, J. R., GIFFEN, H. K., NORRIS, Cl. B., and STROUD, G.: Ectodermal and mesodermal dyplasia with osseous involvement. Arch. Derm. Syph. (Chic.) 41 (1941) 773—778.

COLE, H. N., RAUSCHKOLB, J. E., and TOOMEY, J. A.: Dyskeratosis congenita with pigmentation, dystrophia unguium and leukokeratosis oris. Arch. Derm. Syph. (Chic.) 71 (1955) 451—456.

COLE, H. N., COLE JR., H. N., and LASCHEID, W. P.: Dyskeratosis congenita. Arch. Derm. Syph. (Chic.) 76 (1957) 712—719.

COSTELLO, M. J.: Dyskeratosis congenita with superimposed prickle-cell epithelioma on dorsal aspect of the left hand. Arch. Derm. Syph. (Chic.) 75 (1957) 451.

COSTELLO, M. J., and BUNCKE, C. M.: Dyskeratosis congenita. Arch. Derm. Syph. (Chic.) 73 (1956) 123—132.

COTTINI, G. B.: Sindrome di Ehlers-Danlos clinicamente frusta e istologicamente completa. Gazz. Osp. Clin. (1939).

COTTINI, G. B.: Contributo alla studio distrofie sistemiche del Tessute elastico. G. ital. Derm. 89 (1948) 604—621. Ref. Zbl. Haut- u. Geschl.-Kr. 77 (1951/52) 128.

CROUZON, O.: Dysostose cranio-faciale héréditaire. Bull. Soc. méd. Hôp. (Paris) 33 (1912) 545.

CURTH H. O.: Progeria with erythema on hands and feet, parietal alopecia, congenital coloboma and osteoporosis. Ref. Zbl. ges. Ophthal. 53 (1950) 73.

DANLOS, H. A.: Un cas de cutis laxa avec tumeurs par contusion chronique des coudes et des genoux. Bull. Soc. franç. Derm. Syph. 19 (1908) 70—72.

DUGOIS, P., COUDERC, P., AMBLARD, P., and GALLIEN M.: Une polydysplasie peu connue avec hypoplasie dermique en aires. Presse méd. 76 (1968) 1189—1190.

DURHAM, D. G.: Cutis hyperelastica (Ehlers-Danlos-syndrome) with blue scleras, microcornea and glaucoma. Arch. Ophthalm. (Chic.) 49 (1953) 220.

EHLERS, E.: Cutis laxa, Neigung zu Hämorrhagien in der Haut. Lockerung mehrerer Articulationen. Derm. Z. 8 (1910) 173.

EHRHARDT, S.: Über die Deckfalte am menschlichen Auge. Z. Morph. Anthrop. 43 (1951) 163.

FAGGE, C. H.: Remarks on certain cutaneous affections. Guy's Hosp Rep. 15 (1870) 316.

FISHER, B., ZUCKERMAN, G. H., and DOUGLAS, R. C.: Combined inheritance of purpura simplex and ptosis in four generations of one family. Blood 9 (1954) 1199.

FLECK, F.: Klinische Beobachtungen einer ungewöhnlichen sporadischen Form von ektodermal-mesodermaler Keimblattdysplasie. Derm. Wschr. 132 (1955) 994–1007.

FONTAINE, A., and WELLENS, W.: Syndrome de Jadassohn-Lewandowsky avec polydactylie et alopecie des sourcils. Arch. belges Derm. 14 (1958) 178. Ref. Zbl. Haut- u. Geschl.-Kr. 102 (1958) 123.

FRANCESCHETTI, A., and MAEDER, G.: Cataracte et affections cutanées du type poikilodermie (syndrome de Rothmund) et du type sclérodermie (syndrome de Werner). Schweiz. med. Wschr. 79 (1949) 657. Ophthalmologica (Basel) 117 (1949) 196.

FRANCESCHETTI, A., and KLEIN, O.: Vererbung und Auge. In: Lehrbuch der Augenheilkunde, edited by M. Amsler, A. Brückner, A. Franceschetti et al. Karger, Basel, 1954.

FRANCESCHETTI, A., THIER, C. J.: Über Hornhautdystrophien bei Genodermatosen unter besonderer Berücksichtigung der Palmoplantar-Keratosen v. Graefes Arch. Ophthal. 162 (1961) 610.

FRANÇOIS, J.: Dystrophie dermo-chondro-cornéenne familiale. Ann. Oculist. (Paris) 182 (1949) 409.

FREEMAN JR., C. D.: Congenital ectodermal and mesodermal dysplasia. Arch. Derm. Syph. (Chic.) 71 (1955) 667.

FREUND, J.: Zur Phänogenese des Ehlers-Danlos-Syndroms. Z. Kinderheilk. 77 (1956) 611.

FRIEDERICH, H. C. SEITZ, R.: Über eine Form der ektodermalen Dysplasie unter dem Bilde der Pili torti mit Augenbeteiligung und Störung der Schweiß-Sekretion. Derm. Wschr. 131 (1955) 277–283.

FUCHS: Oligohidrosis hypotrichotica bei ektodermaler Dysplasie. Derm. Wschr. 141 (1960) 448.

FUHS, H.: Über das seltene Syndrom von congenitalen Keratosen an Haut und Cornea. Derm. Z. 53 (1928) 199.

FÜLLING, G., PÜNDER, H.: Augenhintergrundsveränderungen beim Status Bonnevie-Ullrich. Klin. Mbl. Augenheilk. 128 (1956) 724–727.

GARB, J.: Dyskeratosis congenita with pigmentation, dystrophia unguium and leukoplakia oris. Arch. Derm. Syph. (Chic.) 55 (1947) 242–250.

GELDMACHER, M.: Ein Fall von Dystrophia adipo-genitalis mit Cutis laxa und hochgradiger Vulnerabilität der Haut. Inaug.-Diss. Bonn, 1921.

GERTLER, H.: Karzinombildung beim Werner-Syndrom. Derm. Wschr. 150 (1964) 606–616.

GERTLER, W.: Poikiloderma congenitale (Thomson). Derm. Wschr. 130 (1954) 1013–1015.

GILFORD, H.: Progeria: a form of senilism. Practitioner 73 (1904) 198.

GRIMALT, Fr., KORTING G. W.: Anetodermie und Osteopsathyrose (Syndrom von Blegvad-Haxthausen) Z. Haut- u. Geschl.-Kr. 22 (1957) 361–365.

GROENOUW, A.: Beziehungen des Auges zu den Hautkrankheiten. In: Hdb. der Haut- und Geschlechtskrankheiten, Vol. XIV/1, edited by J. Jadassohn. Springer, Berlin, 1930.

GROSS: Der Hypertelorismus. Ophthalmologica (Basel) 131 (1956) 137–156.

GRUPPER, CH., and ZELLER, M.: Poikilodermie cervico-faciale. Bull. Soc. franç. Derm. Syph. 65 (1958) 60–62.

HALLERMANN, W.: Vogelgesicht und Cataracta. Klin. Mbl. Augenheilk. 113 (1948) 115.

HANHART, E.: Über die Kombination von Peromelie mit Mikrognathie, ein neues Syndrom beim Menschen, entsprechend der Akroteriasis congenita von Wriedt und Mohr beim Rinde. Arch. Klaus-Stift Vererb.-Forsch. 25 1950).

HOFMANN, H.: Akuter Keratokonus bei mongoloider Idiotie. Klin. Mbl. Augenheilk. 129 (1956) 756–762.

HOFFMANN, D. H., and SCHIRREN, C.: Über Hornhautveränderungen bei der ektodermalen Dysplasie. Klin. Mbl. Augenheilk. 134 (1959) 413.

JUNG, E. G., and VOGEL, M.: Anhidrotische Ektodermaldysplasie mit Hornhautdystrophie. Schweiz. med. Wschr. 96 (1966) 1477–1483.

KING-LEWIS, F. L., and POLUNIN, J. V.: Two cases of Ehlers-Danlos-syndrome. Arch. Dis. Child. 22 (1947) 170.

KIRMAN, B. H.: Idiocy and ectodermal dysplasia. Brit. J. Derm. 67 (1955) 303–307.

KLEIN, D., and FRANCESCHETTI, A.: Mißbildung und Krankheiten des Auges. In: P. E. Becker: Humangenetik, Vol. VI. Thieme, Stuttgart 1964.

KLINE, A. M., SIDBURY JR., J. B., and RICHTER, C. P.: The occurrence of ectodermal dysplasie and corneal dysplasia in one family. J. Pediat. 55 (1959) 355.

KNOTH, W., BAETHKE, and HOFFMANN, L., Über das Werner-Syndrom. Hautarzt 14 (1963) 145–152, 193–202.

KORTING, G. W.: Poikiloderma congenitum s. infantum Thomson. Derm. Wschr. 134 (1956) 1113.

KORTING, G. W.: Fehlbildungen der Haut und Hautveränderungen bei Fehlbildungssyndromen. In: Hdb. der Haut- und Geschlechtskrankheiten, Vol. III/1, edited by J. Jadassohn. Springer, Berlin 1963.

KORTING, G. W., and GOTTRON, E.: Cutis laxa. Arch. Derm. Syph. (Berl.) 193 (1951/52) 14–33.

KORTING, G. W., and KORINTHENBERG, I.: Hirsutismus als Teilsymptom des abortiven oder tardiven Morbus Pfaundler-Hurler. Z. Haut- u. Geschl.Kr. 37 (1964) 65–70.

KORTING, G. W., and HOLZMANN, H.: Hautveränderungen bei mongoloider Abartung. Med. Welt (Stuttg.) 17 (1966) 2801.

KORTING, G. W., and HOLZMANN, H.: Die Sklerodermie und ihr nahestehende Bindegewebsprobleme. Thieme, Stuttgart 1967

KORTING, G. W., and RUTHER, H.: Ichthyosis vulgaris und akrofaciale Dysostose. Arch. Derm. Syph. (Berl.) 197 (1953/54) 91–104.

LEHMANN, W., und LÖHR K.: Über eine seltene Mehrfachmißbildung der Gliedmaßen und des Urogenitalsystems. Z. mensch. Vererb.- u. Konstitut.-Lehre 33 (1955) 119.

LUDWIG, A., and KORTING, G. W.: Vogt-Koyanagi-ähnliches Syndrom und mandibulofaciale Dysostosis (Franceschetti-Zwahlen) Arch. Derm. Syph. (Berl.) 190 (1950) 307–316.

MIESCHER, G., and STORCK, H.: Morbus Ehlers-Danlos. Dermatologica (Basel) 102 (1951) 381.

MINTZER, I. J., and RUBIN, Z.: Dermatological manifestations of familial autonomic dysfunction (Riley-Day-syndrome). Arch. Derm. Syph. (Chic.) 67 (1953) 561–565.

MULLER, S., and BRUNSTING, L. A.: Cataracts associated with dermatologic disorders. Arch. Derm. Syph. (Chic.) 88 (1963) 330–339.

NAGER, F. R., and DE REYNIER, J. P.: Das Gehörorgan bei der angeborenen Kopfmißbildung. Karger, Basel 1948, zit. n. Hanhart.

OSTER, H.: Die familiäre Dysautonomie. Dtsch. med. Wschr. 82 (1957) 2038–2040.

PACHE, H. D.: Über den angeborenen Schweißdrüsenmangel (Anhidrosis hypotrichotica mit Hypodontia [Siemens]). Münch. med. Wschr. 88 (1941) 1135–1138.

PASTINSKY, I., VÁNKOS, J., and RÁCZ, I.: Ein Beitrag zur Pathologie der "Dyskeratosis congenita" Cole-Rauschkolb-Toomey. Derm. Wschr. 135 (1957) 587–593.

PETERS, A., and HÖVELS, O.: Die Dysostosis maxillo-facialis, eine erbliche typische Fehlbildung des I. Visceralbogens. Z. menschl. Vererb.- u. Konstitut.-Lehre 35 (1960) 434.

PETROHELOS, M. A.: Werner's syndrome. A survey of the cases with a report of the second autopsied. case. Ann. intern. Med. 48 (1958) 1205.

PRETO, J.: Clin. pediat. (Bologna) 30 (1948) 436, zit. n. J. A. Velasco, and A. Prader: Ektodermale Dysplasie vom anhidrotischen Typus. Helv. paediat. Acta 11 (1956) 604.

REDSLOB, E.: Entropion palpébral par malformation des glandes de Meibom. Ann. Oculist. (Paris) 180 (1947) 263.

REMKY, H., and ENGELBRECHT, G.: Dystrophia dermo-chondrocornealis (François). Klin. Mbl. Augenhk. 151 (1967) 319.

REYNAERS, H.: Cutis laxa hyperelastica. Arch. belges Derm. 9 (1953) 29. Ref. Zbl. Haut- u. Geschl.-Kr. 86 (1953/54) 313.

RIEGER, H.: Erbpathologie des Auges. In: Der Augenarzt, Vol. III, edited by K. Velhagen. Leipzig 1960.

ROSSI, E., and ANGST, H.: Das Ehlers-Danlos-Syndrom. Helv. paediat. Acta 6 (1951) 245.

RUSSO, A.: Sindrome genito-sclerodermica e cataracta (morbo di Rothmund). Ann. Ottalm. 62 (1934) 646. Ref. Zbl. Haut- u. Geschl.-Kr. 50 (1935) 124.

SALAMON, T.: Vererbung von Haar- und Nagelkrankheiten. In: Hdb. der Haut- und Geschlechtskrankheiten, Vol. VII, edited by J. Jadassohn. Springer, Berlin 1966.

SCHACHTER, M.: Atrésie congénitale des paupières inférieures, facies de "clown" et malformations squelettiques multiples chez un petit oligophrène, malformations identiques chez le père. Ann. paediat. (Basel) 169 (1947) 345.

SCHÄFER, E.: Zur Lehre von den congenitalen Dyskeratosen. Arch. Derm. Syph. (Berl.) 148 (1925) 425—432.

SCHANDEL, A.: Two cases of progeria complicated by microphthalmus. Acta paediat. (Uppsala) 30 (1942/43) 286.

SCHRECK, E.: Veränderungen des Sehorgans bei Haut- und Geschlechtskrankheiten. In: Dermatologie und Venerologie, Vol. IV, edited by H. A. Gottron and W. Schönfeld. Thieme, Stuttgart, 1960.

SÉZARY, A., and HOROWITZ, A.: Syndrome d'Ehlers-Danlos. Bull. Soc. franç. Derm. Syph. 42 (1935) 1744—1747.

SIEMENS, H. W.: Über Keratosis follicularis. Arch. Derm. Syph. (Berl.) 139 (1922) 62—72.

SKELLER, E., and OSTER, J.: Keratokonus bei Mongolismus. Acta ophthal. (Kobh.) 29 (1951) 149—161.

STOJANOV, P. K., and BAŽDEKOV, B. Případ Wernerova syndromu. Čs. Derm. 35 (1960) 191.

STRACKER, O.: Hypertelorismus. Wien med. Wschr. 101 (1951) 469—472.

STREIFF, E. B.: Dysmorphie mandibulofaciale (tête d'oiseau) et altérations oculaires. Ophthalmologica (Basel) 120 (1950) 79.

SUMMER, K. G.: The Ehlers-Danlos syndrome. Amer. J. Dis. Child. 91 (1956) 419.

THIERS, H., COLOMB, D., CUFFIA, CH., DESCOS, L., and PICOT, C.: Syndrome de Werner. Bull. Soc. franç. Derm. Syph. 71 (1964) 616—617.

THOMAS, CH., NEIMANN, N., CORDIER, J., and ALGAN, B.: Les manifestations oculaires de la maladie d'Ehlers-Danlos. Bull. Soc. franç. Ophthal. (1953) 214. Ref. Zbl. Haut-Geschl.-Kr. 88 (1954) 70.

TOLENTINO, P., and BUCALOSSI A.: Su di una rara sindrome degenerativa multipla ecto-mesodermica. Folia hered. path. (Milano) 2 (1952) 62. Ref. Zbl. Haut- u. Geschl.-Kr. 84 (1953) 385.

TOURAINE, A.: Anidrose avec hypotrichose et anodontie. Bull. Soc. franç. Derm. Syph. 42 (1935) 1529—1539.

TYNDEL, M.: Ein Fall von Riley-Day'schem Syndrom. Wien. med. Wschr. 105 (1955) 189—190.

ULLRICH, O., and FREMERY-DOHNA: Dyscephalie mit Cataracta cogenita und Hypotrochose als typischer Merkmalskomplex. Ophthalmologica (Basel) 125 (1953) 73.

VALERO, A., and GELLEI, B.: Retinitis pigmentosa, hypertension, and uraemia in Werner's syndrome. Brit. med. J. 1960/II, 351.

VILLA, M., STRINGA, S. G., and RAIMONDI, E.: Dysplasia ectodermica hereditaria. Arch. argent. Derm. 4 (1954) 53. Ref. Zbl. Haut- u. Geschl.-Kr. 91 (1955) 317.

VOGEL, M., and KIESSLING, W.: Weiße Hornhautringe in Verbindung mit Aplasia cutis congenita circumscripta des behaarten Kopfes. Ber. dtsch. ophthal. Ges. 66 (1964) 361.

WAGNER, F.: Beitrag zur Frage der Dysostosis multiplex (Pfaundler-Hurler) mit Fehlbildung der Bowman'schen Membran. Z. Kinderheilk. 69 (1951) 179.

WEECH, A. A.: Hereditary ectodermal dysplasia (congenital ectodermal defect). Amer. J. Dis. Child. 37 (1929) 37.

WEYERS, H.: Klinik und Pathologie der Dysostosis mandibulofacialis. Z. Kinderheilk. 69 (1951) 207.

WIEDEMANN, H. R.: Zur Françoisschen Krankheit. Ärztl. Wschr. 13 (1958) 905.

WOLFRAM, G., PRIEGNITZ, F., and WAGNER, H.-J.: Zum Werner-Syndrom. Dtsch. med. Wschr. 84 (1959) 2125—2126.

ZUNIN, C., and MARIOTTI, L.: Le anomalie oculari nello status Bonnevie-Ullrich. Ann. Ottal. 79 (1953) 359—376. Ref. Zbl. ges. Ophthal. 62 (1954) 57.

Keratoses and Dystrophies

Diffuse keratoses (Ichthyoses)

These widespread abnormalities of keratinization of the skin represent either retention hyperkeratosis (ichthyosis vulgaris) or a proliferation hyperkeratosis (ichthyosis congenita). In recent years ichthyosis hystrix, which in the past had been considered a part of ichthyosis vulgaris, has been thought to represent a separate type. The same is true for its variants erythroderma congenitalis ichthyosiformis bullosa, ichthyosis hystrix gravior, and maleformatio ectodermalis generalisata Bäfverstedt.

In ichthyosis vulgaris (simplex, nitida, serpentina, nigricans), 80 per cent of the cases are inherited as an autosomal dominant. The original manifestations begin in the first year of life and the disease progresses until puberty, sparing the flexors. Furthermore, significant improvement occurs in the summer in this diffuse hyperkeratosis. In contrast, in ichthyosis congenita (gravis seu fetalis, mitis and tarda, or érythrodermie ichthyosisforme congénitale) the more marked manifestations occur on the flexor surfaces. Ichthyosis congenita is considered to be inherited as an autosomal recessive.

As far as eye changes in the ichthyoses are concerned, participation of the eyelids is a major finding. SCHRECK in his review reports shrinkage of the eyelids (see also the case of WEBER from the Mainz Clinic) followed by scarring, ectropion, conjunctivitis, symblepharon, and keratitis. He also reports his own observations of punctate white-blue opacities (see also HERMANS; FRANCESCHETTI and SCHLÄPPI) immediately anterior to Descemet's membrane, and other opacities of the cornea. Cases of syndermatotic cataracts have been described in ichthyosis (SIEMENS; JANCKE; PINKERTON et al.). However, this combination is supposed to be accidental

(JAY, BLACH and WELLS). GRAUPER considered that a certain form of corneal change parallels the skin manifestations as "ichthyosis corneae." Furthermore, JUNG and VOGEL reported on the combination of ichthyosis vulgaris with anhidrosis or anhidrotic ectodermal dysplasia in the sense of a retention hyperkeratosis, flat palmar plantar keratoses, and corneal dystrophy.

The combination of "ichthyosis" with hydrophthalmus as well as microcephalus, alteration of the face, deaf and dumbness and abnormalities of the extremities was reported by WESTERLUND. A coincidence of ichthyosis vulgaris with phlebitis retinalis was reported by BÖCK and NIEBAUER. Furthermore, in a case of congenital ichthyosis which was combined with "allergic" (that is, urticarial) manifestations, SAVIN reported changes in the cornea in which nodular thickening in the parenchyma could be seen with a slit lamp. AMALREC, BESSOU, and FARENC noted worsening of the skin and corneal changes in the fall and improvement in the spring in cases of Savin syndrome.

Of those ichthyosis syndromes which can be considered to be entities, Rud syndrome (1927) includes ichthyosis congenita, epilepsy, oligophrenia, hypogenitalismus, partial gigantism, and polyneuritis. In such cases STEWART sometimes noted an association with retinitis pigmentosa.

In the syndrome described by SJÖGREN and LARSSON on the basis of 28 cases, the underlying symptoms consisted of congenital ichthyosis with spasticity and oligophrenia and in 3 of the 28 cases there was degeneration of the macula (see also the case of ROHMER and TEMPÉ).

The Refsum syndrome (described in 1947) consists of hereditary ataxia of cerebellar character beginning in childhood or in adult life and associated with polyneuritides, atrophy of the muscles, and inner ear deafness. In this lipidosis, in which 3, 7, 11, 15-tetramethylhexadecanoic acid is increased, the protein of the liquor of the anterior chamber is also increased without any accumulation of cells. The skin shows ichthyosis vulgaris, while ophthalmologically the symptoms of night blindness or retinitis pigmentosa, as well as anomalies of the pupil, are present.

Palmar and plantar keratoses

The idiopathic palmar plantar keratoses (keratomas, keratodermias) comprise approximately 30 presently known types, many of which are associated with other abnormalities. These are divided into three basic varieties: *Keratosis palmoplantaris diffusa* (THOST 1880, UNNA 1883). This "classic example of regular autosomal dominance" (SCHNYDER) begins as a rule in the first few weeks of life, becomes symmetrically fully developed only with increasing mechanical stress, that is, at the time when walking first begins. The syndrome consists of erythema at the edge of the involved areas, hyperhidrosis and sharply circumscribed flat yellowish-white hyperkeratoses which are crossed by deep painful fissures. Of the special forms only the progressive type of mal de meleda (STULLI, 1826) will be described. This begins in the first few weeks of life, extends beyond the edges of the palms and soles and is not uniformly inherited, since it is either recessive or dominant (GREITHER). Another type is the markedly hyperhidrotic palmoplantar keratosis with periodentopathy described by PAPILLON and LEFÈVRE (1924), which gradually leads to total loss of all teeth.

Hereditary transmission probably also occurs in keratosis palmoplantaris mutilans (PARDO-COSTELLO and MESTRE; VOHWINKEL, 1929), in which spontaneous loss of fingers

and toes develops because of the development of tight bands (Ainhum or dactylolysis spontanea, MESSUM 1821).

Keratosis palmoplantaris insuliformis seu striata. This rare variant described primarily by SIEMENS is associated with very little hyperhidrosis and is manifest only in childhood, not in the newborn. It is frequently associated with other symptoms ("keratosis multiformis").

Keratosis palmoplantaris papulosa. (BUSCHKE and FISCHER, 1906; BRAUER, 1913.) In this apparently regularly autosomal dominant inherited palmar plantar keratosis the initial signs are pinhead- to small pea-sized horny pearls which leave behind typically delled depressions. The manifestation of these keratomas occurs primarily in the second and third decades and only rarely in childhood.

Some of these congenital keratoses are associated with changes in the eye. Not all have stood the test of time (the famous case of MATSUOKA "A Case of Keratosis Corneae with Keratodermia" and criticism by FRANCESCHETTI and SCHNYDER). The case of SCHÄFER (1925) is in retrospect not clear. On the other hand, a combination of the linear palmar and diffuse plantar keratosis with dystrophy of the cornea reported by SPANLANG in 1927 is still carried under the name of Spanlang-Tappeiner syndrome. Furthermore, RICHNER found "herpetoid" changes of the cornea in localized palmar plantar keratoses. Of special note (because of such an association) is the case of HANHART (1947), which was re-examined by FRANCESCHETTI and SCHNYDER. Finally, GRAYSON described a case of keratoma hereditarium dissipatum palmare et plantare of the type Brauer with small grayish intraepithelial alterations of the cornea without inflammatory reaction or vascularization. Histologically this consisted of a circumscribed thickening of the basal membrane with pseudopodialike extensions and, subjectively, photophobia, and epiphora.

Follicular keratoses

Keratosis follicularis lichenoides seu lichen pilaris (WILSON, 1876; T. FOX, 1879). These are presumably autosomal dominant hyperkeratoses of the openings of the follicles. They primarily involve multiple areas of the extensor surfaces of the extremities, are generally more common in the male, and mainly occur before puberty and decrease around the climacteric. There is a peculiar graterlike appearance of the lesions. The individual keratotic plaques are either whitish or reddish and frequently acquire an atrophic character, particularly in the face (kératose pilaire rouge atrophiante de la face).

Another variety of keratosis pilaris rubra which develops primarily in blondes in childhood occurs in the area of the eyebrows as ulerythema ophryogenes (TAENZER, 1889). It probably represents an irregularly dominant genodermatosis (Fig. 67). The clinical appearance is dominated by a pale erythema in bands with follicular keratoses, that develops with time into follicular atrophy. The claimed relationship (MERTENS) to "atopy," that is, to endogenous eczema, cannot be substantiated by us.

Such idiopathic follicular keratoses are mimicked primarily by deficiency of vitamin A (NICHOLLS: phrynoderma) and by deficiency in vitamin C. However, in the presence of avitaminoses hemorrhagic changes appear in the grater- or gooseflesh-like keratoses after some time (Jessner: "lichen scorbuticus").

Keratosis follicularis spinulosa (UNNA, SALINIER). The discontinuous grouped spinulous keratoses are primarily deuteropathic reactions (lichen scrofulosorum, trichophyticus, among others), and in adults there is frequently a "spinolusism" such as occurs in mycosis fungoides or follicular mucinosis. The most important spinulous keratotic genodermatosis that generally appears in the first few weeks of life is "keratosis follicularis spinulosa decalvans" (SIEMENS, 1925). SCHNYDER and KLUNKER consider this disease

a dominant sex-linked disorder because it regularly occurs in families and shows relatively little expression in the heterozygous females in contrast to the males. Shortly after birth the eyelashes in the male are lost, resulting in photophobia, tearing, ectropion of the eyelids, and alteration of the conjunctiva and cornea (keratosis superficialis, punctata, peripheral pannus formation). Furthermore, the eyebrow becomes sparse, particularly in the lateral parts, because of the development of follicular spinous elements. As noted, the eye manifestations are also less pronounced in females. In addition, spontaneous involution occurs at puberty, leaving a residual follicular scarring atrophy of the skin.

Acneiform follicular keratoses. In this very heterogeneous group of disorders, characterized by the existence of inflammatory comedolike horny papules (industrial acne), eye changes are absent as far as we know.

Darier's disease (Keratosis follicularis)

This genodermatosis described in 1889 by DARIER (together with THIBAULT) was originally thought to be due to a microorganism ("psorospermosis"). It occurs sporadically, the first manifestations may begin in the first or second decade or even in early childhood and are characterized by a tendency to grouped, ill-smelling vegetations made up of individual reddish or grayish red, greasy, squamous or crusting, follicular or parafollicular lesions occurring predominantly on the seborrheic intertriginous skin areas (Fig. 68). Under the horny masses one can see the thumbnail-like form of the attached horny lamellae. On the backs of the hands the dominant lesions are flat, warty papules. On the lower part of the back one can frequently see white macules, which probably are the earliest changes of the disease (NEUMANN). Almost always there are accompanying mucosal and nail changes in the form of longitudinal depressions, linear opacities, and crumbling of the nails. They are analogous to the dyskeratoses on the glabrous skin. Very striking interruptions of the papillary patterns are seen in the dermogram. Light seems to worsen the manifestations, as evidenced by the fact that the disease is aggravated in the summer.

From the ophthalmologic point of view, BRÜNAUER reviewed several papers describing the appearance of horny papules in the areas of the outer skin of the lids or the edges of the eyelids which may result in injury to the cornea and even ulceration. In the older literature nodular changes of the cornea were described in Darier's disease (v. BREUCH; JAENS). Furthermore, SPITZER reported on associated amblyopic nystagmus and finally, VON SZILY and GJESSING reported cataract formation in Darier's disease.

Dyskeratosis intraepithelialis benigna hereditaria

This disease has been known as "red eye disease" for about 150 years in a racially isolated population in North Carolina (see V. SALLMANN and PATON). It involves the conjunctiva of the eye and the oral mucosa. For further details see SCHNYDER and KLUNKER. It begins with injection of the conjunctiva within the first few months of life developing into gelatinous proliferating elevations adjacent to the limbus of the cornea that are not attached to the sclera of the eyeball. Both eyes

are affected. In the spring, the cornea can be covered in this fashion by a form of veil, which again disappears in the fall. Dyskeratosis of the oral mucosa also begins early in life and continues until puberty. The mucosal changes in the mouth resemble "white sponge nevus" (CANNON, 1935) and similar alterations can also be found in the vagina and urethra. The histologic picture of both the oral mucosa and the conjunctival epithelium is that of dyskeratosis and acanthosis. A whole variety of other anomalies can coexist with this disease, in particular association with hemoglobin C, coloboma of the iris, retinitis pigmentosa, and others.

Elastosis perforans serpiginosa

This picture was originally described by LUTZ as "keratosis follicularis serpiginosa" and by MIESCHER as "elastoma intrapapillare performans." It begins at an early age on the neck or in the vicinity of the elbows with characteristic circinate serpiginous grouped keratotic or verrucous papules (Fig. 69). Although this dystrophy of the elastic tissue is often associated with certain congenital anomalies in which ocular changes are well known (osteogenesis imperfecta, cutis laxa, pseudoxanthoma elasticum, among others; for review see KORTING, 1965), so far, ophthalmologic changes have not been reported in the pertinent literature. However, in view of the associated findings, they can be expected.

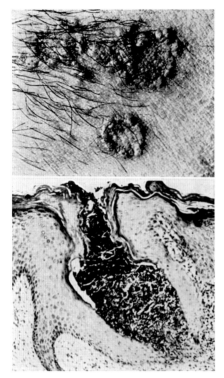

69   Elastosis perforans serpiginosa.

Pseudoxanthoma elasticum

Pseudoxanthoma elasticum was described by DARIER in 1896 histopathologically based on a case of Chauffard. It represents a dysplastic-metabolic, systemic elastosis with certain collagen components (KORTING and HOLZMANN) (Fig. 70). It is apparently inherited in a recessive fashion rather than a dominant one. The cutaneous signs are generally manifest in the first or second decade but may occasionally occur earlier. Clinically, the sites of predilection are the lateral aspects of the neck, the axillary vault, the flexor areas of the major joints and, less commonly, the inner aspects of the upper thighs and the groin area. The lesions consist of ivory white to lemon yellow, linear or grouped papular lesions which frequently are traversed by telangiectatic vessels. Furthermore, the disease is associated with cardiovascular signs (angina pectoris) and alterations of the middle and smaller arteries (asymptomatic occlusion of the ulnar and radial artery, and intermittent claudication). Changes simulating those in the skin may also be found in the gastrointestinal tract (the oral mucosa, and mucosa of the stomach and rectum) (Fig. 71). An important stigma of systemic elastorrhexis is the predilection for calcification which appears to be genetically associated with this disease.

In this connection it is worth pointing out not only that dystrophic elastosis tends toward calcification, but that such calcifications are quite frequent in diffuse connective tissue diseases such as, for example, Thibierge-Weissenbach syndrome (as part of progressive scleroderma) or the calcifications that occur in dermatomyositis. On clinical grounds there is a general relationship of "elastorrhexis systematisata," as pseudoxanthoma elasticum is frequently called today (TOURAINE), particularly to Paget's disease of the bone, to osteopetrosis, to cutis laxa (for which one can suggest that the cause is a genetically determined insufficiency of the normally well-developed texture of collagen), or to Marfan's syndrome. On the other hand, there are also associations, although more rarely, of pseudoxanthoma elasticum with other established disease processes of the elastic tissue. One example is the association of pseudoxanthoma elasticum with elastosis perforans serpiginosa, both of which may occur with cutis laxa (for details see KORTING and HOLZMANN, 1968).

Pseudoxanthoma elasticum occurs with a frequency of 1 to 160,000 to about 1 to 1 million of population (GOODMAN et al.) and is more common in the female sex (64 per cent: GOENINNE, VAN GINNEKENN and BERNARD). The ophthalmologic changes, consisting of linear pigment changes of the retina, have been known since 1889 (DOYNE) and were named "angioid streaks" by KNAPP (1892)

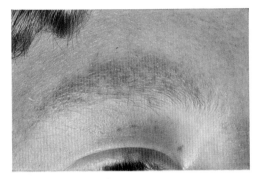

67 Ulerythema ophryogenes.

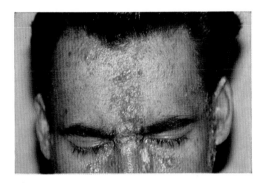

68 Darier's disease.

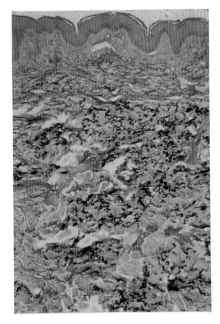

71 Pseudoxanthoma elasticum. Histologic appearance.

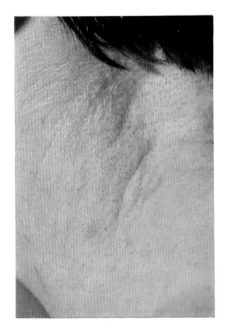

70 Pseudoxanthoma elasticum.

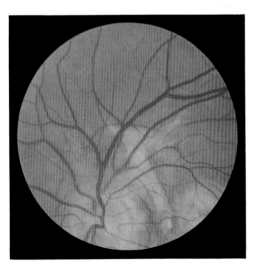

72 Angioid streaks. (Collection of Prof. Nover. Mainz.)

(Fig. 72). The ophthalmologist GRÖNBLAD (1929) and the dermatologist STRANDBERG (1929) associated angoid streaks with Darier's pseudoxanthoma elasticum. For this reason, this combination has been known as the Grönblad-Strandberg syndrome (FRANCES-CHETTI and ROULET, 1936). The first German observation of this kind was made by MARCHESANI and WIRZ. The streaks that simulate blood vessels of the retina are due to the development of plaques and calcification in Bruch's membrane (BÖCK et al., 1938), which in later stages are associated with bleeding into various layers of the retina, particularly arand the macula, and with whitish exudation. The arrangement of these pigmented streaks around the papillary region is apparently due to tension produced at the point of entry of the optic nerve into the eyeball and at first appears without any interference with function, but is later associated with atrophy of the optic nerve (see KLEEMANN). Beyond this, in the further course of the disease, other chorioretinal changes may reduce visual acuity to amaurosis (GOODMAN et al.). MARON (1968) described a tapetoretinal degeneration in addition to angioid streaks with lacerations at the fundus in the vicinity of the papillae. SCHRECK described rare eye changes in pseudoxanthoma elasticum which consisted of atrophy of the optic nerve, exophthalmus due to recurrent orbital hematoma, dislocation of the lens, keratoconus, and cataract. More recent reports concern cloudlike haziness of the cornea (COLLIER).

Secondarily pigmented cracks of the lamina vitrea are partially concentric or radial and usually begin around the optic disc or are located there and may be grossly or finely granular and reddish-brown, surrounded by a pale halo; they can also be found in other diseases, and simulate "angioid streaks." Eight to fifteen per cent of patients with Paget's disease of the bone have such changes (ROWLAND, TERRY, SCHOLZ et al.), occasion-ally associated with pseudoxanthoma elasticum of the skin (SANDBAKA-HOLMSTRÖM, see also McKUSICK). They may also occur in familial hyperphosphatamia (BAKWIN, GOLDEN and FOX; MC PHAUL and ENGEL) and in cases of sickle cell anemia (GEERAETS and GUERRY, PEARCE).

Acanthosis nigricans

This pigmented papillary dystrophy develops classically in the intertriginous skin areas (genital, anal, axillary, submammary, and so on) only after the fortieth year of age (usually after puberty — Translators) and frequently is associated with adenocarcinoma of the gastrointestinal tract, but also in the presence of carcinoma of the ovary, breast, and lung (POLLITZER and JANOVSKY, 1890).

In contrast to the malignant form, there have been described a "benign" form and pseudoacanthosis nigricans (OLLENDORFF-CURTH), which begins prior to puberty in obese persons, perhaps with endocrine dysfunction. The appearance is more discrete, but the localization is the same, namely, a more or less confluent and retiform papillomatosis. Occasionally one can also see the localized appearance of small pendulous fibromas.

From the ophthalmologic point of view, changes of the conjunctivae are quite rare, although they have been described on some occasions (HOLLANDER; MILLER and DAVIS; SCHREUS; HEROLD, KAUFMAN and SMITH; DEGOS and SAINRAPT; HISSARD; DUCUING; ARCHANGELSKIJ, KOPF and LAUSECKER). When changes are present, papillary flesh-colored or rose-red hypertrophies which result eventually in secondary irritation of the cornea predominate (WEISS). GROENOUW likened the conjunctival changes found in this disease with trachoma. CORRADO noted obliteration of the tear ducts by acanthotic papillomas.

References

AMALREC, P., BESSOU, P., and FARENC, M.: Le syndrome de Savin. Bull. Soc. Ophthal. (Paris) 65 (1965) 724–728.

BAKWIN, H., GOLDEN, A., and FOX, S.: Familial osteoectasia with macocranium. Amer. J. Roentgenol. 91 (1964) 609.

BÖCK, J.: Zur Klinik und Anatomie der gefäßähnlichen Streifen im Augenhintergrund. Z. Augenheilk. 95 (1938) 1–50.

BÖCK, J., and NIEBAUER, G.: Über zwei Kranke mit Ichthyosis vulgaris und Netzhautgefäßerkrankung. Wien. klin. Wschr. 76 (1964) 758–760.

BROICH, K.: Psorospermosis follicularis Darier. Verh. dtsch. Derm. Ges. X. Kongr., Frankfurt, 1908.

BRÜNAUER, St. R.: Morbus Darier. In: Hdb. der Haut- und Geschlechtskrankheiten, Bd. VIII/2, hsg. von J. Jadassohn. Springer, Berlin, 1931.

CARLBORG U. EJRUP, B., GRÖNBLAD, E., and LUND, P.: Vascular studies in pseudoxanthoma elasticum and angioid streaks. Acta Med. Scand. Suppl. 350 (1959).

COLLIER, M.: Elastorrhexis systématisée et dystrophies corneennes chez deux soeurs. Bull. Soc. Ophthal. (Paris) 65 (1965) 301–310.

CORRADO, M.: Alterazioni degli annesi oculari da Acanthosis nigricans. Atti Cong. Soc. Oftal. Ital. (1939) 239. Ref. Zbl. Haut- u. Geschlechtskrankheiten 66 (1941) 664.

DARIER, J.: Pseudoxanthoma elasticum. Mh. prakt. Derm. 23 (1896) 609–617.

DAVENPORT, D. D.: Ulerythema ophryogenes. Arch. Derm. Syph. (Chic.) 89 (1964) 74–80.

DEGOS, R., and SAINRAPT, A.: Acanthosis nigricans. Ann. Derm. Syph. (Paris) 7 (1947) 178 zit. n. H.-J. HEITE u. G. VON DER HEYDT.

DOYNE, R. W.: Choroidal and retinal changes, the result of blows on the eye. Trans. Ophthal. Soc. U. K. 9 (1889) 128.

DUCUING, J.: Documents pour l'étude de l'acanthosis nigricans. Bull. Ass. franc. Cancer 25 (1930) 695.

EDDY, D. D., and FARBER, E. M.: Pseudoxanthoma elasticum. Arch. Derm. Syph. (Chic.) 86 (1962) 729–740.

FRANCESCHETTI, A., and GRÖNBLAD, E.: "Angioid streaks" – Pseudoxanthoma elasticum. Der Zusammenhang zwischen diesen gleichzeitig auftretenden Augen- und Hautveränderungen. Acta ophthal. (Kbh.) 10, Suppl. 1, 1–114 (1932).

FRANCESCHETTI, A., and ROULET, E. L.: Le syndrome de Grönblad et Strandberg (stries angioides de la rétine et pseudoxanthome élastique) et ses rapports avec les affections du mesenchyme. Arch. Ophthal. (Paris) 53 (1936) 401–426.

FRANCESCHETTI, A., and SCHLÄPPI, V.: Degenerescence en bandelette et dystrophie de la cornée dans un cas d'ichthyose congénitale. Dermatologica (Basel) 715 (1957) 217–223.

FRANCESCHETTI, A., and SCHNYDER, U. W.: Versuch einer klinisch genetischen Klassifikation der hereditären Palmoplantarkeratosen unter Berücksichtigung assoziierter Symptome. Dermatologica (Basel) 120 (1960) 154–178.

GEERAETS, W. J., and GUERRY, D.: Angioid streaks and sickle cell disease. Amer. J. Ophthal. 49 (1960) 450.

GEERAETS, W. J., and GUERRY, D.: Angioid streaks and sickle cell disease. Amer. J. Ophthal. 50 (1960) 213.

GOENINNE, L., VAN GINNEKEN, E., and BERNARD R.: Syndrome de Groenblad-Strandberg-Touraine. Presse méd. 71 (1963) 2511–2514.

GOODMAN, R. M., SMITH, E. W., PATON, D., BERGMAN, R. A., SIEGEL, Ch. L., OTTESSEN, O. E., SHELLEY, W. M., PUSCH, A. L., and McKUSICK, V. A.: Pseudoxanthoma elasticum: a clinical and histopathological study. Medicine 42 (1963) 279–334.

GRAYSON, M.: Corneal manifestations of keratosis plantaris and palmaris. Amer. J. Ophthal. Ser. 3, 59 (1965) 483–486.

GROENOUW, A.: Beziehungen des Auges zu den Hautkrankheiten. In: Hdb. der Haut- und Geschlechtskrankheiten, Bd. XIV/1, hsg. von J. Jadassohn. Springer, Berlin, 1930.

HANHART, E.: Neue Sonderformen von Keratosis palmoplantaris. Dermatologica (Basel) 94 (1947) 286–308.

HERMANS, R.: Opacité en bandelette de la cornée et ichthyose. Bull. Soc. belge Ophthal. 113 (1956) 316–327.

HEROLD, W. C., KAUFMAN, W. H., and SMITH, D. C.: Acanthosis nigricans. Arch. Derm. Syph. (Chic.) 44 (1941) 789–799.

HISSARD, R.: Un cas d'acanthosis nigricans. Bull. Soc. franç. Derm. Syph. 40 (1933) 595–597.

HOLLANDER, L.: Is it a form of avitaminosis? Arch. Derm. Syph. (Chic.) 48 (1943) 650–655.

JAENSCH, K.: Hornhautbefunde bei Darierscher Dermatose. Klin. Mbl. Augenheilk. 78 (1927) 96.

JANCKE, G.: Cataracta syndermatotica und Ichthyosis congenita. Klin. Mbl. Augenheilk. 117 (1950) 286.

JUNG, E. G., and VOGEL, M.: Anhidrotische Ektodermaldysplasie mit Hornhautdystrophie. Schweiz. med. Wschr. 96 (1966) 1477–1483.

KLEEMAN, W.: Zum Groenblad-Strandberg-Syndrom. Zschr. Haut- und Geschlechtskrankheiten, 41 (1966) 150–155.

KNAPP, H.: On the formation of dark angioid streaks as an unusual metamorphosis of retinal hemorrhage. Arch. Ophthal. (Chic.) 21 1892) 289–292.

KOPF, O., and LAUSECKER, H.: Acanthosis nigricans bei Mensch und Tier. Hautarzt 4 (1953) 250–254.

KORTING, G. W.: Elastosis perforans serpiginosa als ektodermales Randsymptom bei Cutis laxa. Arch. klin. exp. Derm. 224 (1966) 437–446.

KORTING, G. W., and HOLZMANN, H.: Zur Frage humoraler Kollagen-Begleitkomponenten beim Pseudoxanthoma elasticum. Arch. kli. exp. Derm. 231 (1968) 408–414.

KRAUPA, E.: Die Ichthyosis der Hornhaut. Klin. Mbl. Augenheilk. 65 (1920) 903.

KRAUPA, E.: Die familiären degenerativen Hornhautveränderungen (neurotische Dystrophie und Ichthyosis corneae) im System der sog. Dystrophien der Hornhaut. Klin. Mbl. Augenheilk. 70 (1923) 396.

KRAUPA, E.: Familiäre celluläre (ichthyotische) und neurotische Dystrophie der Hornhaut. Klin. Mbl. Augenheilk. 73 (1924) 229.

McKUSICK, V. A.: Heritable disorders of connective tissue, 2d Ed. Mosby, St. Louis, 1960.

McPHAUL, Jr., J. J., and ENGEL, F. L.: Heterotopic calcification, hyperphosphatemia and angioid streaks of the retina. Amer. J. Med. 31 (1961) 488.

MARCHESANI, O., and WIRZ, F.: Die Pigmentstreifenerkankung der Netzhaut, das Pseudoxanthoma elasticum der Haut, eine Systemerkrankung. Arch. Augenheilk. 104 (1931) 522–545.

MARON, H.: Pseudoxanthoma elasticum. Derm. Wschr. 154 (1968) 129–130.

MILLER, T. R., and DAVIS, J.: Acanthosis nigricans occurring in association with squamous carcinoma hypopharynx. N. Y. St. J. Med. 54 (1954) 2333.

MULLER, S. A., and BRUNSTING, L. A.: Cataracts associated with dermatologic disorders. Arch. Derm. Syph. (Chic.) 88 (1963) 330–339.

PEARCE, W. G.: Interpretation of fundus signs in angioid streaks. Trans. Ophthal. Soc. U. K. 85 (1965) 429–435.

PINKERTON, O. D.: Catarakt verbunden mit kongenitaler Ichthyosis. Arch. Ophthal. 60 (1958) 393–396.

RICHNER, H.: Hornhautaffektion bei Keratoma palmare et plantare hereditarium. Klin. Mbl. Augenheilk. 100 (1938) 580.

ROMMER, F., and TEMPÉ, J. D.: À propos du syndrome de Sjögren et Larsson. Rev. Oto-neuro-ophthal. 36 (1964) 179–186.

ROWLAND, D. W.: Angioid streaks. Amer. J. Ophthal. 16 (1933) 61.

SANDBACKA-HOLMSTRÖM, I.: Das Grönblad-Strandbergsche Syndrom. Pseudoxanthoma elasticum, angioid streaks, Gefäßveränderungen. Acta derm.-venereol. (Stockh.) 20 (1939) 684—700.

SAVIN, L. H.: Corneal dystrophy associated with congenital ichthyosis and allergic manifestations in male members of a family. Brit. J. Ophthal. 40 (1956) 82—89.

SCHÄFER, E.: Zur Lehre von den kongenitalen Keratosen. Arch. Derm. Syph. (Berl.) 148 (1925) 425—432.

SCHNYDER, U. W., and KLUNKER, W.: Erbliche Verhornungs-störungen der Haut. In: Hdb. der Haut- und Geschlechts-krankheiten, Bd. VII, hsg. von J. Jadassohn, Springer, Berlin, 1966.

SCHNYDER, U. W., KONRAD, B., SCHREIER, K., NERZ P., and CREFELD, W.: Über Ichthyosen. Dtsch. med. Wschr. 93 (1968) 423—428.

SCHOLZ, R.: Angioid streaks. Arch. Ophthal. (Chic.) 26 (1941) 677—695.

SCHRECK, E.: Veränderungen des Sehorgans bei Haut- und Geschlechtskrankheiten. In: Dermatologie und Venerologie, Bd. IV, hsg. von H. A. Gottron u. W. Schönfeld. Thieme, Stuttgart, 1960.

SCHREUS, TH.: Acanthosis nigricans. Zbl. Haut- u. Geschl.-Kr. 57 (1938) 88.

SIEMENS, H. W.: Über seltenere und kompliziertere Vererbungsmodi bei Hautkrankheiten. Arch. Derm. Syph. (Berl.) 138 (1922) 433—438.

SIEMENS, H. W.: Zur Differentialdiagnose und Prognose der überlebenden Fälle von Ichthyosis congenita. Arch. Derm. Syph. (Berl.) 156 (1928) 624—655.

SPANLANG, H.: Beitrag zur Klinik und Pathologie seltener Hornhauterkrankungen. Z. Augenheilk. 62 (1927) 21.

STRANDBERG, J.: Pseudoxanthoma elasticum. Zbl. Haut- u. Geschl.-Kr. 31 (1929) 689.

TERRY, T. L.: Angioid streaks and osteitis deformans. Trans. Amer. Ophthal. Soc. 32 (1934) 555—573.

WAARDENBURG, P. J., FRANCESCHETTI, A., and KLEIN, D.: Genetics and Ophthalmology, Vol. 1, Charles C. Thomas, Springfield, 1961.

WEBER, G.: Über einen Fall von Ichthyosis congenita mit Lidschrumpfung Derm. Wschr. 128 (1953) 1144—1149.

WEISS, A.: Erkrankungen des Auges bei Acanthosis nigricans. Klin. Mbl. Augenheilk. 78 (1927) 790—796.

WESTERLUND, E.: Clinical and genetic studies of the primary glaucoma disease. Opera ex Domo Biol. Hered. 12. Munksgaard, Kopenhagen, 1947.

Bullous Dermatoses

Pemphigus

While in the past many blistering skin diseases were categorized as "pemphigus" or "pemphigoid" (pemphigus syphiliticus, pemphigus neonatorum, among others), at the present time only one type of disease characterized by acantholytic blister formation is considered true pemphigus of the kind originally postulated by WICHMANN in 1791.

The acute malignant pemphigus of the past century (PERNET and BULLOCH; CROCKER) is represented today by Lyell's syndrome (epidermolysis acuta toxica, KORTING and HOLZMANN) and in infancy is known as dermatitis exfoliativa neonatorum (VON RITTERSHAIN).

The cause was originally considered to be sodium chloride retention or possibly a virus infection, but at present this disease is thought to be immunologically based. It has been shown by means of indirect immunofluorescent staining that the blood serum of patients with pemphigus contains antibodies against the intercellular substance of the epidermis (BEUTNER and JORDAN; WALDORF et al.; CHORZELSKI, VON WEISS and LEVER). In contrast to this, no antibodies localized in this area can be found in the serum of patients with dermatitis herpetiformis (LEVER, 1967).

Pemphigus chronicus can clinically be found in three different varieties, of which pemphigus vulgaris (WICHMANN) is the most common variant (Fig. 73). In this disease there develop initially tense, later flaccid, then collapsing and later cloudy bullae on normal skin, a sign that has been called the Nodet sign. Furthermore, the epithelial defects developing in pemphigus vulgaris frequently show a loose "epidermal collar" around their edge. Striking is the fact that it is possible to move the bullae or the unchanged skin in their vicinity on tangential pressure. This so-called Nikolski

phenomenon is due to a loosening of the material between the epidermal cells, that is, acantholysis (AUSPITZ 1881; CIVATTE 1943). More specifically, the formation of the blisters is suprabasilar in pemphigus vulgaris and subcorneal in pemphigus foliaceus, the disease in which Nikolski first described his phenomenon. The earliest electron microscopic changes show evidence of degeneration of the intercellular substance and the intercellular bridges or desmosomes between and above the basal cells, which generally remain attached to the corium in pemphigus (BRAUN-FALCO and VOGELL).

In almost 50 per cent of patients with pemphigus the disease begins in the oral mucosa where generally only the beginning or the after effects of blister formation are noted as yellowish pale discolorations of the epithelium, moist lesions or erosions.

Pemphigus foliaceus (CAZENAVE, 1844) is a form of pemphigus chronicus which frequently begins as pemphigus vulgaris, that is, in bullous form, but soon changes into an exfoliative erythroderma with intertwined crusted scales. Pemphigus vegetans (NEUMANN, 1896), which also may begin or end as pemphigus vulgaris, is characterized by verrucous or pustulous thickenings, which occur primarily in the intertriginous areas of the skin (the axilla, the inguinal, or submammary regions) and where at the edge of the vegetations remnants of blister tops can be noted. This observation is significant in the differential diagnosis against bromoderma vegetans.

From the ophthalmologic point of view, an occasional manifestation of classical pemphigus vulgaris occurs on the conjunctivae (LEVER and TALBOTT: 16 per cent), which is not surprising considering the extremely frequent involvement of the oral mucosa (Fig. 74). Here, just as in the mouth, the blisters are very transient, so that even after a few hours oozing areas denuded of epithelium are found which characterize the picture of conjunctivitis pseudomembranacea. LEVER points out that conjunctival lesions of pemphigus vulgaris, in contrast to the lesions of benign mucous membrane pemphigus (formerly called pemphigus conjunctivae), heal without the formation of scars, which is to be expected from the intraepithelial location of the blisters. Only in the presence of long-standing purulent infection do scars occur.

In pemphigus foliaceus, the palpebral conjunctiva is frequently reddened or edematous and secretes a seropurulent material without either desquamation or exfoliation.

In Brazilian pemphigus (fogo selvagem, wild fire), which because of its appearance is generally compared to pemphigus foliaceus and which occurs in familial or conjugal fashion, AMENDOLA noted blisters in the vicinity of the eyebrows and entropium and ectropium, trichiasis, and in acute cases even changes of the iris. This was observed in a series of 240 patients. This disease occurs primarily in the state of São Paulo endemically and affects all races. In about 5 per cent of cases of this disease cloudiness of the anterior pole of the lens may occur, similar to the scutelliform cataract occurring in atopic dermatitis. In the Senear-Usher syndrome (pemphigus erythematosus, ORMSBY, 1936) there are some similarities to lupus erythematosus on the face, while other affected areas resemble more either seborrheic dermatitis or pemphigus foliaceus. As in the European pemphigus foliaceus, no changes of the eye have been reported in these diseases.

Bullous pemphigoid

This disease was considered an independent entity by LEVER (1953). It is also known as parapemphigus (PRAKKEN and WOERDEMANN) (Fig. 76) or pemphigus of the aged (STEIG-

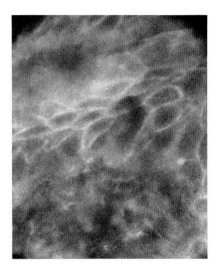

75 Pemphigus vulgaris. Demonstration of fluorescent antibodies against intercellular substance. The light areas fluoresce yellow in the original preparation. (Courtesy Dr. G. Brehm, University Skin Clinic, Mainz)

LEDER). The disease is characterized by the development of many large, tense blisters.

These blisters dominate the clinical picture throughout the entire duration of the disease because they persist for a long time. They can occur on an erythematous and edematous base, in contrast to the blisters of pemphigus vulgaris. Blister formation in this disease is subepidermal, in contrast to the acantholysis of pemphigus vulgaris, and some authors (LA-PIÈRE) believe that although itching is usually absent, this relatively benign bullous pemphigoid represents an exquisitely bullous form of dermatitis herpetiformis. There are, however, important reasons, for instance the absence of peribullous abscesses in pemphigoid (PIERARD and WHIMSTER), to consider it as a disease of its own (C. KRESBACH and HARTWAGNER). From the immunologic point of view, the fixation of autoantibodies to the basal layer appears to be specific for bullous pemphigoid (LEVER). The mucosae are rarely involved, but occasionally the conjunctiva may be the location of blisters.

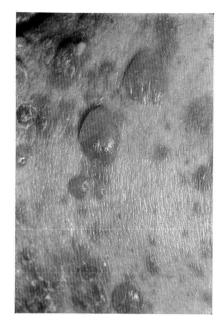

73 Pemphigus vulgaris.

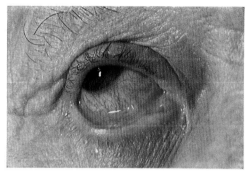

77 Benign mucous membrane pemphigoid.

74 Pemphigus vulgaris. Acantholytic blister formation.

78 Dermatitis herpetiformis. Typical grouped herpes-like arrangement of blisters.

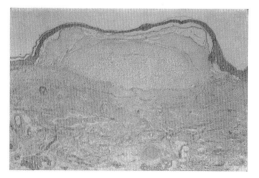

76 Parapemphigus. Subepidermal blister formation.

79 Dermatitis herpetiformis. Histologic appearance.

"Benign" mucous membrane pemphigoid (LEVER)

This designation is an attempt to rename a disease syndrome which was previously described as "pemphigus conjunctivae" or "pemphigus mucosus." To fit the actual observations, it would be better to call the disease "scarring mucous membrane pemphigoid" or "dermatite bulleuse mucosynéchiante et atrophiante" (DEGOS, LORTAT-JACOB and HARDY) (Fig. 77).

As these descriptions suggest, the lesions are primarily located on the mucous membranes, and there is a tendency to scarring and a generally benign course except for the possibility of significant diminution of vision. Pemphigus of the eye is well known to the ophthalmologist under the name "essential shrinkage of the conjunctiva" (VON GRAEFE and also HAGER). Scarring occurs almost always when the conjunctivae as well as the mucosae near the skin are affected by this type of pemphigus. In any case, involvement of the eye occurs almost always in mucous membrane pemphigoid, occasionally only in a solitary fashion. The blisters on the conjunctivae are also transient in this variant of pemphigus and the initial picture appears to be a banal conjunctivitis followed by significant scarring and shrinkage with the usual consequences of synechiae and trichiasis: symblepharon, ankyloblepharon, entropion. The cornea itself is frequently involved (clouding, invasion of new blood vessels, perhaps related to decreased secretion of tears, and so on) until finally the end stage consists of a cornea covered by a horny membranous layer, which is similar to the "blind" eye of a statue. This was the typical outcome prior to the use of corticosteroids.

In general, the disease process can become stationary after a few years, so that some visual acuity remains. However, protracted forms have been reported in which the patient suffers almost all his life. The variability of this disease is so great that, although it is called benign mucous membrane pemphigus, total blindness may result in less than a year's time in spite of modern therapy, as reported recently by KLEINE-NATROP and HAUSTEIN. Some other effects are double vision, which is produced by interference with motion of the eyeball (MAWAS et al.) or perforation following ulcers of the cornea (LINDEMANN). This is a somewhat rare disease. Its incidence varies according to SELBY and PARISI from one in 9,000 to one in 75,000; according to OBERSTE-LEHN as one in 50,000; and according to KANEE as one in 20,000 patients seen for eye disorders Histologic studies of the conjunctiva and cornea in this disease were reported by SOUDAKOFF and WHAILMAN; STREITMAN; SCHRECK; SELBY and PARISI et al.

Familial pemphigus (HAILEY and HAILEY)

This type of pemphigus as a rule manifests itself only in patients more than 30 years of age and may be irregularly dominantly transmitted through several generations. It is characterized by small grouped blisters occurring on only a few areas of the body such as the neck, axilla, and groin, which eventually develop into circular, slightly vegetating, grey-red granulations. As far as is known only oral mucosal lesions have been found in this variant of pemphigus (FISCHER and NIKOLOWSKI).

Dermatitis herpetiformis (DUHRING, 1884)

In this "wide spectrum disease," which may well be the most polymorphous of all dermatoses, serious impairment of health is absent, although itching or a peculiar burning feeling of the skin is usually present. The cutaneous manifestations consist of erythema, urticaria, grouped (herpetiform) vesicles, blisters, pustules, or papules. The lesions are usually symmetric, grouped, and there is predilection for the posterior axillary fold (Figs. 78 and 79). Histologically, one finds in the vicinity of the new lesion multilocular microabscesses and degeneration of the collagen of the papillary tips, and as the lesion progresses, unilocular subepidermal blister formation. The characteristic microabscesses can still be found at the edge of the blisters. In contrast to pemphigus and bullous pemphigoid, no auto-antibodies have been shown in dermatitis herpetiformis (LEVER, 1967). In general, involvement of the mucous membranes is a rare phenomenon in dermatitis herpetiformis, and isolated lesions can hardly be differentiated from mucosal pemphigus. SCHRECK described several eye lesions, including bilateral central chorioretinitis. BONAVOLANTA found eight reports in the world literature up until 1949. The question arises whether in these observations of involvement of the eye, dermatitis herpetiformis or benign mucosal pemphigus was present. It should be pointed out that benign pemphigoid and dermatitis herpetiformis have been considered to be the same disease by a number of French and Belgian authors, for instance Lapiere.

Subcorneal pustular dermatosis (SNEDDON and WILKINSON, 1956)

This disorder is closely related to dermatitis herpetiformis, if it is not identical to it, and has been primarily reported in female patients 20 years or older. Clinically, one sees recurrent crops of small blisters or groups of pustules which change to thinly scaling clustered patches. The head is almost always spared and mucous membrane lesions are almost always absent. Thus it is not surprising that the conjunctiva is not involved.

Hereditary epidermolyses

In this quite variable group of hereditary epidermolyses the underlying defect is a hereditary weakness of the skin structure in the region of the basal cell layer and the adjacent layers. This results in "hereditary predisposition to the formation of blisters" (GOLDSCHEIDER, 1892) and today is mainly diagnosed as "epidermolysis bullosa hereditaria" (KÖBNER, 1896). Inherently we are dealing with a "bullosis mechanica hereditaria" (SIEMENS), that is, the formation of blisters which are provoked primarily by minimal mechanical irritation. Less marked is blister formation in warm weather or warm baths, as described specially by COCKAYNE (1938) as "recurrent blistering eruption of the feet in hot weather" (see the case of KORTING). From these minimal variants, which generally occur later in life on the palms and soles, there are transitional cases leading up to the complete manifestation of epidermolysis bullosa hereditaria simplex. This disease is inherited as an autosomal dominant trait with distinct preference for the male sex. In contrast to other variants, eye and mucosal changes are usually absent.

Epidermolysis bullosa hereditaria dystrophica dominans. In this type, the blisters are subepidermal and one sees as late effects scarring, circumscribed atrophy or formation of keloids (epidermolysis bullosa hyperplastica: TOURAINE). Most of all horny cysts occur at the site of the blisters (milia). Quite frequently in this dominant dystrophic form of epidermolysis (COCKAYNE, TOURAINE) one finds clawlike sclerodactyly, onychodystrophy, and hyperhidrosis of the palms and soles. On the basis of these phenomena, which primarily affect the extremities and may occur independently on a genetic basis, there can develop on the trunk, particularly the lumbosacral area, an "epidermolysis bullosa albopapuloidea" (first described by PASINI, 1928). In this disorder, ivory white lesions may be noted like elevated plaques over the surrounding skin.

Epidermolysis bullosa hereditaria dystrophica recessive (HALLOPEAU; SIEMENS). In this recessive dystrophic variety, which has also been called "epidermolysis bullosa hereditaria dystrophica polydysplastica" (TOURAINE), one finds, in addition to dystrophic changes of the nails, anomalies of the hairs and teeth as well as alterations of the skeleton and mutilation. This disease is of clinical importance because of late manifestations, which because of involvement of the mucosa all the way down to the esophagus, may cause stenosis (see case of KORTING, 1957).

From the ophthalmologic point of view the conjunctiva and cornea are usually not involved in epidermolysis bullosa hereditaria dominans. In the recessive form, the Hallopeau-Siemens form, blisters and erosion may occur on the cornea synchronously with the bullous skin lesions. Symblepharon and scarring of the tear ducts have been noted, similar to that which occurs in mucous membrane pemphigoid (BÄFVERSTEDT and GRANSTRÖM). In the very rare epidermolysis bullosa dystrophica ulcero-vegetans, conjunctivitis with dystrophy of the cornea occurs (SCHNYDER). Finally, in the extremely rare dystrophia bullosa hereditaria, typus maculatus, spotty dystrophies of the cornea have been reported (MENDES DA COSTA and VAN DER VALK, WOERDEMAN).

References

AMÈNDOLA, F.: Catarata no pemfigo foliáceo (nota prévia). Rev. paul. Med. 26 (1945) 286.

AMÈNDOLA F.: Ocular manifestations of pemphigus foliaceus. Amer. 3. Ophthal. 32 (1949) 35—44.

AMÈNDOLA, F.: Ophthal. ibero-amer. 18 (1956) 37.

BÄFVERSTEDT, B., and GRANSTRÖM, K. O.: Fall von Epidermolysis bullosa hereditaria dystrophica mit Augensymptomen. Ref. Zbl. Haut- u. Geschl.-Kr. 65 (1940) 1517.

BEUTNER, E. H., and JORDAN, R. E.: The demonstration of skin antibodies in sera of pemphigus vulgaris-patients by indirect immunofluorescent staining. Proc. Soc. Exp. Biol. (N. Y.) 117 (1964) 505—510.

BONAVOLANTÁ, A.: Dermatite di Duhring-Brocq con manifestazione oculare. G. ital. Oftal. 2 (1949) 169—184.

BRAUN-FALCO, O., and VOGELL, W.: Elektronenmikroskopische Untersuchung zur Dynamik der Akantholyse bei Pemphigus vulgaris, I. Mitt.: Die klinisch normal aussehende Haut in der Umgebung von Blasen mit positivem Nikolski-Phaenomen. Arch. klin. exp. Derm. 223 (1965) 328—346.

BRAUN-FALCO, O., and VOGELL, M.: Elektronenmikroskopische Untersuchung zur Dynamik der Akantholyse bei Pemphigus vulgaris, II. Mitt.: Die akantholytische Blase. Arch. klin. exp. Derm. 223 (1965) 533—550.

CHORZELSKI, T. P., v. WEISS, J. F., and LEVER, W. F.: Clinical significance of autoantibodies in pemphigus. Arch. Derm. Syph. (Chic.) 93 (1966) 570—576.

DEGOS, R., LORTAT-JACOB, E., and HARDY, P.: Dermatite bulleuse mucosynéchante et atrophiante (Pemphigus oculaire). Bull. Soc. franç. Derm. Syph. 63 (1956) 324—325.

DEGOS, R., LORTAT-JACOB, E., and HARDY, P.: La place nosologique du "pemphigus oculaire." Dermatologica (Basel) 115 (1957) 205—217.

FISCHER, H., and NIKOLOWSKI, W.: Die Mundschleimhaut beim Pemphigus benignus familiaris chronicus. Arch. klin. exp. Derm. 214 (1962) 261—273.

HAGER, G.: Augenveränderungen beim Pemphigus. Z. ärztl. Fortbild. 52 (1958) 366—371.

KANEE, B.: Ocular pemphigus with scarring of the skin and mucous membranes. Arch. Derm. Syph. (Chic.) 55 (1947) 37—41.

KLEINE-NATROP, H. E., and HAUSTEIN, F.: "Benignes Schleimhautpemphigoid" mit rascher Erblindung und generalisierten vernarbenden Hautveränderungen. Hautarzt 19 (1968) 8—12.

KORTING, G. W.: Zur Kenntnis der sog. rezidivierenden Blasen-Eruption an den Füßen bei heißem Wetter (Weber-Cockayne). Z. Haut- u. Geschl.-Kr. 17 (1954) 36—40.

KORTING, G. W.: Über Oesophagusstenosen bei Epidermolysis bullosa. Z. Haut- u. Geschl.-Kr. 22 (1957) 282—285.

KORTING, G. W., and HOLZMANN, H.: Universelle Epidermolysis acuta toxica. Arch. klin. exp. Derm. 210 (1960) 1—13.

KRESBACH, H., and HARTWAGNER, A.: Zur Differentialdiagnose zwischen Dermatitis herpetiformis Duhring und bullösem Pemphigoid. Z. Haut.- u. Geschl.-Kr. 43 (1968) 165—176.

LAPIÈRE, S.: La pemphigoide (W. Lever) peut-elle être considérée comme une entite morbide séparée de la dermatite polymorphe de Duhring-Brocq. Dermat-Congr. Budapest, 1965.

LEMMINGSON, W.: Augenbeteilung bei Epidermolysis bullosa. Klin. Mbl. Augenheilk. 122 (1923) 350—353.

LEVER, W. F.: Pemphigus. Pemphigoid. Pemphigus familiaris benignus. In: Hdb.: der Haut- u. Geschlechtskrankheiten, Vol. II/2, edited by J. Jadassohn. Springer, Berlin, 1965.

LEVER, W. F.: Pemphigus and Pemphigoid. Charles C. Thomas, Springfield, Ill., 1965.

LEVER, W. F.: Differentialdiagnose zwischen Pemphigus vulgaris, bullösem Pemphigoid und Dermatitis herpetiformis. Med. Klin. 62 (1967) 1173—1176.

LEVER, W. F.: and TALBOTT, J. H.: Pemphigus. A. clinical analysis and follow-up study of sixty-two patients. Arch. Derm. Syph. (Chic.) 46 (1942) 348—357.

LINDEMANN, M.: Ein Fall von Pemphigus conjunctivae (Demonstration). 34. Tagg. Württemberg. augenärztl. Vereinig. Tübingen, 9. X. 1938. Kl. Mbl. Augenheilk. 102 (1939) 142.

LYELL, A.: Toxic epidermal necrolysis: an eruption resembling scalding of the skin. Brit. J. Derm. 68 (1956) 355.

MENDES DA COSTA, S., and VAN DER VALK, J. W.: Typus maculatus der bullösen hereditären Dystrophie. Arch. Derm. Syph. (Berl.) 91 (1908) 1–8.

MOWAS, J., SIDI, E., MELKI, G., and PACK, J.: Pseudo-pemphigus de l'oeil et le pemphigus cutané. Bull. Soc. Ophtal. Fr. (1953) 205–211.

OBERSTE-LEHN, H.: Bullöse Dermatosen. In: Lehrbuch der Haut- und Geschlechtskrankheiten, 9th ed., edited by H. G. Bode and G. W. Korting. Fischer, Stuttgart, 1962.

PIÉRARD, J., and WHIMSTER, I.: The histological diagnosis of dermatitis herpetiformis, bullous pemphigoid and erythema multiforme. Brit. J. Derm. 73 (1961) 252–266.

PRAKKEN J. R., and WOERDEMAN, M. J.: "Pemphigoid" (parapemphigus): its relationship to other bullous dermatoses. Brit. J. Derm. 67 (1955) 92–97.

SCHNYDER, U. W.: Die hereditären Epidermolysen. In: Hdb. der Haut- und Geschlechtskrankheiten, Vol. VII, edited by J. Jadassohn. Springer, Berlin, 1966.

SCHRECK, E.: Über einander zugeordnete Erkrankungen der Haut, der Schleimhäute und der Deckschicht des Auges (cutaneo-muco-oculoepitheliale Syndrome). Arch. Derm. Syph. (Berl.) 198 (1954) 221–257.

SCHRECK, E.: Veränderungen des Sehorgans bei Haut- und Geschlechtskrankheiten. In: Dermatologie und Venerologie, Vol. IV, edited by H. A. Gottron and W. Schönfeld. Thieme, Stuttgart, 1960.

SELBY, G. D., and PARISI, P. J.: Ocular pemphigus. Report of a case. Arch. Ophthalm. 43 (1950) 238–243.

SOUDAKOFF, P. and WHAILMAN, S.: Ocular pemphigus; Report of a case with the histologic findings in the cornea. Amer. J. Ophthal. 36 (1953) 231–236.

STREITMANN, B.: Benignes Schleimhautpemphigoid. Derm. Wschr. 144 (1961) 780–785.

WALDORF, D. S., SMITH CH. W., and STRAUSS, A. J. L.: Immunofluorescent studies in pemphigus vulgaris. Arch. Derm. Syph. (Chic.) 93 (1966) 28–33.

WOERDEMAN, M. J.: Dystrophia bullosa hereditaria. Typus maculatus. Proc. XI. Internat. Congr. Dermat. Stockholm, Acta derm. venereol. (Stockholm) 3 (1957) 678–686.

Physical and Chemical Injuries to the Skin

Dating back to Fabricius von Hilden, burns (combustiones), scalds (ambustiones), and injuries due to freezing (congelationes) have been divided into three stages: erythema, blisters, necrosis (carbonisatio, dermatitis eschariotica). The extent and prognosis of the injury such as may be obtained by judging the percentage of the surface area of the burned skin can, particularly in burns, only be determined hours to days after the accident. Shortly after thermal trauma, the affected area shows an increased permeability of the cell membrane and together with this, among other things, the burn shock develops under nerve influence, with a tendency to sympathicotonic centralization of the circulation. Finally, because of the marked pain and helped by hypoxia, tissue proteases are activated. Specific "burn poisons" are presently considered nonexistent. In the first three to five days, burn injuries are characterized by a decreased blood volume, which is due to the loss of fluid rich in protein and electrolytes. This is followed by hypoproteinemia and in particular hypoalbuminemia, a reduction in colloidal osmotic pressure, reduction in serum sodium, and increase in potassium. Pathogenetically important are the release of catecholamines and changes in the microcirculation with stasis and clumping of erythrocytes and thrombocytes. With modern therapy, acute renal shutdown (including the myorenal syndrome) has become rare, although coagulation of the muscles in case of high voltage burns almost always results in anuria (KOSLOWSKI, GOERZ and IPPEN) (Fig. 80). In such cases the degree of muscular injury can be followed by measurement of chromoproteinuria and of creatine phosphokinase (CPK). The second, toxemic phase of burn injuries is significantly influenced by the usual bacterial infection of burn wounds (particularly with hospital-derived organisms), while the anemia of burns (such as occurs secondarily by toxic injury of the bone marrow) is one of the late complications.

The clinical appearance of local heat injury is well known. Only in high voltage burns is it necessary to point out that the entry and exit points of the current (5,000 to 25,000 volts) may be discrete or localized in hidden areas in the form of the so-called "current marks." These are characterized as centrally necrotic but peripherally bullous to erythematous, and in the wider periphery sooty-black tissue changes.

From the ophthalmologic point of view, burns and freezing injuries in the vicinity of the eye show the same manifestations as occur on the skin, resulting finally in necrosis. In spite of the frequent wrinkling of eyelids, one can observe defects of tissue resulting in scalf formation as well as ectropion, and ankyloblepharon, among others. Furthermore, SCHÖN-FELD and GROENOUW reported on occasional observations of retinitis hemorrhagica, neuroretinitis or retrobulbar neuritis (THIES) even when the eye was not specifically involved.

In freezing injuries, changes in the vicinity of the visual organ are rare, since the closure of the eyelids and the excellent blood supply of the eyelids provide a certain amount of protection. Such injuries are frequently symmetrical and are related to a variety of factors such as the blood supply. Toxic systemic reactions are rarely noted. Nevertheless, frostbite of the eyelids and cornea may occur in people trapped in a snowstorm or in a war in the high mountains (VON HERRENSCHWAND) or following prolonged application of ice compresses. Finally, extreme cold may result in cloudiness of the lens, as can electrical or lightning injuries. Extensive and deep necrosis may follow introduction of foreign bodies into the eye. The most feared of these are the indelible pencil tissue necroses (ERDHEIM, 1915), which may also occur in the skin. SCHRECK pointed out that parts of indelible pencil have been reported to have been applied to the conjunctiva in artifacts. Severe changes of the eye are also possible following tear gas injury (JAENSCH, MIDTBO), which may also affect the skin (EISSNER and LIEBESKIND). Their description, as far as the eye is concerned, can be found in the ophthalmologic literature.

Otherwise traumatic lesions produced by chemicals are relatively similar: colliquative necrosis may develop following alkali, and sharply limited coagulation necrosis is produced by acids (Figs. 80 and 81). Profuse irrigation of the eye at the time of chemical instillation may save the patient's vision. The dermatologist has to think about these, since eventual late changes, such as secondary glaucoma because of obstruction of the drainage tracts of the fluid of the anterior chamber, require ophthalmologic therapy (ABRAMOWICZ).

References

ABRAMOWICZ, I.: Lime scald of human eye in histopathologic picture. Klin. Poczman 32 (1962) 421–425. Ref. Zbl. ges. Ophthal. 88 (1963) 178.

COLOMBO: Doppelseitige Hornhautveränderung nach Kälteeinwirkung bei einem Flieger. Brit. J. Ophthal. 5 (1921) 553.

EISSNER, H., and LIEBESKIND, H.: Dermatitis durch Chloracetophenon. Z. Haut- u. Geschl.-Kr. 37 (1964) 143–152.

GOERZ, G., and IPPEN, H.: Diagnostische Hinweise bei der Starkstrom-Verbrennung. Dtsch. med. Wschr. 93 (1968) 402–403.

GOHLKE, H., and ULLERICH, K.: Haut- und Augenschäden durch Dichlordiaethylsulfid (Lost). Hautarzt 2 (1951) 404–407.

GROENOUW, A.: Thermische Schädigungen. In: Hdb. der Haut- und Geschlechtskrankheiten, Vol. XIV/1, edited by J. Jadassohn. Springer, Berlin, 1930.

V. HERRENSCHWAND: Über Schädigungen der Hornhaut im Hochgebirgskriege. Zbl. prakt. Augenheilk. (1916).

JAENSCH, zit. n. SCHRECK, E.: Hautveränderungen durch äußere Einwirkungen. In: Dermatologie und Venerologie, edited by H. A. Gottron and W. Schönfeld. Thieme, Stuttgart, 1960, Vol. IV.

KOSLOWSKI, L.: Die Verbrennungskrankheit. Dtsch. med. Wschr. 88 (1963) 233–239.

MIDTBØ, A.: Eye injury from tear gas. Acta ophthal. (Kbh.) 42 (1964) 672–679. Ref. Zbl. ges. Ophthal. 93 (1965) 209.

SCHMITZ, R.: Nekrosen, Gangrän, Geschwüre. In: Hdb. der Haut- und Geschlechtskrankheiten, Vol. III/1, edited by J. Jadassohn. Springer, Berlin, 1963.

SCHÖNFELD, W.: Dermatologie für Augenärzte. Thieme, Stuttgart, 1947.

SCHRECK, E.: Veränderungen des Sehorgans bei Haut- und Geschlechtskrankheiten. In: Dermatologie und Venerologie, Vol. IV, edited by H. A. Gottron and W. Schönfeld. Thieme, Stuttgart, 1960.

SNELL, A. C.: Metastatic infiltration of cornea (Ringabscess). Amer. J. Ophthal. 4 (1921) 419.

THIES, O.: Doppelseitige schwere Neuritis retrobulbaris bei Hautverbrennung. Klin. Mbl. Augenheilk. 72 (1924) 391–394.

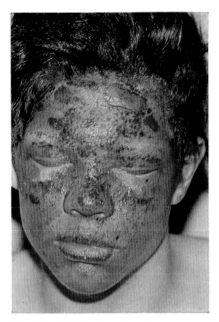

80   Electrical high voltage burn.

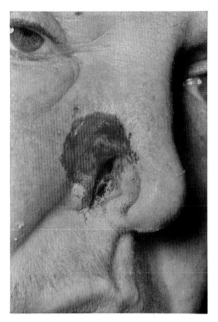

82   Trophic ulcer of the nose secondary to electrocoagulation of the semilunar ganglion for neuralgia of the trigeminus.

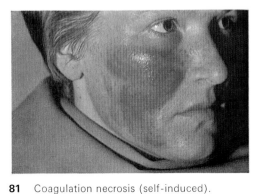

81   Coagulation necrosis (self-induced).

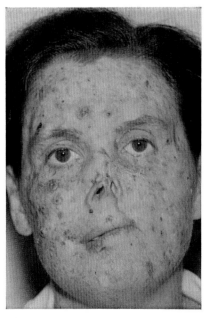

83   Xeroderma pigmentosum

Diseases Due to Light (Including Radiation Reactions)

Normal individuals develop similar injuries of the skin and eye, as for instance in the case of the ordinary sunburn (dermatitis solaris acuta). The epithelium of the conjunctiva has no particular protective function, as has been shown by the studies of MIESCHER (see also THIES). In contrast to this, THIES points out that its pigment content protects the iris to some extent, as has also been shown by KRAUPA. In the dermatologic literature there have been described in addition to "the red sunscreen" of the erythrocytes (FINSEN), the "brown sunscreen" of the pigment (MEYER) and the "cloudy sunscreen" (SCHALL and ALIUS), which is due to the thickening of the horny layer secondary to irradiation by sunshine (the "light callus" of GUILLAUME, 1926, and MIESCHER 1931). These light screens are natural protective measures. In recent years an additional physiological protection of the epidermis against light has been found in the formation of urocaninic acid. Under normal circumstances, the erythema of the skin and the epithelial cell damage is not due to the visible component of light or the long wave ultraviolet light, but rather to the emission of UVB and UVC (ULLERICH, WULF, and WISKEMANN). The erythema action spectrum shows two peaks, the first of which is located in the UVC between the wavelengths of 230 and 260 mμ, the second peak is located in the UVB at around 300 mμ. Erythema of the skin and the usual solar conjunctivitis are thus produced by the same wavelength of the sun rays.

Chronic changes due to light generally occur only in the corium. Here changes in the normal connective tissue structure develop, similar to the end stage of senile elastosis, cutis rhomboidalis nuchea, or an abortive xeroderma pigmentosum. These changes consist of degeneration of the ground substance with increase in acid mucopolysaccharides, the development of abnormal elastic tissue and

changes in the wall and increase in size and dilation of the blood vessels of the cutis (see IPPEN).

Xeroderma pigmentosum

Of the various actinic dermatoses (from which the porphyrias were separated and discussed under the metabolic diseases), the striking clinical picture of xeroderma pigmentosum (KAPOSI, 1870) will be described first (Fig. 83). In this hereditary disease, which is generally autosomal recessive and only rarely autosomal dominant ("light-shrunk skin" of E. HOFFMANN), there develop lentigo or frecklelike hyperpigmentations based on congenital hypersensitivity to sunlight radiation due to inability of the skin to protect itself against ultraviolet radiation. Later, not only do these changes develop into an atrophic poikiloderma but there develop warty keratoses followed by squamous cell carcinomas, keratoacanthomas, basal cell carcinomas, and occasionally even melanomas—the hallmark of the premature aging due to xeroderma pigmentosum (POPOFF: ephelidosis maligna). Recently, such patients have been shown to have elevated copper and decreased glutathion levels in the blood serum and aminoaciduria (LANGHOF). The frequent association of mental defects has been described in the past (xerodermatic idiocy: DE SANCTIS and CACCHIONI; ELSÄSSER; ELSÄSSER, FREUSBERG and THEML). Presumably the hypersensitivity of the skin to the effects of light in xeroderma pigmentosum is based on a defective capability to repair ultraviolet-induced injury to DNA.

From the ophthalmologic point of view, the common formation of ectropion of the eyelids is associated with other mutilating changes; associated eye changes are not at all rare (LARMANDE and TIMSIT: in 18 of 20 patients). Furthermore, the eyelids, particularly the lower eyelids, frequently show ulcers as well as the already noted tumor formations (REESE and WILBER). THIEL and others described a carcinoma of the cornea. The association with changes of the cornea in xeroderma pigmentosum was frequently reported (literature in KLEIN and FRANCESCHETTI). EL-HEFNAWI and MORTADA describe the following as typical corneal changes: edema, round cell infiltration, vascularization, and ulceration. HUERKAMP noted a similar appearance of the iris (further description of this phenomenon in LEBAS).

Light urticaria

This is a rare form of physical allergy, which can be elicited by a variety of wavelengths of light. Most patients react to radiation with wavelengths of less than 3700 Å. More recently, familial protoporphyrinimic light urticaria was described (LANGHOF, MÜLLER and RIETSCHEL: Amino-aciduria, sideropenia, and abnormalities of blood clotting, pellagra-like changes, and skin degeneration simulating colloid milium).

Eczema solare (UNNA, VEIEL, WOLTERS) and chronic polymorphic light eruption (HAXTHAUSEN)

The clinical appearance, which affects primarily girls or women in the spring or early summer on the uncovered areas of the skin and represents a form of hydroa aestivalis or hydroa vacciniforme (BAZIN), must always be differentiated from congenital porphyria. In eczema solare, which has a predilection for lichenification of the areas of the skin exposed to light, as occurs in chronic polymorphous light eruption (erythematous, erythematoid or pruriginous), hardening develops in the

course of the spring and summer so that patients become less sensitive to light and may even develop spontaneous remissions in the fall or winter. According to WISKEMANN and WULF the pathogenic effect of radiation is due primarily to long wavelengths in the ultraviolet range between 320 and 400 mµ. Of pathogenetic importance furthermore is the appearance of breakdown products of protein in the urine that are derivatives of tryptophane metabolism (KIMMIG: "Lichtbandstoffe," i.e. indolyl-acryloyl-glycine).

From the point of view of the ophthalmologist, hypersensitivity of the conjunctiva to light is frequently associated and may persist for many years synchronous with the skin changes, manifesting itself periodically (see the case of URBANEK). In the skin changes of the form of hydroa, which is associated with porphyrin excretion in the urine, associated changes in the fundus have been frequently reported (papilledema, optic neuritis, retinitis, edema of the retina, retinal bleeding, and chorioretinitis; literature in ULLERICH, WULF and WISKEMANN). KOCHS and LIBOWITZKY observed a 13-year-old girl with a chronic polymorphous light eruption, who showed inflammation of the conjunctiva and bilateral subcapsular anterior cataract without any porphyrinuria (increased excretion of normal porphyrins) or porphyria (excretion of normally absent porphyrins). (For the older literature, see also LEBAS, p. 744.)

Springtime pernio

This photodermatosis, which has been described by HELLER (1907), KEINING (1940) and BURCKHARDT (1942), occurs primarily in children and causes papular or papular vesicular lesions on the rim of the ears. It usually recurs year after year. In spite of phenomenologic relationship to the diseases classed as hydroa, simultaneous involvement of the eye has not yet been reported.

Photodynamic reactions

The phototraumatic reaction due to an overdose of short-wave ultraviolet light is known as dermatitis solaris acuta (ZENNER and BEUTNAGEL). In this disorder, commonly known as "sunburn," a spotty follicular erythema forms at first which then develops into large flat red areas. In addition to blister formation, transient generalized symptoms such as headache, vomiting, and so on may occur. Painful keratitis or keratoconjunctivitis and headaches are reported not infrequently by people who have been exposed to artificial UV lamps. Such injury is usually not serious (HORNBERGER). The description of keratoconjunctivitis photoelectrica, keratitis punctata superficialis following intensive exposure to sunlamps, welding arcs, or Klieg lights, belongs primarily to the ophthalmologic literature (see SCHRECK).

Phototoxic reactions are the result of specific absorption of UV radiation and photoallergic reactions are due to the development of sensitizing substances (EPSTEIN, BURCK-HARDT). Phototoxicity originally is primarily due to a substance contained in plants, e.g. the grass dermatitis (dermatitis bullosa striata pratensis (OPPENHEIM) and to a certain extent also berloque dermatitis (FREUND-ROSEN-THAL) which is produced by oil of bergamot and furocoumarines (KUSKE). Photoallergies are primarily produced by sulfonamides (including oral antidiabetic agents), hydro-chlorothiazides, declomycin, optical brighteners, certain antimycotics, and phenothiazines. ULLERICH, WULF, and WISKEMANN reported on a severe keratoconjunctivitis occurring after the administration of the latter, which could only be brought under control by suturing both eyelids. Wavelengths of light which normally cause no injury were the cause in this patient. By irradiation of skin tests, the presence of photoallergy against one of the common phenothiazine derivatives could be established. Symptomatically, the use of Schott filters in special eyeglasses which removed the irritating wavelengths had the desired protective effect.

Radiation reactions (roentgen, laser, and others)

As is well known, the effect of x-rays is not observed by the patient at the time when they are applied. Most of the presently used experimental systems for demonstrating damage also show the ill effects of radiation only after some time. Immediate reactions to x-rays can, however, be determined with an instrument which measures skin resistance (SCHUPPLI and WAGENER).

The after-effects of injury by ionizing irradiation extend from acute to chronic to late radiation damage. The late changes, for instance the ulcers that occur as late as several years later, are usually not only due to radiation damage but to a combination of injuries (FLASKAMP: for further details see STEIN). Of clinical importance is the fact that, in general, depigmentation and telangiectasia occur even after relatively mild radiation, and slight atrophy and checkered pigmentation occur in general only after more serious injury (KLOSTERMANN). From the pathologic and anatomic point of view, the radiation changes are characterized by fibrous sclerosis of the connective tissue, and a depopulation of cells of the connective tissue (GAHLEN), a change which is different from the usual dermatologic scars. On the other hand, as far as the acid mucopolysaccharides are concerned, no difference between radiation sclerosis and, for instance, the sclerosis of the corium in circumscribed scleroderma can be found (ISHIKAWA, THIES, SCHUMACHER and KLASCHKA). Radiation sclerosis generally is progressive and eventually goes on via necrobiosis to late-developing ulcers which, because they heal so slowly, are clinically worrisome. However, the tissue reaction to x-rays in a general way follows certain individual laws (for details see NOSKO). This increased knowledge of the late changes of x-rays (KOPP and REYMANN, among others) has recently been extended to changes due to ionizing radiation (SCHIRREN).

The dermatologist who uses x-rays is well aware of the fact that the eye, more so than other sensory organs, can be damaged by ionizing radiation. The dermatologist using x-rays in the face area is particularly aware of the possibility of cataract formation. In man, the dose necessary to develop cloudiness of the lens is of the order of 400 r (details in RAJEWSKY, HOBITZ and HARDER). The radiation cataract is a classical late change, which develops several months to two to three years after radiation. Neutron radiation is much more capable of causing cataracts than x-rays or gamma radiation. It is therefore absolutely necessary that the eyes be covered by lead shields during any radiation therapy, and that in the treatment of tumors of the cornea very soft, rapidly absorbed radiation is used (WACHSMANN). A number of reports on the effect of beta radiation on the eye are available (review in SCHRECK). Beta irradiation causes telangiectasia and keratoses of the conjunctiva, atrophy of the sclera, keratitis punctata superficialis, vascularizing pannus of the cornea, iridocyclitis, and cataract, usually after a latent period of many years. The smallest dose to produce cataracts in man by radium was found to be 200 r if given in a single dose. If the radiation was distributed over three weeks to three months, the dose needed was 400 r (MERRIAM). The latent period to cataract formation after a single radium dose to one eye was four and a half to eight and a half years, depending on the dose. By this and other published reports the older results of ROHR-SCHNEIDER (1930) were confirmed, showing that the lens is the most radiation-sensitive part of the eye and that the amount of radiation necessary to injure it is reasonably similar to the amount necessary for the epilation of hair, and that when the dose is divided, a larger dose is needed. High doses of gamma radiation cause ulcerative keratitis, panophthalmia, secondary glaucoma, atrophy of the optic nerve, and rarely, cataracts (CLAUS, DIETHELM and CULLMANN).

The injuries secondary to atomic explosions are due to both heat and radiation effects. Here, skin changes are only a part of the general systemic effects. Direct effects on skin and eye are similar to lightning burns. Although the distance from the explosion is usually rather far and the effect only lasts for a short period of time, third degree burns of the skin are possible and are often followed by particularly severe, bluish-colored keloid formation. From the point of view of the ophthalmologist, blinding and burns of the cornea, retina and chorioid are of importance. In radiation sickness, just as in leukemias or anemia, bleeding into the fundus or exudation may occur. A significant late effect of radioactive injury is the development of radiation cataract, which has been found as far as four kilometers from the explosion center more in young than in old people after a latent period of at least ten months, and with a dose of about 400–600 roentgens (JAMMET, DOLLFUS and HAYE). Laser radiation (laser an acronym for "light amplification by stimulated emission of radiation") is primarily absorbed by pigment in the skin, i.e. by the melanin of the epidermis and the hemoglobin of the red cells, but may also occur after application of indelible ink. Histologically, the effect of repeated laser radiation consists of acanthosis, alterations of the nuclei, changes in the appendages of the skin, thickening of the blood vessels, perivascular infiltration and hemorrhages. Attention must be called to appropriate protection not only of the skin, but particularly of the eye in those who work in laser laboratories (GOLDMAN and RICHFIELD).

References

BURCKHARDT, W.: Zur Frage der photosensibilisierenden Wirkung des Teers. Schweiz. med. Wschr. 69 (1939) 83.

BURCKHARDT, W.: Über eine im Frühling, besonders an den Ohren auftretende Lichtdermatose. Dermatologica (Basel) 86 (1942) 85—91.

EL-HEFNAWI, H., and MORTADA, A.: Ocular manifestations of xeroderma pigmentosum. Brit. J. Derm. 77 (1965) 261—276.

ELSÄSSER, G., FREUSBERG, O., and THEML, F.: Das Xeroderma pigmentosum und die "xerodermische Idiotie." Arch. Derm. Syph. (Berl.) 188 (1950) 651—655.

FLASKAMP, W.: "Über Röntgenschäden." Sonderband Strahlentherapie 12 (1930).

GAHLEN, W.: Histologische Veränderungen röntgenbestrahlten Gewebes. Aesthet. Med. 15 (1966) 193—202.

GOLDMAN, L., BLANEY, D. J., FREEMOND, A., and HORNBY P.: The biochemical aspects of Lasers. J. Amer. Med. Ass. 188 (1964) 304—306.

GOLDMAN, L., and RICHFIELD, D. F.: The effect of repeated exposures to laser beams. Acta derm.-venereol. (Stockh.) 44 (1964) 264—268.

HELLER, J.: Über das gehäufte Vorkommen einer eigenartigen Affektion der Haut der Ohrmuscheln bei Schülern einer Schule. Dermatitis pustularis aurium. Med. Klin. 3 (1907) 1131—1132.

HORNBERGER, W.: Schädigung durch neuere Lichtquellen. Dtsch. Med. Wschr. 75 (1950) 1441—1443.

HUERKAMP, B.: Irisschwund bei Xeroderma pigmentosum. Klin. Mbl. Augenheilk. 119 (1951) 286—292.

IPPEN, H.: Chronische Hautveränderungen durch Lichteinwirkung. Strahlentherapie 123 (1964) 622—631.

ISHIKAWA, H., THIES, W., SCHUMACHER, W., and KLASCHKA, F.: Vergleichende Untersuchungen über das Verhalten der sauren Mucopolysaccharide bei Sklerose der Haut nach Betatron-Bestrahlung bei Sklerodermie und einigen sklerosierenden Dermatosen. Hautarzt 18 (1967) 174—180.

JAMMET, H., DOLLFUS, M. A., and HAYE, Chr.: Les effets oculaires des explosions atomiques. Ann. Oculist. (Paris) 196 (1963) 329—361.

KEINING, E.: Die Frühlingsperniosis zum Unterschied von der Herbstperniosis. Derm. Wschr. 110 (1940) 26—35.

KLOSTERMANN, G. F.: Über die Dosisabhängigkeit verschiedener Röntgen-Spätsymptome der Haut. Z. Haut- u. Geschl.-Kr. 41 (1966) 132—140.

KOCHS, A. G., and LIBOWITZKY, H.: Chronisch polymorpher Lichtausschlag mit Frühjahrskatarrh der Bindehäute und Linsenschädigung. Derm. Wschr. 114 (1942) 438—444.

KOPP, H., and REYMANN, F.: Fortgesetzte Untersuchungen von Spätschäden nach dermatologischer Strahlentherapie. Z. Haut- u. Geschl.-Kr. 42 (1967) 349—354.

KORTING, G. W.: Haut- und Nervensystem in Klinischer Sicht. Münch. med. Wschr. 109 (1967) 1544—1555.

KUSKE, H.: Experimentelle Untersuchungen zur Photosensibilisierung der Haut durch pflanzliche Wirkstoffe. I. Mitt. Lichtsensibilisierung durch Furokumarine als Ursache verschiedener phytogener Dermatosen. Arch. Derm. Syph. (Berl.) 178 (1939) 112—123.

KUSKE, H.: Über die Lokalisation verschiedenartiger Hautausschläge im Sternoclaviculardreieck. Gleichzeitig ein Beitrag zur Resistenzverminderung des Integumentes durch Lichteinflüsse. Dermatologica (Basel) 80 (1939) 6—17

KUSKE, H.: Aktinische Dermatosen. In: Hdb. der Haut- und Geschlechtskrankheiten, Bd. II/2, hsg. von J. Jadassohn. Springer, Berlin, 1965.

LANGHOF, H., MÜLLER, H., and RIETSCHEL, L.: Untersuchungen zur familiären, protoporphyrinaemischen Lichturticaria. Arch. klin. exp. Derm. 212 (1960/61) 506—518.

LARMANDE, A., and TIMSIT, E.: À propos de 20 cas de Xeroderma pigmentosum. Algérie méd. 59 (1955) 557—562. Ref. Zbl. ges. Ophthal. 68 (1956) 171.

MERRIAM, G. R.: Late effects of beta radiation on the eye. Arch. Ophthal. (Chic.) 53 (1955) 708—717.

MIESCHER, G.: Die Schutzfunktionen der Haut gegenüber Lichtstrahlen. Strahlentherapie 39 (1931) 601—618.

NOSKO, L.: Zur Entstehung und Klinik des Röntgenulcus. Z. Haut- u. Geschl.-Kr. 14 (1953) 251—255.

RAJEWSKI, B., HOBITZ, H., and HARDER, D.: Biologische Grundlagen der Röntgentherapie. In: Hdb. der Haut- und Geschlechtskrankheiten, Bd. V/2, spez. S. 133, hsg. von J. Jadassohn. Springer, 1959.

ROHRSCHNEIDER, W.: Schädigungen des Sehorgans bei therapeutischer Anwendung von Röntgen- und Radiumstrahlen. Zbl. ges. Ophthal. 23 (1930) 289.

ROHRSCHNEIDER, W.: Über die Wirkung der Röntgenstrahlen auf das Auge. Strahlentherapie 38 (1930) 665—683.

REESE, A. B., and WILBER, I.: The eye manifestations of xeroderma pigmentosum. Amer. J. Ophthal. 26 (1943) 901—911.

SANCTIS, de C., and CACCHIONE, A.: L'idocia xerodermica. Riv. sper. Freniat 56 (1932) 269—292. Ref. Zbl. Haut- u. Geschl.-Kr 43 (1933) 448—449.

SCHIRREN, C. G.: Ist die Anwendung von Thorium-X-Lack in der dermatologischen Praxis noch vertretbar? Hautarzt 12 (1961) 65—68.

SCHRECK, E.: Veränderungen des Sehorgans bei Haut- und Geschlechtskrankheiten. In: Dermatologie und Venerologie, Bd. IV, spez. S. 880 f., hsg. von H. A. Gottron u. W. Schönfeld. Thieme, Stuttgart, 1960.

SCHUPPLI R., and WAGENER, H. H.: Über ein neues Verfahren zur Erfassung der Sofortwirkung von Röntgenstrahlen. Strahlentherapie 112 (1960) 561—566.

SCHUPPLI, R., and WAGENER, H. H.: Untersuchungen zur Frage des Sofort-Nachweises biologischer Wirkungen von Röntgenstrahlen. II. Mitt. Dermatologica (Basel) 122 (1961) 30—36.

STEIN, G.: Röntgenfolgezustände im Bereich der Haut. Strahlentherapie 121 (1963) 247—258.

THIEL, R.: a) Xeroderma pigmentosum, b) Opticus-Tumoren. Kong. 26.—28. IX. 1949 in Heidelberg. Ber. dtsch. ophthal. Ges. 55 (1949) 346. Ref. Zbl. ges. Ophthal. 51 (1949/50) 165.

THIES, O.: Schädigungen des Auges durch Licht. Derm. Wschr. 85 (1927) 1024—1027.

ULLERICH, K., WULF, K., and WISKEMANN, A.: Augenaffektionen infolge Lichtsensibilisierung durch eine photodynamisch wirksame Substanz. Klin. Mbl. Augenheilk. 131 (1957) 30—46.

URBANEK, J.: Lichtdermatitits und eine Lichterkrankung der Conjunctiva. Z. Augenheilk. 61 (1927) 66—70.

WACHSMANN, F.: Allgemeine Methodik der Röntgentherapie von Hautkrankheiten. In: Hdb. der Haut- und Geschlechtskrankheiten, Bd. V/2, spez. S. 211, hsg. von J. Jadassohn. Springer, Berlin, 1959.

WENDT, G. G.: Allgemeine Humangenetik. In: Hdb. der Haut- und Geschlechtskrankheiten, Bd. VII, hsg. von J. Jadassohn. Springer, Berlin, 1966.

ZENNER, B., and BEUTNAGEL, J.: Dermatitis solaris acuta als besondere Erscheinungsform hochgradiger Lichtüberempfindlichkeit. Strahlentherapie 81 (1950) 617—622.

Disturbances of Pigmentation

Melanogenesis is primarily influenced directly by the cells, but in addition to this there are controlling factors superimposed. These controlling factors consist either of positive ones (primarily the melanocyte stimulating hormone [MSH] which is formed in the pars intermedia of the hypophysis) and negative ones (melatonin, synthesized in the pineal gland, LERNER). Relatively colorless chromogens can be directly oxidized by ultraviolet light. Indirect tanning occurs secondary to ultraviolet-induced erythema. Furthermore, melanogenesis depends to a certain extent on additional factors such as heavy metals and ascorbic acid, among others (KORTING and UHLMANN).

Melanodermas

Trivial examples of melanodermas are the ephelides which, even when they involve most of the face, usually spare the eyelid area since they require insolation for their development ("summer freckles"). Other pigmented areas, such as local accumulations of melanin without the formation of nevus cells (such as nevi spili and lentigines) are also most commonly found in the facial area. These may also represent a partial expression of certain phakomatoses.

One of the various kinds of chloasma (which today is seen primarily following intake of contraceptive drugs) is the "masque biliaire" of the French authors, also called "chloasma hepaticum" (KAUFMANN), which involves the periorbital region. In this relatively sharply demarcated, periocular hyperpigmentation, which HEBRA already assumed to be related to disturbances of the ovaries, one finds dark to light coffee- or dirty nutbrown discolorations, which extend close to the eyebrows and medial to the root of the nose, lateral to the edge of the orbit and caudal to the zygoma. They occur particularly in pyknic females. Such melanodermas should not be mistaken for "rings around the eyes" since these are not due to increased pigmentation, but rather increased dryness of the tissue and loss of the muscle tonus in this area. The "liver masque" certainly does not occur in all women who have bile duct disease and certainly all women who have periocular hyperpigmentation do not have disease of the gall bladder either (PHLEPS, Fig. 84). Similar melanin pigmentation which, however, is darker at the inner canthi, can, according to the author's observation, be found in morbus Basedow, and when present in this disease, is known as the Jellinek sign. Pigmentation which is marked in the periocular area also sometimes occurs as a paradoxical effect following the long-term use of mercury-containing ointments used for the removal of freckles (literature in LÜDERS, FISCHER and HENSEL, Fig. 85). Depigmentation of the eyelids was reported as a side effect of a dermatitis due to eserin used in the treatment of glaucoma (JACKLIN). In younger patients with long-standing allergic rhinitis, the orbital palpebral groove may become bluish-black, probably secondary to stasis of the marginal veins due to edema of the mucosa (Marks, Fig. 86). Among other types of chloasma in the periorbital region is the brown forehead ring described by VOLHARD in 1903 in a 17-year-old male who had left-sided hemiatrophy of the face. This "linea fusca" (ANDERSEN and WERNOE, HAXTHAUSEN; for further literature see MICHEL) occurs primarily in patients who have or had chronic encephalitis. The cardinal example of a hormonally dependent hyperpigmentation is of course that occurring in Addison's disease, where yellowish to very deep brown-bronze discoloration occurs preferentially in the creases of the palms, the nipples, the genital-anal area, and spottily in the oral mucosa.

Pigmentary changes of Addison's disease not infrequently also involve the conjunctiva and particularly the edge of the eyelids. Occasionally, pigment deposition has been reported at the limbus of the cornea, and BITTORF pointed out the darkish discoloration of the fundus (see KAUFMANN). As far as discoloration of the eyeball itself is concerned the reader is referred to the review by SCHRADER.

Nevus fuscocaeruleus ophthalmo-maxillaris of Ota. In 1916, PUSEY reported on the combination of facial and scleral hyperpigmentation in a young Chinese student. OTA in 1930 in Japan named the syndrome as above and FITZPATRICK and his collaborators have renamed it "ocular and dermal melanocytosis." (Figs. 87 and 88). DORSEY and MONTGOMERY believe that the nevus of Ota is a particular localization of the Mongolian spot, i.e. an "atavistic" diffuse mesodermal pigmentation which usually is found extensively on the sacral area and over the buttocks. The nevus of Ota has been reported most commonly in Japan; TANINO recorded 26 cases in 1940 and ITO and YOSHIDA reported on 110 cases. Cases have also been reported in Negro children (HELMICK and PRINGLE). Macroscopically, the sclera is markedly blue and the homolateral vicinity of the eye, upper and

lower lid, the forehead and the cheeks also show a bluish discoloration. Histologically, there are ectopic melanocytes in the sclera. Just as in the blue nevus and the Mongolian spot, rare occurrence of malignant melanoma with metastasis has been described in the nevus of Ota (DORSEY and MONTGOMERY, ALBERT and SCHEIE). Pigmentation of the posterior aspect of the cornea was noted by KRAUS and JUNG.

Incontinentia pigmenti (BLOCH-SULZBERGER). This characteristic and ophthalmologically important pigmented dermatosis occurs more commonly in girls, usually soon after birth or in the first few months of life (Fig. 89). In its earlier stages it is associated with marked eosinophilia of blood and tissue but otherwise with little impairment of well-being. The skin manifestations occur in crops, beginning with lentil-sized urticarial, papulopustular, or even linearly arranged vesicular skin eruptions, all of which resemble dermatitis herpetiformis (DUHRING). In contrast to typical dermatitis herpetiformis, there develops (in connection with the skin manifestations) a characteristic pigmentation in form of a spattered or networklike, sometimes deer-antlerlike, slate gray or chocolate brown figures which may be grouped and associated with lichenoid or

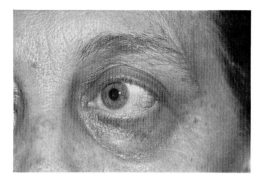

84 "Liver mask."

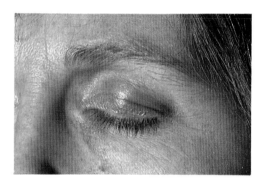

85 Silver deposition in the skin.

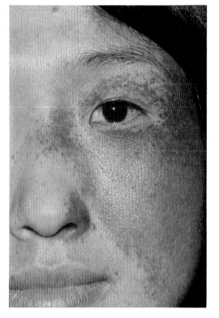

87 Nevus of Ota.

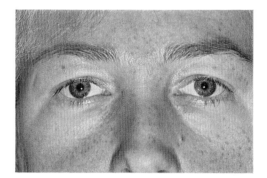

86 Bluish discoloration of the orbital skin secondary to venous stasis because of swelling of the mucosa in the nose.

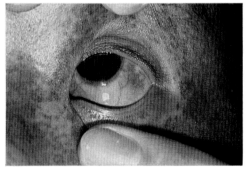

88 Nevus of Ota.

verrucous lesions. Other changes consist of anomalies of the teeth and alopecia of the pseudopelade type.

In contrast to the "familial chromatophore nevus" type of FRANCESCHETTI and W. JADASSOHN, which consists of an increased reticular pigmentation and shows no eye changes, variable abnormalities of the eye are part of the complete picture of incontinentia pigmenti of BLOCH-SULZBERGER. FRANCESCHETTI and JADASSOHN found an incidence of 27.4 per cent and WOLLENSAK found 46 cases with ophthalmologic changes among 175 reported ones (LEBAS: 25 to 35 per cent). The changes are dominated by posterior uveitis, optic atrophy (characteristically of postneurotic form), pseudogliomas (usually secondary to a fetal uveitis), and, not infrequently, strabismus (eight per cent). In the author's own observations in Macedonia (UEBEL, LUDWIG and KORTING) retrolental fibroplasia and the picture of dysgenesis mesodermalis corneae and iridis (RIEGER) have been found and this has been confirmed by FINDLAY.

SCIALDONE and ARTIFONI reported on a rare pigmented dermatosis, which is not incontinentia pigmenti but consists of diffuse areas of achromia separated by larger pigmented areas and associated with retinopathia pigmentosa, cataract, congenital deafness, oligophrenia, and cerebellar ataxia — a syndrome similar to that of Hallgren.

Leukopathias

The genetically determined disease albinism is today considered to be primarily due to blocking of melanocytic tyrosinase and not to a congenital lack of melanocytes in the skin, the hair, and the pigmented layers of the eye (Fig. 90). Albinism may occur in a complete or incomplete form (albinoidism) or may be circumscribed, such as in the development of a white forelock (poliosis circumscripta). Complete universal albinism is genetically usually a simple recessive disease in which, from the ophthalmologists' point of view, marked photophobia, horizontal nystagmus, decrease of visual acuity, and hypoplasia of the macula lutea are found. Other anomalies of the eye consist of alterations of refraction, malformation of the lens, retinitis pigmentosa and astigmatism. Color vision and adaptation to the dark are usually unaffected. Circumscribed partial albinism (leukism) expresses itself as an irregular alternation of pigmented and pigmentless skin areas. The already noted white forelock is usually dominant and occipital poliosis has also been reported.

Albinism of the eye presents either as albinism solum bulbi or albinism solum fundi. The syndrome, described by KLEIN and WAARDENBURG, consists of circumscribed albinism, deaf mutism, blepharophimosis, dystopy of the lower tear punctae, and hypoplasia of the iris with or without heterochromia. This disease is inherited with variable penetrance in the dominant fashion (KLEIN). Syntropy of albinism with congenital miosis was noted by AGOSTON and GROF. In the Chediak-Higashi syndrome, an analog of which also occurs in certain kinds of mink (WINDHORST, ZELICKSON and GORD), there is found, in addition to partial albinism of the skin and eye, granulation of leukocytes, hepatosplenomegaly, general enlargement of lymph nodes and hyperhidrosis. This disease is inherited in a recessive fashion. As far as concerns other pigment changes of the skin associated with deaf mutism, reference is made to the review by REED, STONE, BODER, and ZIPRKOWSKI.

Vogt-Koyanagi-Harada Syndrome. In 1873 SCHENKL described a poliosis circumscripta of the eyelids with sympathetic ophthalmia (Fig. 91). In 1883 NETTLESHIP and in 1892 HUTCHINSON reported similar cases associated with nontraumatic uveitis. Finally, VOGT and KOYANAGI in 1906 and 1929 described the syndrome as a separate entity consisting of "dysacusis, alopecia, and poliosis with severe

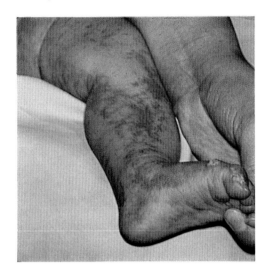

89 Incontinentia pigmenti.

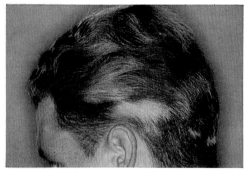

91 Vogt-Koyanagi syndrome.

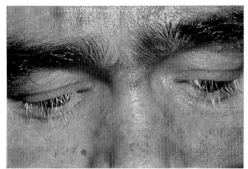

92 Poliosis of the eyelashes.

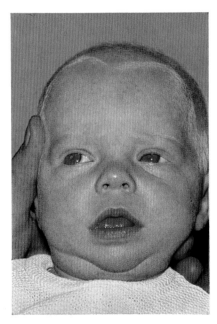

90 Albinism.

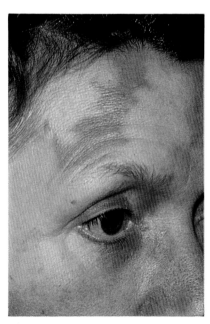

93 Vitiligo.

uveitis of nontraumatic origin." This has been compared with the Harada syndrome (1962) as "the various expressions of an inflammatory process in the eye" (SCHRECK). From the pathogenic point of view, it is thought that either an allergic mechanism (autoimmune reaction against the uveal pigment) or an infection with a virus (case of LUDWIG and KORTING, see also for other cases) is the cause (JOHNSON).

From the point of view of skin changes, the signs of canities or poliosis occur primarily on the eyebrows and eyelashes (Fig. 92) but may also involve the hairs of the head, the axilla, and the mons pubis. Furthermore, changes resembling vitiligo, including sparse hair which sometimes cannot really be differentiated from alopecia areata, may occur (LEWIS and ESPLIN: 62 per cent of cases). In the author's own experience, the changes consist of confluent, bizarre, hairless areas, which are different from the circular areas of alopecia areata. Finally, JESS reported on a case of poliosis, vitiligo, and severe bilateral iridocyclitis in the presence of dystrophia adiposogenitalis.

Nevus of Sutton. A particular variant of circumscribed albinism, i.e., the nevus depigmentosus, is the nevus of Sutton (leukoderma acquisitum centrifugum). This may be isolated or multiple, and may represent a perineval vitiligo usually surrounding a nevus

cell nevus (HALTER). Against the idea that it is vitiligo surrounding a nevus is the clinical observation that, in contrast to typical vitiligo, the pale area surrounding the nevus usually shows no hyperpigmentation at the edge.

Vitiligo. Vitiligo is an acquired acroleukopathia which may be idiopathic or primary, (in that it does not occur following some other changes). It is usually sharply circumscribed, frequently begins in the perianal or genital area, is often symmetrical and characterized by hyperpigmentation of the border of the depigmented area (Fig. 93). Its development is slow; there is no relationship between its localization and any nerve or blood vessel distribution. PINKUS believes that in vitiligo the tyrosinase of the melanocytes located at the junction loses its function, while LERNER assumes that a substance similar to melatonin is secreted by the peripheral nerve endings. This lightens the pigment cells and interferes with the new formation of melanin. One frequently has the impression that the disease is inherited in a dominant fashion. In 20 to 30 per cent of cases a familial distribution is noted. According to Lerner, vitiligo occurs before the 20th year of life in more than 50 per cent of patients. From the ophthalmologic point of view, JÜTTE noted increased tortuosity of the blood vessels of the retina and HOFF the possibility of the simultaneous depigmentation of the iris (see also WEXLER).

References

AGOSTON, I., and GROF, P.: La miose congénitale et l'albinisme. Ophthalmologica Bd. 155, Nr. 5, S. 399 (1968).

ALBERT, D. M., and SCHEIE, H. G.: Nevus of Ota with malignant melanoma of the choroid. Arch. ophthal. (Chic.) 69 (1963) 774–777.

ANDERSEN, O., and WERNDL, T. B.: Der braune Stirnring. Ugeskr. Laeg. 2 (1930) 817.

DORSEY, C. S., and MONTGOMERY, H.: Blue nevus and its distinction from mongolian spot and the nevus of Ota. J. Invest. Derm. 22 (1954) 225–236.

FINDLAY, G. H.: On the pathogenesis of incontinentia pigmenti. Brit. J. Derm. 64 (1952) 141–146.

FITZPATRICK, T. B.: Albinism. In: Metabolic basis of inherited disease. STONBURY, J. B. et al. (eds.) 2d. Edition. McGraw-Hill, New York, 1960.

FITZPATRICK, T. B.: Albinism: some thoughts on the color problem and integration of dermatology and medicine. J. Invest. Derm. 35 (1960) 209–214.

FITZPATRICK, T. B., and LERNER, A. B.: Biochemical basis of human melanin pigmentation. Arch. Derm. Syph. (Chic.) 69 (1954) 133–149.

FITZPATRICK, T. B., ZELLER, R., KUKITA, A., and KITAMURA, H.: Ocular and dermal melanocytosis. Arch. Ophthal. (Chic.) 56 (1956) 830–832.

FRANCHESCHETTI, A., and JADASSOHN, W.: À propos de "l'incontinentia pigmenti," délimitation de deux syndromes différents figurant sous le même terme. Dermatologica (Basel) 108 (1954) 1–28.

HALLGREN, B.: Retinitis pigmentosa in combination with congenital deafness and vestibulocerebellar ataxia. Acta genet. (Basel) 8 (1958) 97.

HARADA, E.: Beitrag zur klinischen Kenntnis von nicht eitriger Choroiditis. Nippon Ganka Gakkai Zasshi 30 (1926) 356–378.

HARADA, E.: Clinical study of non suppurative choroiditis. A report of acute diffuse choroiditis. Acta Soc. Ophthal. Japon. 30 (1926) 356, zit. n. B. Leiber u. G. Olbrich: Die klinischen Syndrome, Bd. I, IV. Ed. Urban & Schwarzenberg, München, 1966.

HAXTHAUSEN, H.: Skin changes in chronic encephalitis. Acta derm.-venereol. (Stockh.) 8 (1932) 408–416.

HELMICK, E. D., and PRINGLE, R. W.: Oculocutaneous melanosis or nevus of Ota. Arch. Ophthal. 56 (1956) 833–838.

HOFF, F.: Akuter totaler Pigmentverlust. Dtsch. Med. Wschr. 79 (1954) 284–287.

HUTCHINSON, J.: A case of blanched eyelashes. Arch. Surg. (Lond.) 4 (1892/93) 357–358.

ITO, M.: Studies on melanin. Tohoku J. exp. Med. Suppl. 55 (1952) 42.

JACKLIN, H. N.: Depigmentation of the eyelids in eserine allergy. Amer. J. Ophthal. 59, 3 (1965) 89–92.

JELLINEK, S.: Ein bisher nicht beachtetes Symptom der Basedow'schen Krankheit. Wien. klin. Wschr. 17 (1904) 1145.

JESS, A.: Demonstration eines Falles von Poliosis, Vitiligo und schwerster doppelseitiger Iridocyclitis bei Dystrophia adiposo-genitalis. Klin. Wschr. 11 (1932) 2165.

JOHNSON, W. C.: Vogt-Koyanagi-Harada syndrome. Arch. Derm. Syph. (Chic.) 88 (1963) 146–149.

JÜTTE, A.: Über Augenhintergrundsveränderungen beim akuten Lupus erythematodes visceralis. Klin. Mbl. Augenheilk. 137 (1960) 765–772.

KAUFMANN, E.: Die pathologischen Pigmentierungen der Haut in innerer Medizin, Neurologie und Psychiatrie. In: Hdb. der Haut- und Geschlechtskrankheiten, Bd. IV/2, hsg. von J. Jadassohn, 1027–1028, Springer, Berlin, 1933.

KLEIN, D.: Les diverses formes héréditaires de l'albinisme. Bull. schweiz. Akad. med. Wiss. 17 (1961) 351–364.

KLEIN, D., and FRANCHESCHETTI, A.: Mißbildungen und Krankheiten des Auges. In: Humangenetik, Bd. IV, hsg. von P. E. BECKER. Thieme, Stuttgart, 1964.

KORTING, G. W., and UHLMANN, W. J.: Untersuchungen über die Wirkung von Ergänzungsfaktoren am Modell der Tyrosin-Tyrosinase-Reaktion. Derm. Wschr. 122 (1950) 1149–1154.

KOYANAGI, Y.: Dysakusis, Alopecia und Poliosis bei schwerer Uveitis nicht traumatischen Ursprungs. Klin. Mbl. Augenheilk. 81 (1929) 194–211.

KRAUS, E., and JUNG, E.: Pigmentierungen der Hornhautrückfläche bei Naevus fusco-caeruleus ophthalmo-maxillaris (Ota). Ber. dtsch. ophthal. Ges. (1967) 442.

LEBAS, P.: Les Syndromes oculo-cutanés, Bruxelles, Imprim. Médicale et Scientifique, 1959, S. 581.

LERNER, A. B.: Vitiligo. J. Invest. Derm. 32 (1959) 285–310.

LUDWIG, A., and KORTING, G.: Vogt-Koyanagi-ähnliches Syndrom und mandibulofaciale Dysostosis (Francheschetti-Zwahlen). Arch. Derm. Syph. (Berl.) 190 (1950) 307–316.

LÜDERS, G., FISCHER, H., and HENSEL, U.: Hydrargyrosis cutis mit allgemeinen Vergiftungserscheinungen nach langdauernder Anwendung quecksilberhaltiger Kosmetica. Hautarzt 19 (1968) 61–65.

MARKS, M. B.: Significance of discoloration in the lower orbitopalpebral grooves in allergic children (allergic shiners). Ann. Allergy 21 (1963) 26–32.

MICHEL, P.-J.: La "Linea Fusca." (brown forehead ring d'Andersen, Werner et Haxthausen). Ann. Derm. Syph. (Paris) 85 (1958) 509–521.

NETTLESHIP, E.: A case of sympathetic ophthalmitis with whitening of the eyelashes. Trans. ophthal. Soc. U. K. 4 (1883/84) 83–84.

OTA, P.: Naevus fusco-caeruleus ophthalmo-maxillaris. Tokyo Jji Shinski (1930) 1243.

PHLEPS, R.: Die klinische Bedeutung der perioculären Hyperpigmentierung ("Pigmentlarve"). Münch. med. Wschr. 96 (1954) 1186–1188.

PUSEY, W. A.: Facial pigmented naevus involving the sclera. Ophthal. Rec. 25 (1916) 618.

REED, W. B., STONE, V. M., BODER, E., and ZIPRKOWSKI, L.: Pigmentary disorders in association with congenital deafness. Arch. Derm. Syph. (Chic.) 95 (1967) 176–186.

SCHENKL, A.: Ein Fall von plötzlich aufgetretener Poliosis circumscripta der Wimpern. Arch. Derm. Syph. (Prag) 5 1873) 136–139.

SCHRADER, K.: Auge und Allgemeinleiden. Schattauer, Stuttgart, 1966.

SCHRECK, E.: Veränderungen des Sehorgans bei Haut- und Geschlechtskrankheiten. In: Dermatologie und Venerologie, Bd. IV, hsg. von H. A. GOTTRON und W. SCHÖNFELD. Thieme, Stuttgart, 1960.

SCIALDONE, D., and ARTIFONI, E.: Un insolito quadro di dermatopatia pigmentaria associata a retinite pigmentosa, cataratta e sordita congenite oligofrenia ed atassia cerebellare. G. ital. Oftal. 17 (1964) 49–60. Ref. Zbl. ges. Ophthal. 96 (1966) 149.

TAMINO, H.: Über eine in Japan häufig vorkommende Nävusform: "Naevus fusco-caeruleus ophthalmo-maxillaris Ota": Beobachtungen über die Augenmelanose als Komplikation dieses Naevus und seine klinische Erscheinung. Jap. J. Dermat. Urol. 47 (1940) 51.

UEBEL, H., LUDWIG, A., and KORTING, G.: Zur Kenntnis der Incontinentia pigmenti Bloch-Sulzberger. Arch. Derm. Syph. (Berl.) 190 (1950) 114–124.

VOGT, A.: Frühzeitiges Ergrauen der Cilien und Bemerkungen über den sogenannten plötzlichen Eintritt dieser Veränderung. Klin. Mbl. Augenheilk. 44 (1906) 228–242.

VOLHARD, F.: Über chronische Dystrophien und Trophoneurosen der Haut im Anschluß an kasuistische Mitteilungen. Münch. med. Wschr. 50 (1903) 1108–1111.

WEXLER: Ocular depigmentation accompanying generalized vitiligo. Arch. Ophthal. 57 (1928) 393.

WINDHORST, D. B., ZELICKSON, A. S., and GOOD, R. A.: A human pigmentary dilution based on a heritable subcellular structural defect. J. Invest. Derm. 50 (1968) 9–18.

WOLLENSAK, J.: Charakteristische Augenbefunde beim Syndroma Bloch-Sulzberger (incontinentia pigmenti). Klin. Mbl. Augenheilk. 134 (1959) 692–706.

YOSHIDA, K.: Nevus fusco-caeruleus-ophthalmo-maxillaris Ota. Tohoku J. Exp. Med. 55, Suppl. 1 (1952) 34.

Circulatory and Vascular Disorders

The angioneuropathias and angiolopathias which are common disorders for the dermatologist (acrocyanosis, erythrocyanosis, perniosis, livedo reticularis s. cutis marmorata) are of very little importance for the periorbital region. Occasionally, livedo reticularis may occur in connection with cerebral vascular changes, for instance, together with a homonymous hemianopsia (SNEDDON). In the area of the cheeks and nose occasionally a purely vasomotor permanent redness occurs on a constitutional basis, which has been called either the "farmer type" (MONCORPS) or simply noted as a constitutional facial mask (OHNSORGE and KEHRER, FISCHER-BRIGGE, BLAICH and ENGELHARDT). The essential telangiectasias, which are sometimes of central origin, may resemble these vegetative permanent erythemas. However, in their presence, one should explore the possibility of anoxia, such as strangulation during delivery or during later life (BRAUN-FALCO and MARGHESCU). The slowly progressive symmetrical telangiectasias of the facial skin (and particularly of the conjunctiva), which are part of the Louis-Bar syndrome, will be described in more detail under the angiophakomatoses. However, very frequently generalized essential telangiectasia of the skin is observed without a particular reason for it (WEBER and ROTH).

It should be noted, however, that angiitis of the retinal vessels found by the ophthalmologist should not in general be considered part of an angiitis obliterans, since according to RATSCHOW, involvement of the eye is very rare in comparison with the frequency of that disease. Altogether, there are hardly any other primarily vascular skin manifestations of the periorbital region.

Endangiitis obliterans

Although, as has been pointed out, ophthalmologic findings are by no means common in angiitis obliterans (of the peripheral occlusive type, phlebitis migrans, livedo racemosa, and so on), there are exceptions which will be mentioned.

In this connection it is necessary to describe the Cogan syndrome, a nonsyphilitic interstitial keratitis with functional disturbances of the statoacusticus nerve, although the sensory abnormalities occurring in this syndrome cannot definitely be ascribed to vascular changes. Furthermore, eye changes occur in the syndrome in only 20 per cent of cases and are (even when histologically determined) of nonspecific nature (FISHER and HELLSTROM).

In thromboangiitis cutaneo-intestinalis disseminata, described originally in 1941 by KÖHLMEIER as "multiple necrosis of the skin in thromboangiitis obliterans" and in 1942 considered to be an independent disease by DEGOS, we find an obliterating vasculitis of the small blood vessels of the skin. DEGOS, DELORT, and TRICOT in 1942 named this entity "dermatite papulo-squameuse atrophiante" and in 1948 renamed it "papulose atrophiante maligne." Some 25 cases have now been reported. The disease manifests itself clinically in the extensive development of crops of pale, edematous papules with central necrosis, which spare the face, palms, and soles although they are otherwise widespread. As the disease progresses, it is characterized by weakness, lack of appetite, and loss of weight, and the gastrointestinal tract is involved with multiple infarcts of the wall of the gut. Equivalent manifestations of these systemic obliterations of blood vessels are found in the eye either as circumscribed areas of ischemia, as scleritis, episcleritis and end-arteriolitis obliterans of the subconjunctiva (EICHENBERGER and collaborators), micro-aneurysms of the conjunctiva bulbi, and chorioiditis or visual disturbances with fixation of the pupil. MAWAS and SIDI noted the onset of a "papulose strophiante maligne" in the eye.

Periarteritis nodosa

The entity originally described by KUSS-MAUL and MAIER of Freiburg in 1866 under the name "aneurysma verminosum hominis" and then described in detail in the same first volume of the "German Archiv for Clinical Medicine" as "periarteritis nodosa" has been known for over 100 years and described in detail in at least 1000 cases and several monographs. Since GRUBER we have known its tendency to extend irregularly to multiple areas and about the allergic genesis of this systemic blood vessel disease. Through the reports of ZEEK, WEGENER, CHURG and STRAUSS and others, we have known about the existence of a variety of special forms, such as hypersensitivity angiitis, giant cell angiitis, and others. The disease occurs primarily in males between the ages of 20 and 45 years. Often, it is preceded by allergic disease (bronchial asthma, urticaria, drug allergy, and the administration of sera). The clinical signs are extensive, with involvement of the kidneys, heart, liver, the perirenal and gastrointestinal areas, the peripheral nerves, and, of course, the skin. The extent of the involvement of the skin varies between 15 and 64 per cent of cases (KOR-TING). There may be a variety of cutaneous manifestations, which may present as multiple neurotic gangrene of the skin, apoplexia cutanea (FREUND), or multiple polymorphous erythemas of the skin, and occasionally may consist only of Schönlein-pur-pura or livedo racemosa (GOLDSCHLAG and CHWALIBOGOWSKI).

As can be seen from the review by SCHRECK, the eye is involved in periarteritis nodosa in the form of disturbances of vision in about 25 per cent of cases, and as objective findings of the eye in 10 to 20 per cent of cases. The eyelids and conjunctiva may show, although only rarely, the classical nodular lesions. BERN-STEIN described thrombosis of the central artery, SANNICANDRO atrophy of the optic nerve, while KERNOHAN and WOLTMAN noticed chorioiditis.

DEJAJANNE reported on retinitis pigmentosa. VOGEL found in the literature reports on exophthalmus, corneal ulcers, intraocular bleeding and amotio retinae. The clinical diagnosis of eye changes in periarteritis nodosa is thus made difficult because of the polymorphous manifestations (see LEBAS).

Wegener's granulomatosis

This "rhinogenous granulomatosis" consists of a giant cell granulomatous angiitis involving the nose, sinuses, lungs, and spleen. It is also associated with changes in the kidney without hypertension or edema. It was probably described as early as 1896 by McBRIDE and particularly in 1931 by KLIN-GER (see MILLS). In 1947 it was called "Wegener's granulomatosis" by RINGERTZ. It is most prevalent in the third and fifth decades of life and occurs primarily in men, as does periarteritis nodosa. It is not easy to classify Wegener's granulomatosis clearly, and it has been considered by many authors as a real tumor (for instance, a reticulum cell sarcoma) or as identical with granuloma gangraenescens (KRAUSS, 1929). Nevertheless, this giant cell granulomatous angiitis has been separated as a special respiratory-renal variety of periarteritis nodosa. As far as the skin changes are concerned, one finds most commonly hemorrhagic macular or nodose changes, which may become ulcerated (KUNTZ, BENEKE and

KNOTH). According to the author's own observations, a clinical picture simulating SCHÖNLEIN's purpura or a papulo-necrotic picture ("arteriolitis hyperergica") may develop (KORTING and LACHNER).

The eye changes which have been noted were conjunctivitis (KESSELRING and ZOLLINGER), iritis, and scleritis or episcleritis (BOCK). KRAUSS, VORTEL and associates described also ulcers at the edge of the cornea and we ourselves have seen infiltrates near the limbus. BÖKE and REICH point out that the cornea is "apparently only involved in the region of the limbus." Most of the ocular signs such as exophthalmus, ptosis, disturbances of the eye muscles, dacryocystitis, episcleritis, iritis, bleeding of the retina or cotton-wool areas are nonspecific. However, the eye may also be involved directly in Wegener's granulomatosis by the disease process from the sinus into the orbit or by the hematogenous route. BÖKE and REICH reported necrotizing scleritis and ulcus rodens as part of Wegener's granulomatosis (MOOREN, see also in LEBAS). Unilateral or bilateral exophthalmus, sometimes with optic neuritis and blindness, were also described by KLINGER in 1931, FAHEY and associates — (Case 7) and LAPP.

Arteritis temporalis
(HORTAN, MAGATH and BROWN)

In this giant cell arteritis (GILMOUR), which was probably described in 1890 by HUTCHINSON, the terminal branches of the external and internal carotid and of the temporal or ophthalmic arteries are primarily involved. This has been considered a failure of the vasa vasorum because of age by PILLAT. Symptomatically, severe unilateral attacks of headache occur. These usually awaken the patient at night, one to two hours after he has fallen asleep, because the unendurable orbital or temporal pain does not allow him any further rest. The pain is associated with epiphora,

chemosis, and rhinorrhea. Changes in vision occur in 10 to 40 per cent of the cases (KINMONT and McCALLUM). Such ocular symptoms can in untreated cases lead very rapidly to amaurosis within a few hours because of occlusion of the central artery with or without ischemic edema and colliquative necrosis of the optic nerve near the papilla, or because of unilateral (50 per cent) or bilateral (25 per cent) retrobulbar neuritis (VALENTI). Paresis of the oculomotor muscles has also been noted. On the other hand, eye changes may be relatively minor even though there are severe skin changes (necrosis of the scalp) due to temporal giant cell arteritis (case of LUGER and WUKETICH). Furthermore, the vascular changes may also involve the coronary arteries and the supra-aortal arteries, producing the picture of the pulseless disease (TAKAYASU). Histologically, the findings consist of narrowing of the lumen of the blood vessels and splitting of the elastic fibers which, as shown by HAMPERL, may act as foreign bodies. In such an area there develops a tissue full of lymphocytes and fibroblasts, with a striking content of epithelioid and giant cells within the granuloma.

Arteriolitis "allergica" cutis

In 1957, RUITER included a series of clinical manifestations (maladie trisymptomatique de Gougerot, allergides nodulaires dermiques, etc.) under this title because of similar pathologic changes in the small blood vessels of the corium. These primary vasculitides of the skin, which are characterized by fibrinous changes and inflammatory reaction, are clinically very polymorphous. Thus, one can find either urticarial, hemorrhagic, papulonecrotic, or nodular appearances. In the author's experience, no ophthalmic changes have been found associated with this leukocytoclastic vasculitis. WILKINSON reported eye changes as well as renal and arthritic components of importance.

References

BERNSTEIN, A.: Periarteritis nodosa without peripheral nodules diagnosed ante mortem. Amer. J. Med. Sci. 190 (1935) 317.

BLAICH, W., and ENGELHARDT, H.: Zur Frage der Entstehung der essentiellen Teleangiektasien, der "vasomotorischen Dauerrötung" und ähnlicher Gefäßveränderungen. Hautarzt 5 (1954) 357—362.

BOCK, H. E.: Die hyperergischen Gefäßerkrankungen. In: Angiologie, hgs. von M. Ratschow, Thieme, Stuttgart, 1959.

BÖKE, W., and REICH, H.: Augenbeteiligung bei der Wegenerschen Granulomatose. Klin. Mbl. Augenheilk. 151 (1967) 802—822.

BRAUN-FALCO, O., and MARGHESCU, S.: Essentielle Teleangiektasien nach Verbrühung und Strangulation. Derm. Wschr. 153 (1967) 553—558.

COGAN, D. G.: Syndrome of nonsyphilitic interstitial keratitis and vestibuloauditory symptoms. Arch. Ophthal. 33 (1945) 144—149.

DEGOS, R.: Malignant atrophic papulosis: a fatal cutaneointestinal syndrome. Brit. J. Derm. 66 (1954) 304—307.

DEGOS, R., DELORT, J., and TRICOT, R.: Dermatite papulosquameuse atrophiante. Ann. Derm. Syph. (Paris) 49 (1942) 148—150, 281.

DEGOS, R., DELORT, J., and TRICOT, R.: Bull. Soc. Méd. Hôp. Paris 64 (1948) 803, zit. n. Eichenberger, Landolt u. Wegmann.

DEJEAU, VIALLEFONT, CHAMPION: Périartérite noueuse et rétinite pigmentaire. Bull. Soc. Ophthal. Fr. (1953) 252.

EICHENBERGER, H., LANDOLT, E., and WEGMANN, W.: Papulose atrophiante maligne Degos. Schweiz. med. Wschr. 97 (1967) 1639—1649.

FAHEY, J. L., LEONARD, E., CHURG, J., and GODMAN, G.: Wegener's granulomatosis. Amer. J. Med. 17 (1954) 168.

FARRERAS VALENTI, P.: Die Arteriitis temporalis. Münch. med. Wschr. 107 (1965) 1859—1864.

FISHER, E. R., and HELLSTROM, H. R.: Cogan's syndrome and systemic vascular disease. Arch. Path. 72 (1961) 572—592.

GOLDSCHLAG, F., and CHWALIBOGOWKI, A.: Periarteriitis nodosa. Lemberger Dermat. Ges. Sitzung 26. IV. 1934, Ref. Zbl. Haut- u. Geschl.-Kr. 49 (1935) 1.

HAMPERL, A.: Elastische Fasern als Fremdkörper. Bemerkungen zur sog. Arteriits temporalis. Virchows Arch. path. Anat. 323 (1953) 591.

HORTON, B. T., MAGATH, T. B., and BROWN, G. E.: Arteritis of the temporal vessels. Arch. intern. Med. 53 (1934) 400—409.

HORTON, B. T., MAGATH, T. B., and BROWN, G. E.: Arteriitis temporalis oder cranialis. In: Angiologie, hsg. von M. Ratschow. Thieme, Stuttgart, 1959.

KERNOHAN, J., and WOLTMAN, H. W.: Periarteritis nodosa. A clinico-pathologic study with special reference to the nervous system. Arch. Neurol. Psychiat. (Chic.) 39 (1938) 655.

KESSELRING, Fr., and ZOLLINGER, H. U.: Die Wegenersche Granulomatose. In: Ergebn. inn. Med. Kinderheilk. 16 (1961) 41—78.

KINMONT, P. D. C., and McCALLUM, D. I.: Skin manifestations of giant-cell arteritis. Brit. J. Derm. 76 (1964) 299—308.

KLINGER, H.: Grenzformen der Periarteriitis nodosa. Frankfurt. Z. Path. 42 (1931) 455.

KÖHLMEIER, W.: Multiple Hautnekrose bei Thrombangiitis obliterans. Arch. Derm. Syph. (Berl.) 181 (1941) 783—792.

KORTING, G. W.: Über cutane Periarteriitis nodosa unter besonderer Berücksichtigung begleitender Leberstörungen und der sogenannten Thrombophlebitis migrans. Arch. Derm. Syph. (Berl.) 199 (1955) 332—349.

KORTING, G. W., and LACHNER, H.: Arteriolitis hyperergica als Hauterscheinungsbild bei Wegenerscher Granulomatose. Die Med. Welt 1968, 2115.

KRAUS, Z., VORTEL, V., FINGERLAND, A., SALAVEC, M., and KRCH, V.: Unusual cutaneous manifestations in Wegener's granulomatosis. Acta derm. venereol. (Stockh.) 45 (1965) 288—294.

KUNTZ, E., BENEKE, G., and KNOTH, W.: Die Wegenersche Granulomatose. Med. Welt. (Stuttg.) 18 (1967) 285—304.

LAPP, H.: Pathologisch-anatomischer Beitrag zur Pathogenese und nosologischen Stellung des malignen Granuloms der Nase. Virchows Arch. path. Anat. 331 (1958) 487.

LEBAS, P.: Les Syndromes oculo-cutanés, Bruxelles, Imprimerie Médicale et Scientifique, 1959, 233.

LUGER, A., and WUKETICH, St.: Kopfschwartennekrose bei temporaler Riesenzellarteriitis. Derm. Wschr. 153 (1967) 89—98.

MAWAS, H., and SIDI, H.: Première localisation oculaire de la papulose atrophiante maligne. Bull. Soc. Ophthal. Fr. (1961) 70—74.

MILLS, C. P.: Wegener's granulomatosis. Brit. J. Derm. 77 (1965) 203—206.

PILLAT, A.: Über allergisch-hyperergische Gefäßerkrankungen an der Netzhaut. Ber. Dtsch. Ophthalm. Ges. 61 (1968) 66.

RATSCHOW, M.: Angioorganopathien. In: Angiologie, hsg. von M. Ratschow. Thieme, Stuttgart, 1959.

RUITER, M.: Über die sogenannte Arteriolotis (Vasculitis) allergica cutis. Hautarzt 8 (1957) 293—301.

RUITER, M., and OSWALD, F. H.: Weiterer Beitrag zur Kenntnis der Arteriolitis (Vasculitis) "Allergica" cutis (Purpura Schoenlein, leukoklastische Mikrobide, anaphylactoide Purpura, maladie trisymptomatique de Gougerot, allergides nodulaires dermiques usw.). Hautarzt 14 (1963) 6—18.

SANNICANDRO, G.: Periarterite nodulare dell'arto inferiore sinistro. Arch. ital. Derm. 9 (1933) 70.

SCHRECK, E.: Veränderungen des Sehorgans bei Haut- und Geschlechtskrankheiten. In: Dermatologie und Venerologie, Bd. IV, hsg. von H. A. Gottron u. W. Schönfeld. Thieme, Stuttgart, 1960.

SCHUPPLI, R.: Periarteriitis nodosa. In: Hdb. der Haut- und Geschlechtskrankheiten, Bd. II/2, hsg. von J. Jadassohn. Springer, Berlin, 1965.

SNEDDON, I. B.: Cerebro-vascular lesions and livedo reticularis. Brit. J. Derm. 77 (1965) 180—185.

VOGEL, K.-H.: Die Periarteriitis nodosa und ihre Verlaufsformen. I. Die Periarteriitis nodosa. Med. Welt (Stuttgart) 1961, 2328—2337.

WEBER, G., and ROTH, W. G.: Generalisierte essentielle Teleangiektasien an Haut und Conjunctiven. Z. Haut- u. Geschl.-Kr. 42 (1967) 655—658.

WEGENER, F.: Über eine eigenartige rhinogene Granulomatose mit besonderer Beteiligung des Arteriensystems und der Nieren. Beitr. path. Anat. 102 (1939) 36—38.

WILKINSON, D. S.: Some clinical manifestations and associations of "allergic" vasculitis. Brit. J. Derm. 77 (1965) 186—192.

Hemorrhagic Diatheses

VIRCHOW (1854) defined hemorrhagic diatheses as the propensity for abnormal bleeding from the openings of the body or into the skin, mucosa or inner organs, including alterations of clotting. The cutaneous and extracutaneous manifestations of the hemorrhagic diatheses are at the present time classified into vascular, thrombocytic, or coagulopathic varieties.

Angiopathies

Osler's disease

A number of earlier observations are now classified as part of Osler's disease (FRANK, 1840, LEGG, 1876, etc.) (Fig. 94). Among these, the description by RENDU (1896) is particularly important, since he pointed out the frequent nose bleeds, sparing of the extremities, and involvement of the mucosa (lips, cheeks, tongue, soft palate) with pinhead to lentil size "real hemangiomas." Furthermore, he noted that these are due to a dilatation of the superficial vessels of the skin. RENDU particularly attempted to differentiate his observations from real hemophilia. Later, in addition to the classic manifestations of the disease such as dominant inheritance and bleeding phenomena, some less characteristic and only occasionally appearing dilatations of the terminal blood vessels were described, that OSLER in 1901 characterized as "telangiectasias" rather than as angiomatosis. In this connection it must be pointed out that the concept "angioma" in the past has been used without clear definition of the type of proliferation of blood vessels, without dividing sharply between ectasia or new formation (that is, "angioma" formation).

In the modern concept of Osler's disease there are telangiectasias in circumscribed small areas on otherwise normal skin, in some areas simulating spider nevi, in other areas appearing more like angiomas. These lesions are localized primarily on the flexor areas of the fingers, also under the nails, then in the middle portion of the face and, according to the author's observations, also on the toes. In contrast to this, the trunk, except the upper chest, is almost always free of lesions. Of particular clinical importance (in addition to the primary sign of nose bleeds) are the manifestations of Osler's disease on the mucosa of the stomach, intestinal tract, bladder, liver, and lung (arteriovenous lung fistula). These latter manifestations may be of prime importance in individual patients.

As far as the pathologic anatomy of the more angiomatous lesions (found in the phase where nose bleed is a prime symptom) is concerned, and the more telangiectatic lesions that are found later in life, early reports describe a "particularly striking dilatation of the blood vessels in the area of the affected skin, to which not only the papillary capillaries but also the deeper-lying blood vessels contribute" (GOTTRON). In contrast to this, a reduction of the perivascular elastic fibers and edema of the surrounding collagen tissue is considered to be of less pathognomonic importance (MEMMESHEIMER). In the histologic description of Osler's disease both a cavernous and a thrombotic character of the dilatation of the papillaries also have been reported (FINGERLAND and JANOUSEK). Later, the knowledge of the characteristic histopathologic changes in Osler's disease was significantly expanded by the demonstration of arteriovenous anastomoses and occlusion of arteries (KINDLER and TIEDEMANN; NÖDL; GOTTRON and KORTING). Underlying the completely developed telangiectasia in Osler's disease may be a convolution of latent dysplastic blood vessels or an alteration of the

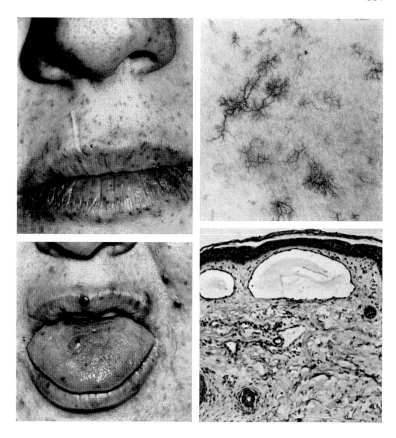

94   Osler's disease.

peripheral course of the blood vessels that can result in a quantitative and qualitatively increased reactive development, particularly in response to irritation mediated by hormonal labile phases. According to the author's own observation this does not have to occur in every case. Based on the author's own extensive observations, the pathologic and anatomic basis of Osler's disease, even in serial sections, may only be "simple" telangiectases, so that even on histologic grounds there is a certain polymorphism (KORTING).

As can be expected, there are also, in this hereditary vasculopathy, changes in the eye area of the terminal vessels associated with increased weakness of the vascular wall and fragility of the blood vessels. In his review, SCHRECK points out "bloody tears," similar to nose bleeds (see also FRANÇOIS), which are caused by lesions of Osler's disease of the eyelids, the orbit, and the conjunctiva. (For further description see PITTER; WITTMER; BERGMANN and WIEDEMANN; GARNER and GROSSMANN.) A variety of changes have also been noted, such as miliary aneurysms, streaklike bleeding, subretinal bleeding with changes similar to retinitis exudativa externa or hemorrhagic glaucoma. In the presence of pure punctate bleeding in the area of the palpebral or bulbar conjunctiva, one has to consider in the differential diagnosis the possibility of a fat embolus, particularly when the biopsy shows sudanophilic material in the lumen of the dermal blood vessels (STEPHENS and FRED).

Observations of essential telangiectasia of the skin with cloudiness of the lens, occasionally combined with defects of the stroma of the iris and malformations of the heart and aorta (NIEDEN; PETERSEN) are related to classical Rendu-Osler disease. Furthermore, the occurrence of melanomas of the chorioid has occasionally been observed in Osler's disease (DEKKING; ORAVISTO). This observation, similar to the observations of cataract formation, probably should be considered only coincidental (LEBAS).

Schönlein's Purpura

Schönlein's purpura, which, in the presence of renal and abdominal involvement, has also been termed Henoch's purpura, occurs primarily in young people. It has a marked predilection for the lower parts of the body ("purpura orthostatica"). In the beginning, the lesions are urticarial, appearing as bright red macules which only under pressure (or when the initial exudative component subsides) show intrafocal bleeding points which simulate flea bites. Only in older or recurrent exanthems are the purely petechial lesions prevalent, as originally described by SCHÖNLEIN. From the histologic point of view this purpura really presents a leukocytoclastic-hemorrhagic microbid which develops according to allergic reactions to infection, food, or drugs, gradually becoming "rheumatoid". Later on, particularly in protracted cases, autoimmune reaction mechanisms may appear, giving rise to the picture of the so-called Waldenström purpura hyperglobulinemica (KORTING and ADAM). As far as the periorbital region is concerned, SCHÖNFELD reported bleeding into the skin of the eyelids, the conjunctiva, and the vitreous. LÖWENSTEIN noted red patchlike lesions of the iris that probably represented areas of bleeding.

Progressive Pigmentary Purpura

The clinical manifestations of rust brown, pigmented, often more or less annular, cutaneous manifestations were classified under the above title by LEVER and by KALKOFF. Dermatologists, based on some clinical differences, consider this purpura variously as purpura Majocchi, Schamberg's disease, dermatite lichénoide purpurique et pigmentée (GOUGEROT and BLUM), purpura arciformis (TOURAINE), eczematidlike purpura (DOUKAS and KAPETANAKIS), itching purpura (LOWENTHAL), ADALIN purpura (LOEB as well as MULZER and HABERMANN, and others). All these diseases are characterized by an orthostatic arrangement and rarely extend further up than the belt line, so that appearance in the area of the eyelids is possible, but extremely unlikely.

Thrombocytopathies

Morbus maculosus Werlhof

The classical signs of thrombocytopenia originally described by WERLHOF in 1735 on clinical grounds (which symptomatically simulate thrombasthenia of GLANZMANN) are characterized by a triad of bleeding from all body openings (in the absence of bleeding into the joints), petechiae and suffusions; in other words, the appearance side by side of punctate and flat ecchymoses (Fig. 95). Cooperating causes are trauma (Fig. 96) or certain endocrine influences (premenstruum). Laboratory examination shows thrombocytopenia without any significant anemia or leukopenia, immature megacariocytosis of the bone marrow, prolongation of the bleeding time, positive Coombs test and prolonged recalcification time; further, pathologic clot retraction, and positive Rumpel—Leede test (in the absence of splenomegaly).

In children, some sort of infection usually precedes the exanthematous-appearing acute Werlhof disease by several days, while the diminution of blood platelets in adults usually due to drug allergy quinine frequently appears after several hours (Fig. 97). Chronic Werlhof's disease occurs primarily in girls or young women.

In the syndrome of thrombotic thrombocytopenic purpura (Moschcowitz, 1925) the classic disease manifestations include hemolytic anemia and transient nerve disturbances.

The syndrome of Kasabach and Merritt, in which splenogenic anemia in the presence of giant hemangiomas temporarily leads to thrombocytopenic purpura, will be discussed in the chapter on angiomas (see p. 149).

From the ophthalmologic point of view, thrombocytopenic hemorrhages not infrequently involve the skin of the eyelids or the conjunctiva, as can be seen in Fig. 95. Schönfeld also mentions extravasations into the vitreous or the retina, as well as exophthalmus secondary to bleeding. Schreck points out the appearance of cuplike hemorrhages between the retina and choroid with suffusion of the choroid, and mentions the observations of Schall on dilatation and edema of the blood vessels and edema of the papilla as being part of Werlhof's disease.

Coagulopathies

The basis of these hereditary hemorrhagic diatheses is usually a defect in protein formation or an enzymopathy. The best known of these is the recessive, sex-linked hereditary disease, hemophilia A, described by Schoenlein in 1828, which is due to diminution of factor VIII. Following the least trauma there develops in the first few days of life or later at the first attempts at walking, bleeding into the skin or muscle or joints (hemarthrosis) of various degrees.

Hemophilia B (Christmas disease, PTC deficiency), also a recessive, X-chromosome-linked disease, is clinically identical, except the cause is a deficiency of factor IX.

Parahemophilia is due to a lack of factor V and is clinically hardly separable from hemophilia.

In the eye area, Schrader noted bleeding into the eyelid, the conjunctiva, the anterior chamber, the vitreous, and the retina. In the presence of retrobulbar bleeding, exophthalmus and lagophthalmus may develop. Compression of the optic nerve and of the central artery may cause blindness. Schreck pointed out the possibility of fatal hemorrhages to the outside following bleeding injuries of the eyelid or the conjunctiva, or uncontrollable bleeding from the inner eye itself.

References

Bergmann, G., and Wiedemann, E.: Beobachtung in Sippen und Teleangiectasia hereditaria haemorrhagica. Dtsch. Arch. klin. Med. 202 (1955) 26—51.

Dekking, H. M.: Presentation of cases: Phacomatosis (?), teleangiectases of the face and a choroideal melanoma. Ophthalmologica (Basel) 118 (1948) 1034.

Garner, L. L., and Grossmann, F. E.: Hereditary hemorrhagic teleangiectasis. Amer. J. Ophthal. 41 (1956) 672—679.

Korting, G. W., and Adam, W.: Purpura Schönlein und Leberzirrhose in ihrer Abgrenzung von der "Purpura hyperglobulinaemica." Derm. Wschr. 131 (1955) 121—128.

Korting, G. W., and Denk, R.: Über die klinischen Unterschiede zwischen Fabry-Krankheit und M. Osler (hier weitere Lit.) Med. Welt (Stuttg.) 17 (1966) 851—855.

Lebas, P.: Les Syndromes oculo-cutanés. Bruxelles, Imprimerie Médicale et Scientifique, 1959, 124.

Löwenstein, A.: Roseolanähnliche Affektion der Regenbogenhaut neben punktförmigen Bindehautblutungen bei hämorrhagischer Diathese. Klin. Mbl. Augenheilk. 59 (1917) 583—588.

Nieden, A.: Cataractbildung bei teleangiektatischer Ausdehnung der Capillaren der ganzen Gesichtshaut. Zbl. prakt. Augenheilk. 11 (1887) 353—357.

Oravisto, T.: Hereditary hemorrhagic teleangiectasis as an eye disease. Acta ophthal. (Kbh.) 30 (1952) 447—452.

Petersen, H. P.: Teleangiectasis and cataract. Acta ophthal. (Kbh.) 32 (1951) 565—571.

PITTER, J.: Eine seltene Lokalisation der Osler'schen Krank-
   heit unter der Bindehaut. Klin. Mbl. Augenheilk. 107
   (1941) 76–80.
SCHÖNFELD, W.: Dermatologie für Augenärzte. Thieme,
   Stuttgart, 1947.
SCHRADER, K. W., and RAUH, W.: Auge und Allgemein-
   leiden, Schattauer, Stuttgart, 1966.
SCHRECK, E.: Veränderungen des Sehorgans bei Haut- und

Geschlechtskrankheiten. In: Dermatologie und Venerolo-
   gie, Bd. IV, hsg. von H. A. Gottron u. W. Schönfeld.
   Thieme, Stuttgart, 1960.
STEPHENS, J. H., and FRED, H. L.: Petechiae associated with
   systemic fat embolism. Arch. Derm. Syph. (Chic.) 86
   (1962) 515–517.
WITTMER, R.: Conjunctivalveränderungen beim Morbus
   Osler. Ophthalmologica (Basel) 121 (1951) 158–159.

Metabolic and Storage Diseases

Porphyrias

Porphyrinurias are characterized by abnor-
mally high excretion of normal porphyrins
(HOPPE-SEYLER). Porphyrias, in contrast, are
characterized by the excretion of porphyrins
which normally do not occur and are primary,
genetically determined abnormalities of the
synthesis of heme. As far as the clinical
appearance is concerned, depending on the
porphyrin disease present, gastrointestinal,
central nervous system, and particularly dermatologic changes dominate the picture.
Eruptions consisting of vesicles and bullae
(at first clear as water and then cloudy) and
crusting eruptions leading to scarring occur
in the early years of life, sometimes as late as
puberty. The areas of the body which are
exposed to light are affected, and such eruptions are called hydroa aestivale or summer
prurigo (HUTCHINSON) or, if the disease is
more serious and forms scars, hydroa vacciniforme (BAZIN). Both these forms of hydroa
may be associated with porphyrinuria or may
occur without it. Most of these entities
which in the past have been included under
the heading of hydroa probably belong either
to the chronic polymorphic light eruption
or to other porphyrias.

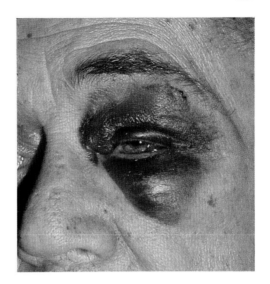

95   Werlhof's disease.

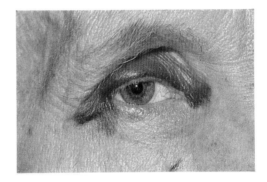

96   Ordinary hematoma.

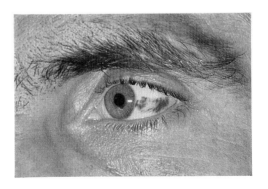

97   Ecchymosis of bulbar conjunctivitis.

Hepatic porphyria

Porphyria hepatica acutc intermittens. This disease is generally transmitted as a hereditary dominant trait on an endogenous toxic basis and is basically an abnormality of enzymes of the liver manifested by the excretion in the urine of precursors of porphyrin, such as porphobilinogen and delta-amino-levulinic acid. As far as the skin is concerned, hyperpigmentation is present, and abdominal pain simulating an "acute abdomen," as well as polyneuritis or central nervous system syndromes, is found.

Porphyria cutanea tarda (WALDENSTRÖM) *s. hepatica* (WATSON). This porphyria, which generally begins late in life (i.e., rarely before the 50th or 60th year) is characterized by the development of bullae produced by trauma or light or both, a tendency to diabetes, and increased serum iron associated with lack of transferrin (BREHM and HOLZMANN). Milia may follow blister formation; hypertrichosis of the eyebrows (bushy eyebrows) is found frequently, and diffuse hyperpigmentation and sclerovitiliginous areas may develop (Fig. 98). It occurs primarily in males with outdoor occupations and frequently in the presence of excessive intake of alcohol.

South African form of porphyria (porphyria variegata). As far as the skin is concerned, this disease shows changes similar to those of cutaneous porphyria, but is associated with symptoms of acute intermittent porphyria. Its manifestations develop primarily at the time of puberty.

Erythropoietic porphyria

Porphyria congenita (GÜNTHER). Up to this time there have been less than 100 cases of this disease reported. It is due to the formation of porphyrin type I, presumably based on a recessively inherited trait, which expresses itself early in childhood. Characteristic features are the burgundy red urine and extreme light sensitivity which leads to formation of blisters, hydroa aestivale. This is followed by ulcers and scarring which affect primarily the nose, the top of the ears, and the fingers (sclerodactylia) Additional symptoms are loss of hair, a dirty-gray coloration of the skin, atrophy as well as loss of nails, and erythrodontia. Some of the red blood cells may fluoresce bright red ("porphyrocytes") and occasionally hemolytic anemia develops.

Familial protoporphyrinemic photodermatosis (urticaria due to light). In this light-induced skin disorder, which was originally described by KOSENOW and TREIBS (1953), and LANGHOF (1960), reddish brown wheals or blisters develop in early childhood following sunlight exposure. These are preceded by early erythema and eventually result in the formation of colloid milia. A second form of this type of erythropoietic protoporphyria presents itself as a light-induced urticaria of the cholinergic type. In contrast to Günther's disease, there are no excretion of uroporphyrin type I isomer, and no mutilations. Characteristic is the bright-red fluorescence of red blood cells in tissue sections due to increased formation of protoporphyrins. Furthermore, occasionally abnormal amino acid excretion in the urine occurs.

Congenital erythropoietic coproporphyria. In this light dermatosis, which was first described by HEILMEYER and CLOTTEN and is characterized by itching and swelling of the skin following exposure to light without permanent skin changes or changes of urine or erythrodontia, the primary abnormality is an increase of coproporphyrin III in the erythrocytes. It is not present in the urine or the feces.

In Hartnup's syndrome, which was observed in 1956 by Dent in a London family named Hartnup, an error in enzyme formation of the tryptophane metabolism is the underlying cause. Based on a recessively inherited gene, mental retardation and marked cerebellar stigmata develop and are associated with an increased excretion of certain amino acids and indoles but not of porphyrins.

From the *ophthalmologic* point of view, mutilation may occur in the eye area similar to the changes on fingers, nose, and ears in porphyria congenita. These changes may result in ectropion, symblepharon, leukoma, or ulcerations of the sclera. In acute porphyria BARNES and BOSHOFF described round retinal edemas simulating cottonwool exudates, which eventually developed into spotty pigmentations. Occasionally, deep bleeding points could be seen. JAFFE observed hemorrhages in acute porphyria anterior to the nerve fiber layer and chorioid and edema of the papilla. In porphyria cutanea tarda, Barnes and Boshoff noted progressive dislike or ring-shaped choroidal pigmentations, which eventually involved the entire eyeground up to the area of the macula and were usually preceded by edema. TOPPEL also noted circular pigmentation, hemorrhages, cottonwool exudates, papilledema, and other changes of the blood vessels of the retina in porphyria cutanea tarda. In fifteen patients with this disorder seen by the author, however, such pigmentation of the fundus could be noted in only one patient. CULLMANN, DENK, and HOLZMANN were able to show disturbances of color vision, particularly a deuteroanomaly in five of fifteen male patients, (in considerable excess over that found in the normal population).

Lipoidoses

The lipid storage diseases are based on genetic enzyme defects, resulting in a pile-up and storage of metabolic components or intermediary products, which are formed before the point of blockage of the enzyme.

Hypercholesterinemic xanthomatosis. In this disease there occur plane papular or tuberous xanthomas of the skin whose characteristic yellowish red color is due to a mixture with carotinoids. The blood serum remains clear since the neutral fats are only slightly increased. In general, the genesis of xanthomata involves primarily mesenchymal tissues which are rich in mucopolysaccharides, thus explaining the frequent appearance of xanthomas of the tendons. The dangerous character of hypercholesterinemic xanthomatosis (which is generally transmitted in a dominant fashion) is expressed by the high percentage of coronary arteriosclerosis (MONTGOMERY: about 25 per cent), which may even cause instant cardiac death in young patients.

From the *ophthalmologic* point of view, xanthelasmas are found on the eyelids. Xanthelasmas of the eyelids rarely develop before the end of the second decade and in general as late as the fourth decade of life. Presence of xanthelasma palpebrarum should always be a reason for an investigation of the serum cholesterol level and less often suggests the presence of diabetes mellitus and hypertension. However, there are many patients with xanthelasma palpebrarum who themselves, as well as their siblings, have normal serum cholesterol (Figs. 99 and 100). In general, these patients should be considered as carriers of primary hypercholesterinemic xanthomatosis. Eyelid xanthelasma occurs twice as often in women as in men (EPSTEIN, ROSENMAN, and GOFMAN) and is particularly common in Jews (ROBINSON). The flat, ivory to lemon-yellow xanthelasmas develop at the inner canthus and extend outward on the upper lid. The lower lid is rarely affected. The development is usually symmetrical and slow until it reaches a stationary phase. Spontaneous partial involution, however, occurs only rarely. In older persons, xanthelasma of the eyelids may show comedones or cysts (xanthelasma sebaceum, HUTCHINSON; for details see SALFELD). This is understandable, since the periorbital region in any case tends to the development of grouped late comedones even without the development of the complete picture of *élasteidose cutanée nodulaire avec kystes et comédons, Favre-Racouchot* (Fig. 101). In patients showing such grouped periorbital comedones, AGIUS found a frequent association with hypertension and diabetes mellitus.

In association with xanthelasmas of the eyelid, one frequently finds arcus senilis which generally is also due to the deposition of cholesterol. As has been pointed out by RINTELEN, this "arcus lipoides" develops at some distance from the edge of the limbus, first in the area of Descemet's and then that of Bowman's membrane. In patients over 70 years of age, this gerontoxon occurs regularly, and in patients in their forties it occurs in about 15 per cent of cases. In young patients such a deposition of cholesterol, phospholipids, and neutral fats (as already pointed out under the heading of *Xanthelasma*) should always make one suspicious of an underlying idiopathic hypercholesterinemia, but in addition local tissue changes in the vicinity of the limbus are probably necessary. In any case, Rintelen does not believe that the presence of an arcus lipoides should be considered an alarming symptom as far as arteriosclerosis or reduced life expectancy is concerned. KLEIN and FRANCESCHETTI pointed out that ring-shaped, cloudy areas (which simulate arcus senilis) may be congenital, lie in the superficial areas of the parenchyma, and be separated from the limbus by a clear area. Such changes may be associated with megalocornea, blue sclerae and other abnormalities.

In contrast to this, the Kayser-Fleischer corneal ring is a pathognomonic early sign of Wilson's disease, which is due to a congenital enzymatic abnormality in which, because of lack of ceruloplasmin, serum copper becomes freed and may be deposited in the brain stem and the liver. In contrast to arcus senilis, this brownish green or greenish gray corneal "copper" ring is located directly at the edge of the limbus in the deep layers of the cornea, i.e., in Descemet's membrane, where in early cases it can be detected only with the slit lamp. Furthermore, not infrequently a "sunflower pseudocataract" may develop, as occurs after intraocular injury with copper chips.

Hyperlipemic xanthomatosis. In this form one sees almost entirely papular and tuberous xanthomas in the presence of milky or creamy serum. The papular xanthomas arise suddenly in eruptive form, and may also disappear spontaneously. The major type which occurs in children was described and is associated with hepatosplenomegaly by BÜRGER and GRÜTZ (1932). Hyperlipemia may also develop subsequently to glycogen storage disease, diabetes mellitus, pancreatitis, and nephrosis.

Necrobiosis lipoidica (OPPENHEIM, 1929, URBACH, 1932). This is found primarily on the leg, generally as single lesions which are thickened, somewhat depressed in the center and either ivory-white or billiard-ball yellow. Diabetic women between the ages of 20 and 40 are primarily affected. Occasionally, lesions may be found on other parts of the body where they are then difficult to differentiate from granuloma annulare (Fig. 102).

Nevoxanthoendothelioma (MAC DONAGH, 1912). This is the most common form of xanthoma occurring in children (Fig. 104). From the histopathologic point of view, it is a form of histiocytoma, but nevoxanthoendothelioma, in contrast to histiocytoma, frequently involutes spontaneously. Clinically, the lesions consist of single or multiple, dome-shaped, whitish or yellowish nodules, about 1 cm. in diameter; they do not ulcerate. The prime location is the edge of the scalp. The disease occurs at any time up to the second year of life. Occasionally, internal lesions such as those in the testes or lungs have been noted. They therefore must be differentiated from Hand-Schüller-Christian disease. Occasionally, nodules have been reported on the iris (BLANK, EGLICK, and BEERMAN; MAUMENEE; NEWELL; HELWIG and HECKNEYN). Secondary glaucoma (BLANK and associates; NEWELL; LEVER) and bleeding into the anterior parts of the eye were also noted. WILK-WILCZYNKSA and WOŽNIAK described uveitis and hydrophthalmus which resulted in enucleation of the affected eye; occasionally enucleation was performed because of suspicion of sarcoma or melanoma.

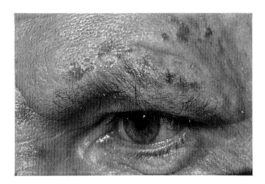

98 Porphyria cutanea tarda

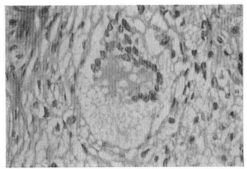

100 Xanthoma. Histologic section with Touton giant cell.

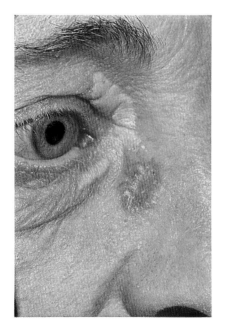

99 Xanthelasma palpebrarum.

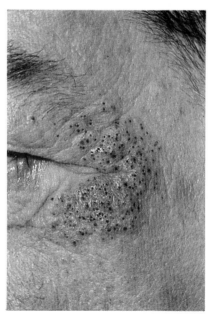

101 Grouped periorbital comedones.

Amyloidoses

In the amyloidoses (VIRCHOW, 1853) the infiltrate consists of paraplastic mixtures of materials consisting of carbohydrates, protein and lipids; these mixtures appear greasy, waxy or sago grainlike and stain with iodine just like starch. In particular, the material stains well with congo red (BENNHOLD, 1923). The deposits are grouped around either collagenous or reticular structures of connective tissue (MISSMAHL). Development of amyloidosis leads to chronic exhaustion of the organs and eventually to death. In principle, this dystrophic disease affects the reticulohistiocytic system (BRUNS) and is on a genetic basis; it may be acquired or is "idiopathic." Demonstration of the disorder is based either on polarization-microscopic or fluorescence-microscopic methods using congo red or thioflavine as stains. Acquired pericollagenous amyloidoses occur primarily in the presence of plasmacytoma or in macroglobulinemia. In the differential diagnosis, familial amyloidosis must be separated from nephro-, neuro-, or myocardiopathies.

Skin changes occur in generalized amyloidosis in about 25 per cent of patients. Individual lesions are either petechial or papular, or lichenoid eruptions, subcutaneous infiltrates, or sclerodermalike or poikilodermic manifestations (GOTTRON). The localized skin para-amyloidosis can resemble neurodermatitis circumscripta or prurigo, particularly when located on the legs ("lichen amyloidosus").

In the periorbital region, one finds in systemic amyloidosis either bleeding into the skin of the eyelids or densely grouped nodules, and, less commonly, flat or tumorlike masses (Fig. 105). In one of the author's cases, which recently was published by Hoede, there were many individual, milialike skin-colored or brownish pinhead-sized papules, which only occasionally showed a more reddish tint (Fig. 105).

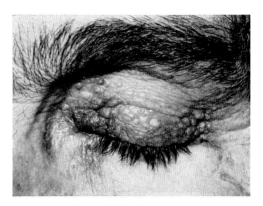

105   Systemic amyloidosis.

Hyalinosis cutis et mucosae (WIETHE, 1924), Lipoid proteinosis (URBACH)

In this disease, which is somewhat similar to para-amyloidosis, a PAS-positive substance consisting primarily of phospholipids and glycoproteins, probably infiltrates from the blood stream, are deposited (WEYHBRECHT and KORTING). In the earliest months of life of the patient one notes hoarseness, which is due to such deposits (Fig. 107). This is followed by slowly progressive, at times even reversible, skin changes on the face, scalp, axilla and other parts of the body; they consist of nodular, hyperkeratotic, whitish, yellow to brown, verrucous nodules associated with white atrophic areas. In this disease, due to the deposit of lipid material, macroglossia with loss of the papillae of the tongue may occur. The associated involvement of the esophagus and stomach, as well as intracranial calcification are important manifestations. This disease is apparently autosomally recessively inherited.

From the ophthalmologic point of view, J. BACSKULIN and E. BACSKULIN noted that eye involvement in hyalinosis cutis and mucosae is primarily expressed at the edges of the eyelids. There it may show a variety of pictures, which do not always parallel the

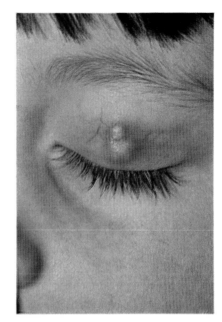

102 Granuloma annulare.

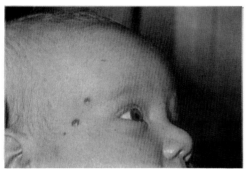

103 Nevoxanthoendothelioma.

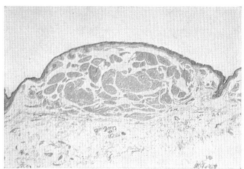

106a Histologic appearance of an amyloid nodule (congo-red stain).

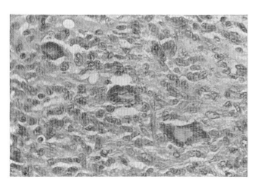

104 Nevoxanthoendothelioma. Histologic appearance.

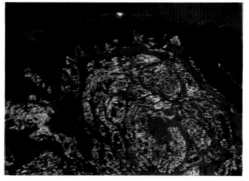

106b Histologic appearance of an amyloid nodule (congo-red fluorescence technique).

changes on the mucosa or skin (Fig. 108). SCHRECK saw glassy, pale nodules, which were arranged like a string of pearls along the edges of the eyelids. Furthermore, there was abnormal growth of the eyelashes, which stood on individual little elevations in several irregular rows. GERTH and FLEGEL noted the following in a 33-year-old patient: multiple, "drusen"-like, fine, small, round, grayish-white nodules with an unsharply determined

tween the fourth and fifth decades but also in children as cretinism, the abortive and localized lesions of mucinosis (circumscribed pretibial myxedema) (RICHTER; KEINING) (Fig. 109), and the euthyroid myxedemas, such as lichen myxedematosus (NEUMANN, 1935) (Figs. 110 and 111) or scleromyxedema complicated by paraproteinemia (ARNDT-GOTTRON), are found only in the older age groups.

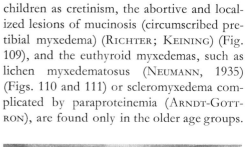

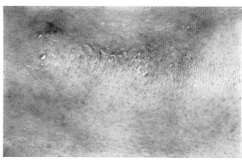

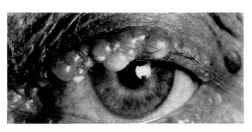

108 Hyalinosis cutis et mucosae. String of pearl appearance of nodules along the edge of the eyelid.

109 Localized myxedema. Orange peel skin.

border and without pigment. They primarily involve the lamina elastica interna. These same nodules also are present at the periphery, particularly on the lower part of the conjunctiva. HOLTZ, together with SCHULZE, reported "drusen"-like changes of the macula, which were found because the patient had an acute attack of glaucoma and which manifested themselves as small, darkly pigmented patches next to yellowish areas. RODERMUND reported dystrophic areas on one side in the vicinity of the macula as "an expression of the deposit of hyaline substances." Opacity of the cornea is rare according to FRANÇOIS, BACSKULIN, and FOLLMANN.

Mucinosis

In contrast to diffuse myxedema (GULL; ORD), which is seen primarily in women be-

The tuberonodular or circumscript myxedema is due either to an attempt at treatment or to an exacerbation of *hyperthyroidism* (details in GOTTRON and KORTING). Much rarer is localized myxedema due to obliterating vascular changes (KORTING and WEBER). By electron microscopy such myxedema shows two different types of cells: on the one hand, the stellar-shaped "mucoblast," and on the other, the usual fibroblast (KORTING, NÜRNBERGER, and MÜLLER).

The progressive *protrusion of the eyeball*, which, however, occurs more frequently without localized myxedema, appears in the latter *after* treatment of hyperthyroidism.

SCHRADER mentions in connection with myxedema loss of eyelashes, strabismus, errors of refraction, dystrophy of the cornea, papillary edema, and optic atrophy. The exophthalmos in Hand-Schüller-Christian disease is in most cases due to the destruction of the orbit; exceptionally, exophthalmos may

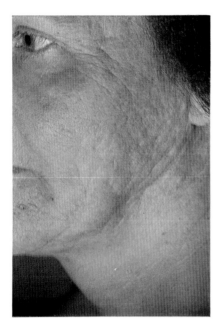

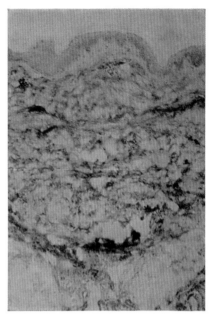

107 Hyalinosis cutis et mucosae. Trabecular wrinkling of the skin.

111 Lichen myxedematosus. Histologic appearance (PAS Alcian blue stain).

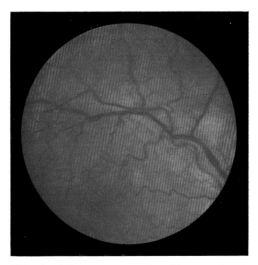

114 Angiokeratoma corporis diffusum *(Fabry)*. Eyeground. (Collection of Prof. Hager, Berlin.)

occur with eosinophilic granuloma of the bone. The pathogenesis of the so-called *malignant endocrine exophthalmos* of SAUTTER depends upon a failure of the regulation between the hypophysis and the thyroid, especially when the thyrotropic hormone of the anterior part of the hypophysis and the exophthalmos factor, which apparently are chemically different, are secreted in increasing amounts due to a sudden stoppage of the thyroid function. HORST and ULLERICH arrive at the explanation "that the endocrine ophthalmopathy is triggered by the hypophysis via the nervous system." The severity of a high grade protrusion of the eyeball manifests

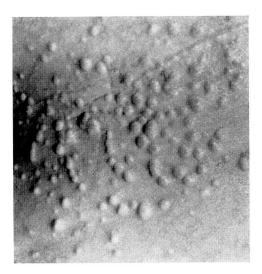

110 Lichen myxedematosus.

itself clinically in stasis and strangulation of the blood vessels of the conjunctiva, insufficient closure of the eyelids with subsequent defects of the cornea, disturbance of the motility of the eye with subjective double vision, retrobulbar neuritis with central scotoma, and sometimes secondary glaucoma. MICHEL saw a very hypertensive patient with scleromyxedema, who presented "sclerotic" chorioiditis.

Sphingolipidoses

While cutaneous lesions are absent in Tay-Sachs disease, Niemann-Pick disease presents melanin pigmentation, occasionally resembling mongolian spots and, less frequently, xanthomas and erysipelaslike or granulomatous infiltrates. The cutaneous manifestations of Gaucher's disease consist of melanin pigmentation, purpuric exanthema, or uncharacteristic ecthymas.

The phosphatid lipoidosis of NIEMANN (1914) and PICK (1927) has been observed in about 100 cases, mostly in small children. Their ophthalmologic examination showed in the region of the macula a cherry-red spot surrounded by a white zone; this resembles the "red spot" found in Tay-Sachs idiocy (for details see PFÄNDER). Gaucher's disease, described in 1892 by the French dermatologist GAUCHER as "primary epithelioma of the spleen," shows storage of the cerebroside kerasin in the reticuloendothelial system of liver, spleen, lymph nodes and bone marrow; occasionally, the conjunctiva adjacent to the cornea shows brownish-yellow spots resembling pingueculae (EAST and SAVIN).

Cutaneous gout

Primary chronic cutaneous gout (usually in the form of butter-cream colored tophi on the rims of the ears or the tips of the fingers) affects as a rule males after the age of 40. With regard to *ocular* involvement, GOTTRON and KORTING saw a 22-year-old butcher with a serum uric acid level of 16.6 mg. per cent presenting somewhat above and nasally to the macula of the left eye a glistening point, which according to HARMS represented a crystalline deposit similar to the observation of WAGEMAN (1897). In addition, SCHRECK reports ocular signs of gout such as tophi of the eyelids, conjunctivitis, or episcleritis nodosa.

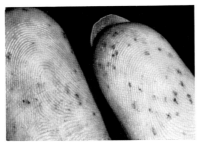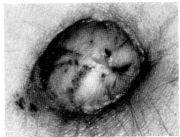

112 Angiokeratoma corporis diffusum *(Fabry).*

Calcinosis cutis

Deposition of calcium salts in the skin or subcutaneous tissue may appear *"in metastatic form"* (VIRCHOW), that is, in systemic disturbances of calcium and phosphorus metabolism (vitamin D intoxication, and the milk and alkali syndrome) or in *metabolic* form (with normal values of blood calcium and phosphorus in scleroderma (THIBIÈRGE and WEISSENBACH, 1911), or dermatomyositis (RUDOLPH, 1934). Localized disturbances of circulation also seem to be responsible for calcium deposits (findings of I. BREHM and HEINZEL of the Mainz Clinic). The term *dystrophic calcinosis* is used to describe calcification of previously damaged tissue (for example, in pseudoxanthoma elasticum, in Malherbe's calcifying epithelioma, or in the Ehlers-Danlos syndrome).

tions by CROCKER, FOX, and others (Figs. 112 and 113). Subsequently, the cutaneous concept was extracutaneously enlarged by the addition of a *cardio-vaso-renal syndrome* by RUITER et al. (1939, 1947, 1954, etc.), which in retrospect makes us understand the findings of Fabry and Anderson. They reported in their first cases associated proteinuria followed by edema of the legs and eyelids. SCRIBA (1950) called the disorder a *phosphatid storage disease*; additional reports by HORNBOSTEL, SPIER, KOCH, and SCRIBA but also by RUITER are generally known today.

The *cutaneous lesions* of this metabolic disease appear as purple or dark blue to blackish macules; they are in places slightly elevated, but appear only rarely as keratotic, superficial papules. These manifestations measure a few millimeters in diameter (STEINER and VOERNER: "angiomatosis miliaris"); they are cir-

Angiokeratoma corporis diffusum (FABRY)

Our knowledge of the cutaneous manifestations of angiokeratoma corporis diffusum is primarily based on two reports which appeared in 1898; one by FABRY, citing HEBRA, speaks of "purpura papulosa hemorrhagica." However, in 1916, FABRY, reviewing his original description, changed the name to the one currently used. The other report by ANDERSON ("a case of angiokeratoma") also refers to probably analogous earlier observa-

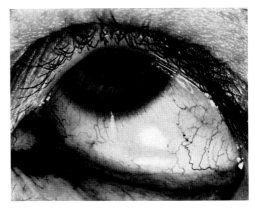

113 Angiokeratoma corporis diffusum *(Fabry).* Tortuosity of the blood vessels of the conjunctiva.

cular or hemispherical. Very important for the diagnosis of Fabry's disease as a storage disease is the presence of double-refractive lipoids seen as fine or larger droplets in muscles and also in the endothelium of cutaneous blood vessels. By electron-microscopy large lysosomes can be seen. The more or less uniform cutaneous lesions of Fabry's disease favor as an exanthema the central portions of the trunk, especially the umbilical region, and appear rather densely by the hundreds or more, in somewhat symmetrical distribution (see further details in KORTING and DENK).

Ophthalmologically, in about 130 cases, deposits of trihexose-ceramide (a glycolipid) have been observed (see KREMER and DENK), together with a periorbital edema; in his original observation Fabry himself had described this edema. The fundus showed tortuosity of the vessels; aneurysmatic dilatation of the vessels of the conjunctiva, papillary edema, and, especially, corneal opacities were present. The latter are, according to SPAETH and FROST, so characteristic that their presence should raise the possibility of this diagnosis. In addition, these authors emphasize the frequent development of posterior capsular cataracts. The opacities of the cornea presumably first described by WEICKSEL (1925) may be the only ocular sign in females, according to RAHMAN. Lately, the stigma of "cornea verticillata" (whorl-like—Translators) (see GRUBER) has been associated with Fabry's syndrome (FRANCESCHETTI; RINTELEN).

Ochronosis

Alkaptonuria described in a 44-year-old diabetic in 1859 by BOEDEKER—although there had been previous reports (see HENNEMANN and KRAUSE)—was the first known example of recessive inheritance in man. In 1891, WOLKOW and BAUMANN found that homogentisic acid was the chemical substrate of the disorder. VIRCHOW, however, had

spoken as early as 1866 of "ochronosis of the cartilage and bonelike structures."

Alkaptonuria is caused by an enzymatic defect in the metabolism of phenylalanine and tyrosine, and can be diagnosed by examining the urine by TROMMER's or FEHLING's tests, paper chromatography, or by the addition of ferric chloride, which produces an evanescent blue or green discoloration.

Pseudoalkaptonuria has been observed following external application of phenol, pyrogallol, or beechwood tar.

The conspicuous brownish or bluish deposits of pigment on the ears or ochronosis of the entire cartilaginous system occur around the thirtieth or fortieth year of life, at a time when homogentisic acid is not any longer completely excreted in the urine (see PÄRTAN and THUMB). The patient suffering from ochronosis may present cicatrizing alopecia (FLECK) and diminished intelligence. Death may be due to ochronotic nephrosis (COOPER and MORAN).

Ophthalmologically, one frequently sees on the sclera within the palpebral fissure grayish-blue spots similar to the ones seen on the skin or the cartilage of the ear. Scleral ochronosis develops somewhat simultaneously with the pigmentation of the ears (SMITH). FRIDERICH and NIKOLOWSKI observed that within the conjunctiva the pigment was distinctly arranged around the smaller vessels, whereas at the limbus of the cornea the pigment in the epithelium was grouped in clumps. The fundus of the eye offered no unusual findings. LAYMON observed a 71-year-old man suffering from ochronosis showing slate gray sclerae while the superficial part of the cornea on both sides of the limbus presented dark blue pigmented spots. LEBAS mentions reports of optic neuritis and glaucoma in ochronosis.

Phenylketonuria

In 1934, FÖLLING first described phenylpyruvic oligophrenia, a disease which becomes manifest shortly after birth. The cutaneous changes are not too characteristic: delicate skin, tendency to diminished pigmentation, eczematous changes, follicular keratoses, hyperhidrosis, photosensitivity and others. The disease is transmitted by an autosomal recessive gene. The diagnosis depends on the dark blue green color that develops when 10 per cent ferric chloride solution is added to the acidified urine, or on the "phenistix test" (for further details see Bickel and Cleve).

Ophthalmologically, the preponderance of blue-eyed individuals with blond hair is emphasized (Jervis). This characteristic was observed even in a mulatto and a Negro and was thought to be due to the inhibition of melanogenesis by the phenol bodies (Danois and Balis).

Carotinosis

The young child with carotinosis presents a distinctive yellow discoloration of the palms and soles, and occasionally systemic involvement (sparing the conjunctivae). It is caused by eating food containing carotinoids. In the adult, however, carotinosis is also decisively determined by the level of the serum lipids. "Aurantiasis" may develop, therefore, in adults even without excessive intake of carotene in the presence of diabetes, hypothyroidism or nephrosis.

Argyria

This irreversible silver impregnation of the skin comes about through long-continued absorption of silver nitrate by several types of application; this results in the storage of silver in the skin for life, at first in a bluish discoloration of the face, neck, and hands, but later, without any systemic ill effects, a slate gray color develops. Sometimes this manifestation appears quite suddenly, for instance, after intense sun exposure (Mehnert). As far as the eye is concerned, the mouse-gray discoloration—for instance, following application of eyedrops containing silver nitrate—affects preponderantly the lower half of the bulbus oculi. Schreck mentions also dustlike grayish deposits in the lens similar to those seen in chrysiasis (due to parenteral use of gold salts —Translators) or mercury deposits in the lens observed in workers in thermometer factories.

References

Agius, J. R. G.: Grouped periorbital comedones. Brit. J. Dermat. 76 (1964) 158–164.

Bacskulin, J., and Bacskulin, E.: Beitrag zur Klinik der Hyalinosis cutis et mucosae (Lipoid-Gluko-Proteinose Urbach-Wiethe). Acta ophthal. (Kbh.) 43 (1965) 610–628.

Barnes, H. D., and Boshoff, P. H.: Ocular lesions in patients with porphyria. Arch. Ophthal. 48 (1952) 567–580.

Bickel, H., and Cleve, H.: Metabolische Schwachsinnsformen. In: Humangenetik, Bd. V/2, edited by P. E. Becker. 208, Thieme, Stuttgart, 1967.

Blank, H., Eglick, P. G., and Beerman, H.: Nevoxanthoendothelioma with ocular involvement. Pediatrics 4 (1953) 202.

Brehm, G., and Holzmann, H.: Quantitative Bestimmungen der Bluteiweiße, speziell des Transferrins und des Haptoglobins bei der Porphyria cutanea tarda. Klin. Wschr. 42 (1964) 283–286.

Brehm, I., and Heinzel, J.: Calcinosis circumscripta als Folge einer umschriebenen Durchblutungsstörung. Derm. Wschr. 149 (1964) 10–15.

Bruns, G.: Die Amyloidosen. Klin. Wschr. 45 (1967) 868–874.

Cooper, J. A., and Moran, T. J.: Studies on ochronosis. Arch. Path. 64 (1957) 46–53.

Cullmann, B., Denk, R., and Holzmann, H.: Zur Häufung von Farbsinnstörungen bei der Porphyria cutanea tarda (Waldenström). v. Graefes Arch. klin. exp. Ophthal. 170 (1966) 201–208.

Dancis, J., and Balis, M. E.: A possible mechanism for disturbance in tyrosine metabolism in phenylpyruvic oligophrenia. Pediatrics 15 (1955) 63.

East, T., and Savin, L. H.: A case of Gaucher's disease with biopsy of the typical pinguecula. Brit. J. Ophthal. 24 (1940) 611.

Epstein, N. N., Roseman, R. H., and Gofman, J. W.: Serum lipoproteins and cholesterol metabolism in xanthelasma. Arch. Derm. Syph. (Chic.) 65 (1952) 70–81.

FLECK, F.: Zur Symptomatik und Entstehung der endogenen Ochronose. Derm. Wschr. 134 (1956) 1317–1326.

FRANCESCHETTI, A.: Personal communication.

FRANÇOIS, J., BACSKULIN, I., and FOLLMANN, P.: Manifestations oculaires du syndrome d'Urbach-Wiethe. Ophthalmologica 155, 433–448 (1968).

FRIDERICH, C. H., and NIKOLOWSKI, W.: Endogene Ochronose. Arch. Derm. Syph. (Berl.) 192 (1951) 273–289.

GERTH, H., and FLEGEL, H.: Hyalinosis cutis et mucosae. Derm. Wschr. 133 (1956) 10–19.

GOTTRON, H. A., and KORTING, G. W.: Zur Pathogenese des Myxoedema circumscriptum tuberosum. Arch. Derm. Syph. (Berl.) 195 (1953) 625–649.

GOTTRON, H. A., and KORTING, G. W.: Chronische Hautgicht. Arch. klin. exp. Derm. 204 (1957) 483–499.

GRUBER, M.: "Cornea verticillata" (Eine einfache dominante Variante der Hornhaut des menschlichen Auges). Ophthalmologica (Basel) 111 (1946) 120–129.

HELWIG, E. B., and HACKNEY, V. C.: Juvenile xanthogranuloma (nevoxanthoendothelioma). Amer. J. Path. 30 (1954) 625.

HENNEMANN, H. H., and KRAUSE, H.: Alkaptonurie mit Ochronose. Geschichte einer Krankheit, in der sich die Entwicklung der naturwissenschaftlichen Medizin widerspiegelt. Med. Welt (Stuttg.) 1964, 1549–1553.

HOEDE, N.: Beschreibung einer Systemamyloidose. Derm. Wschr. 154 (1968) 145–153.

HOLZMANN, H., DENK, R., and CULLMANN, B.: Porphyria cutanea tarda und Farbenblindheit. Klin. Wschr. 44 (1966) 592–593.

HORST, W., and ULLERICH, K.: Hypophysen-Schilddrüsen-Erkrankungen und endokrine Ophthalmopathie. Bücherei des Augenarztes, Enke, Stuttgart, 1958.

JAFFE, U. S.: Acute porphyria associated with retinal hemorrhages and bilateral oculomotor nerve palsy. Amer. J. Ophthal. 33 (1950) 470–472.

JERVIS, G. A.: Phenylpyruvic oligophrenia (phenylketonuria) Assoc. Resp. Nerv. Dis. Proc. 33 (1954) 259.

KAUFMANN, E.: Die pathologischen Pigmentierungen der Haut in innerer Medizin, Neurologie und Psychiatrie. In: Hdb. der Haut- und Geschlechtskrankheiten, Bd. IV/2, edited by J. Jadassohn. 1011, Springer, Berlin, 1933.

KORTING, G. W.: Die Beziehungen zwischen Haut und Leber mit ihren diagnostischen und therapeutischen Möglichkeiten. Therapiewoche 12 (1962) 19.

KORTING, G. W., and DENK, R.: Über die klinischen Unterschiede zwischen Fabry-Krankheit und M. Osler. Med. Welt (Stuttg.) 17 (1966) 851–855.

KORTING, G. W., NÜRNBERGER, F., and MÜLLER, G.: Zur Ultrastruktur der Bindegewebszellen beim Myxoedema circumscriptum praetibiale. Arch. klin. exp. Derm. 229 (1967) 381–389.

KORTING, G. W., and WEBER, G.: Bericht über eine solitäre tuberöse Mycodermie des linken Handrückens. Arch. klin. exp. Derm. 216 (1963) 354–364.

KREMER, G. J., and DENK, R.: Angiokeratoma corporis diffusum (Fabry). Lipoidchemische Untersuchungen des Harnsediments. Klin. Wschr. 46 (1968) 24–26.

LAYMON, C. W.: Ochronosis. Arch. Derm. Syph. (Chic.) 67 (1953) 553–559.

LEBAS, P.: Les Syndromes oculo-cutanés, Bruxelles, Imprimerie Médicale et Scientifique, 1959, 776.

LEVER, W. F.: Histiocytosis. Arch. Derm. Syph. (Chic.) 79 (1959) 608.

MAUMENEE, A. E.: Ocular lesions of nevoxantho-endothelioma (infantile xanthoma disseminatum). Trans. Amer. Acad. Ophthal. Otolaryang. 56 (1956) 401.

McDONAGH, J. E. R.: A contribution to our knowledge of the naevo-xantho-endotheliomata. Brit. J. Dermat. 24 (1912) 85–99.

MEHNERT, E.: Argyrose – eine seltene, irreversible Arzneimittelschädigung. Dtsch. Gesundheitswesen 23 (1968) 228–231.

MICHEL, P. J.: Generalis. Mucinosis papulosa Arndt-Gottron mit Sklero-Myxödem. Bull. Soc. franç. Derm. Syph. 73 (1966) 146.

MISSMAHL, H. P.: Klinik und Diagnostik der verschiedenen Amyloidosen. Münch. med. Wschr. 107 (1965) 846–850.

MONTGOMERY, H.: In der Diskussion zu N. N. Epstein, R. H. Roseman u. J. W. Gofman. Arch. Derm. Syph. (Chic.) 65 (1952) 79–80.

NEWELL, F. W.: Nevoxanthoendothelioma with ocular involvement. Arch. Ophthal. 58 (1957) 321.

PÄRTAN, J., and THUMB, N.: Ochronose und Alkaptonurie. Wien. Z. inn. Med. 38 (1957) 385–388.

PFÄNDLER, U.: Stoffwechselkrankheiten. In: Humangenetik, Bd. III/1, edited by P. E. Becker. Thieme, Stuttgart, 1964.

RAHMAN, A. N.: The ocular manifestations of hereditary dystropic lipoidosis. Arch. Ophthal 69 (1963) 708–716.

RINTELEN, F.: Cornealpathologie im Dienste allgemeinmedizinischer Diagnostik. Med. Welt (Stuttg.) 19 (1968) 991–1082.

ROBINSON, R. V. C.: Comparative incidence of xanthelasmata in Jews and Gentiles. Arch. Derm. Syph. (Chic.) 70 (1954) 662–663.

RODERMUND, O. E.: Zur Hyalinosis cutis et mucosae (Urbach-Wiethe). Z. Haut- u. Geschl.-Kr. 43 (1968) 493–503.

SALFELD, K.: Beitrag zum Xanthelasma sebaceum und zum Hutchinson-Syndrom. Hautarzt 15 (1964) 565–567.

SAUTTER, H.: Wie behandelt man den malignen Exophthalmus? Dtsch. med. Wschr. 87 (1962) 261–262.

SCHRECK, E.: Veränderungen des Sehorgans bei Haut- und Geschlechtskrankheiten. In: Dermatologie und Venerologie, Bd. IV, edited by H. A. Gottron and W. Schönfeld. Thieme, Stuttgart, 1960.

SMITH, J. W.: Ochronosis of the sclera and cornea complicating alkaptonuria. J. Amer. Med. Ass. 120 (1942) 1282.

SPAETH, G. L., and FROST, P.: Fabry's disease. Arch. Ophthal. 74 (1965) 760–769.

TOPPEL, L.: Veränderungen des Augenhintergrundes bei Porphyria cutanea tarda. Münch. med. Wschr. 107 (1965) 933–935.

WEICKSEL, J.: Angiomatosis bzw. Angiokeratosis universalis (eine sehr seltene Haut- und Gefäßkrankheit). Dtsch. med. Wschr. 51 (1925) 898–900.

WEYHBRECHT, H., and KORTING, G. W.: Zur Pathogenese der Hyalinosis cutis et mucosae. Arch. Derm. Syph. (Berl.) 197 (1954) 459–478.

WILK-WILCZYŃSKA, M., and WOŹNIAK, L.: Changes in the eyeball in xanthogranuloma juvenile (naevoxantho-endothelioma). Klin. oczna 35 (1965) 473–480.

Collagenosis Group (Generalized Connective Tissue Diseases)

In 1942, KLEMPERER described the characteristic histologic systemic change of the intercellular substance of the connective tissue of this group—especially of the type of fibrinoid alteration—seen in acute lupus erythematosus, dermatomyositis, and scleroderma, but only in extremely heterogeneous variations. The three diseases show very little macro- or microscopic similarities and these so-called collagenoses cannot be considered a group possessing abnormal autoimmune reactions. On the other hand, these diseases do have in common some hetero- or polytopic localization in certain organs, as well as dysproteinemias, and sometimes fever, chronic progression, and so on.

Lupus erythematosus affects joints, serous membranes, kidneys, and the heart, characteristically affecting the muscles of the heart and skeleton; progressive scleroderma involves the upper gastrointestinal tract, lungs, and kidneys, and also the heart. Within the same organ the three diseases may show marked differences.

For instance, in visceral lupus, the heart presents endocarditis (LIBMAN-SACKS); in dermatomyositis, changes of the myocardium proper occur; in progressive scleroderma, fibrosis of the myocardial tissue above all is present. Clinically, we observe in scleroderma cardiac insufficiency not responding to digitalis, a fact which is easily understood. Absence of edema is due to the sclerosis of the skin or to left or right hypertrophy as a result of kidney changes or pulmonary hypertension; acute lupus erythematosus presents pericarditis and tachycardia, and dermatomyositis shows irregular pulse or long-lasting tachycardia.

Genetically, nothing definite can be said about the influence of hereditary factors in dermatomyositis. In progressive scleroderma the few reported cases with a genetic background do not offer a sufficient basis; however, in acute lupus erythematosus the present tendency toward the assumption of a hereditary disposition would point to a lupus erythematosus diathesis. The hypothesis of such a diathesis has been supported by the presence of a number of antibodies in clinically healthy relatives of patients suffering from lupus erythematosus; to this hypothesis experimental similarities in animals can be added. Scleroderma presents *far fewer* and only suggestive hints that an autoaggressive process is underlying it. Consequently, scleroderma remains primarily a neurovascular affliction with pathogenetically decisive alteration of several collagen fractions. This puts scleroderma again in definite opposition to dermatomyositis, which immunologically is hardly conspicuous; diagnostically, however, a pathologic mechanism of enzymes is important. The simultaneous occurrence of dermatomyositis and malignancies should, by the way, lead in each case to an energetic search for a tumor. (For further details about classification of the collagenoses see KORTING, 1967.)

Scleroderma

Circumscribed scleroderma (morphea) is a disease with spontaneous involution, while the diffuse kind of scleroderma (better described as "progressive") is a systemic fatal disease of the blood vessels and the connective tissue.

Clinically, a circumscribed patch of scleroderma starts with subacute erythema which fades in the center; it later shows a lilac ring around the edge, and then follows a waxy- to boardlike induration, which may progress to atrophy (with pigmentation). According to the extension one differentiates between scleroderma "en plaque", a linear type or "en bandes," or a variation showing small areas of involvement.

Progressive diffuse scleroderma starts with the well-known change between syncope and acroasphyxia of Raynaud's disease. The indurative stage progresses to a waxy- to boardlike consistency of the upper cutaneous layers, which cling more and more to the underlying bone; the end result is acrosclerosis, sclerodactylia, microstomia and similar conditions. In addition, scleroderma presents the full range of a fatal systemic disease (myocardial and pulmonary sclerosis, gastrointestinal slowing of passage with constipation up to the terminal stage of paralytic ileus or complete renal failure shortly before death [for details see KORTING and HOLZMANN]).

In circumscribed scleroderma the eye may show morphea of the lids (upper eyelid: WITTELS, SCHUBERT; lower eyelid: GROEGER, AYRES) with sclerosis and atrophy and occasionally discoloration of eyelashes and eyebrows. SCHRECK cites in this connection hypesthesia of the same side of the cornea, anisocoria, and periarteritic changes of the retina (case of MEUNIER and TOUSSAINT). ECOLI and LEPRI saw localization at the tear ducts; COYLE observed association with superficial keratitis.

In progressive scleroderma torpid swellings of the lids, diminished secretion of tears, and later damage to the cornea caused by imperfect closure of the eyelids (GRÜTTE) (Fig. 115) are present. When the clinical picture is fully developed the lids are thickened and hard; only very late will atrophy with well-marked thinning of the skin be observed. Quite often progression toward Sjögren's keratoconjunctivitis sicca occurs. Much rarer is paresis of the muscles of the eye due to myosclerosis or retinitis caused by hypertension (LEINWAND, DURYEE, and RICHTER). Some reports mention retinal degeneration (MANSCHOT), arterial thrombosis of the retina (MYERSON and STOUT), or bilateral subluxation of the lens (SACKNER). The most frequent ocular involvement in progressive scleroderma is a cataract, which according to SAUTTER occurs always bilaterally and mainly during the fourth decade of life, sometimes even before the cutaneous changes. This progressive type of cataract resembling radial presenile circular cataract requires reevaluation of the diagnosis because of the possibility of Werner's syndrome (see page 62). According to our present knowledge cataracts in progressive scleroderma are rather rare (according to LEINWAND et al., only *one* such instance in 150 cases). "Cottonwool spots" of the retina are a terminal sign and occur with sclerodermatous involvement of the kidney. Among 63 patients with progressive scleroderma, FALCK and ZABEL found two cases with glaucoma (the author could confirm these findings) and in one case clouding of the vitreous was observed. Eight out of ten patients with progressive scleroderma observed by the author showed premature destruction of the vitreous going beyond what is usually seen in older age groups (ultrasonic documentation of GÄRTNER, LÖPPING, and HOLZMANN).

Scleredema adultorum. In many respects this disease closely resembles Hardy's (1877) "edematous scleroderma." ABRAHAM BUSCHKE established this clinical picture in 1900 as a disease entity; it represents a mucopolysaccharide storage disease. Observation of the 230 patients reported so far reveals that the disorder tends to occur slightly more frequently in females and in patients under the age of 20. Characteristic is the onset of the disorder after an infectious disease, the absence of manifestations on the distal parts of the extremities, especially the hands, and the definite tendency to spontaneous involution, chiefly in adolescents. Cutaneous changes of waxy or paraffinlike consistency are present; in general, these changes are more palpable than visible. The surface of the skin may be nodular, wavily trabecular, or furrowed.

Ocular involvement in scleredema is characterized by painful swelling of the eyes (WATRIN and MICHON) or edema of the lids and conjunctiva (BREININ), transitory constriction of visual fields (MERENLENDER and ZAND), and development of glaucoma in "edematous scleroderma" (STREIFF).

Lupus erythematosus

Chronic discoid or cutaneous lupus erythematosus presents on the face follicular erythema, follicular hyperkeratosis (carpet-tack phenomenon), follicular, and, finally, diffuse atrophy, primarily in butterfly distribution (BATEMAN). There is a whole chain of manifestations extending from "congestive seborrhea" (VON HEBRA, 1845; CAZENAVE, 1851) to the sometimes mutilating "ulerythema centrifugum" (UNNA) (Fig. 116). Sometimes, the pinna of the ears and parts of the extremities are affected as in the so-called *"Chilblain lupus"* (HUTCHINSON); with further multiple and rapid focal spreading (for instance, on the back) we speak of "lupus erythematodes chronicus cum exacerbatione acuta" (EHRMANN).

In widespread visceral acute (subacute) lupus erythematosus there are discrete cutaneous changes with petechial intrafocal bleeding, especially on mucous membranes. Simultaneously, marked constitutional signs are present; they help to establish the diagnosis. There is involvement of the lymph nodes, joints, serous membranes such as polyserositis and non-bacterial, verrucous inflammation of the valves and the walls of the endocardium (KAPOSI, 1872; LIBMAN and SACKS, 1924), and of the kidneys with histologically characteristic wireloop lesions and deposition of a homogeneous eosinophilic substance in the thickened basal membrane of the glomeruli or the nervous system.

The eye, according to the findings of ROHRSCHNEIDER and EHRING, shows chronic discoid changes of lupus erythematosus by spreading of the chronic discoid changes to the lids, conjunctiva or eyeball, but according to KLAUDER and DELONG this happens very rarely. By histologic examination an analogy with cutaneous changes (VILANOVA, CARDENAL, and COPDEVILA) can be shown. A third of the patients suffering from chronic lupus erythematosus present, in the periphery of the fundus, sharply circumscribed pigmented foci (frequently only unilaterally), which resemble chorioiditis disseminata. The fundus changes in acute lupus erythematosus (Fig. 117) correspond to those seen in other generalized septic diseases ("retinitis septica"; see HOTZ). P.A. MIESCHER, CLUSKEY, ROTHFIELD, and A. MIESCHER describe in one-fifth of their lupus erythematosus patients the presence of small, circumscribed, whitish foci in the fundus ("cytoid bodies"; see also BRIHAYE VAN GERTRUYDEN and DANIS; CLIFTON and GREER); also, more rarely, bleeding and papillary edema. HAUSER emphasizes the frequency of ocular vascular changes in acute lupus erythematosus such as dilatations, tortuosities, differences in calibre of vessels, crossing phenomena, phlebitis, as well as hemorrhages, subretinal edema, and above all cottonwool-like exudates, especially near the macula, which BERGMEISTER (1929) had first described but had misinterpreted as tuberculous nodes. HAUSER, relying on his own observations, stresses the appearance of "pseudoneuritis," and as another change of the fundus, thrombosis of the central vein. Such changes usually parallel the stage of the disease, but later, due to cardiorenal involvement, fundus changes caused by hypertension may be superimposed. Lately, BÖKE and BÄUMER, in addition to the above, mentioned cottonwool areas and intraretinal bleeding with transition into exudative detachment of the retina and pigmentary anomalies, and called attention to whitish-gray, bandlike infiltrates in the deeper layers of the cornea (HALMAY and LUDWIG) and to corneal opacities stainable with fluorescin in acute lupus erythematosus (SPAETH). Patients complain about photophobia, abnormal flow of tears, foreign body sensation, and similar matters. SPAETH saw depigmentation of the retina which was different from that caused by chloroquine.

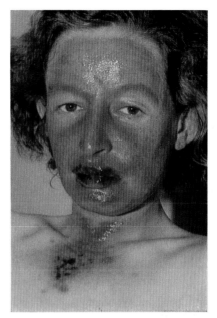

117 Subacute lupus erythematosus.

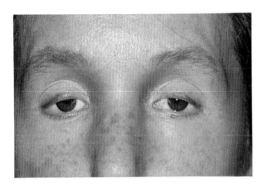

115 Systemic sclerosis.

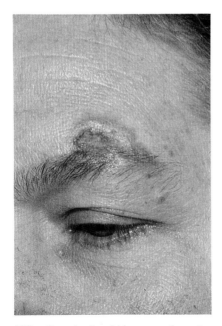

116 Chronic discoid lupus erythematosus.

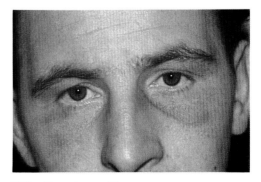

118 Dermatomyositis.

Dermatomyositis

It is important in dermatomyositis (Fig. 118), as mentioned before, to consider a possible underlying disease (malignant tumor, "rheumatic" disease, or tuberculosis), light sensitivity, temperature changes, or other external injurious agents. The disease was already known in the last century (WAGNER, 1863; HEPP, 1887 and UNVERRICHT, 1891); if acute, it presents peculiar, wine to violet-red, later telangiectatic or speckled, rarely porcelain-colored hues, which especially in the nuchal area may be interspersed with lichenoid papules. After healing, the skin presents a speckled appearance composed of atrophy, pigmentation, and telangiectasia. JACOBI described in 1906 this triad as "poikilodermia atrophicans vasculare," which can in rare instances be congenital or can appear in early childhood. In such cases it is frequently accompanied by various defects. Such congenital-poikilodermic syndromes are striking. The diagnosis of dermatomyositis has today been facilitated beyond performing a skin-muscle biopsy and determination of the creatine excretion in the urine (0.1—0.4 gram per day) by the electromyogram and the determination of enzymes.

Such "muscle ferments" compromise the enzymes SGOT (10–40 units per ml. WROB-LEWSKI), SGPT (5–35 units per ml. WROB-LEWSKI), aldolase (1.8–4.9 units per ml. WROB-LEWSKI), *creatine-phosphokinase* (CPK up to 72 units per ml.) and myokinase (MK 2–15 units per ml.). These values do not go strictly parallel (only to a large degree) with the activity of the disease and also do not follow exactly its slowing up by cortisone therapy.

Ophthalmologically, one sees the same changes as with the other diffuse connective tissue diseases or with serum sickness, i.e., alteration of the choroid and retina (LISMAN), cottonwool-like exudates (MAUMENEE, and DE KRIES), streaklike bleedings into the nerve fiber layer or papillary edema. This is discussed in connection with the dysoric fundamental disorder, which underlies dermatomyositis. In any case, the so-called "cytoid bodies or the dysoric nodules" (THOMAS, CORDIER, and DUPREZ) containing hyaline corpuscles and invaded by neurofibrils are not entirely characteristic of a specific collagenosis (PASCHER, WALSH). Beyond this it is not surprising that cases of dermatomyositis with polymyositis as their basis show malfunction of the muscles of the eye (case 16 of WALTON and ADAMS). Occasionally, the eyelids can be affected by the same disease process (DE GRACIANSKY and QUÉRAN).

References

AYRES, S.: Discussion. In: A case for diagnosis (elephantiasis) by P. Anderson. Arch. Derm. Syph. (Chic) 50 (1944) 441. Ref. Zbl. Haut- u. Geschl.-Kr. 84 (1953) 122.

BÖKE, W., and BÄUMER, A.: Klinische und histopathologische Augenbefunde beim akuten Lupus erythematodes. Klin. Mbl. Augenheilk. 146 (1965) 175–178.

BREININ, G. M.: Scleredema adultorum. Ocular manifestations. Arch. Ophthal. 50 (1953) 155–162.

BRIHAYE VAN GERTRUYDEN, and DANIS, M.: Fundus lesions with disseminated lupus erythematosus. Arch. Ophthal. 51 (1954) 799.

CLIFTON, F. C., and GREER, H.: Ocular changes in acute systemic lupus erythematosus. Brit. J. Ophthal. 39 (1955) 1.

COYLE, E. F.: Scleroderma of the cornea. Brit. J. Ophthal. 40 (1956) 239.

ECOLI, G., and LEPRI, G.: Dacryosialoadenopathia atrophicans and sclerodermia. Ophthalmologica (Basel) 124 (1952) 129.

FALCK, I., and ZABEL, R.: Die Beziehungen klinisch unterschiedlicher Bilder der Sklerodermie zur Inneren Medizin. Derm. Wschr. 152 (1966) 593–764.

GÄRTNER, J., LÖPPING, B., and HOLZMANN, H.: Über Glaskörperveränderungen bei Sklerodermie. Untersuchungen mit Ultraschall. Arch. klin. exp. Derm. 229 (1967) 110–116.

GODER, G.: Vasculäre Augenveränderungen bei der generalisierten Sklerodermie. Klin. Mbl. Augenheilk. 144 (1964) 370–383.

DE GRACIANSKY, P., and QUÉRAN, J.: Études de deux cas de dermatomyosite. Bull. Soc. franç. Derm. Syph. 55 (1949) 407–409.

GROEGER: Unna-Tagg, 24.–26. IX. 1948, Hamburg, Kranken-Demonstration. Derm. Wschr. 122 (1950) 844. Arch. Derm. Syph. (Berl.) 189 (1949) 454.

GRÜTTE: 66. Tgg. Vereinig. Südwestdtsch. Dermatol. 10.–11. VI. 1939 in Freiburg/Brsg. Zbl. Haut- u. Geschl.-Kr. 63 (1940) 473.

HALMAY, O., and LUDWIG, K.: Bilateral bandshaped deep keratitis and iridocyclitis in systemic lupus erythematosus. Brit. J. Ophthal. 48 (1964) 558–562.

HARAYAMA, R.: A case of scleroderma diffusum with fundus involvement. Jap. J. clin. Ophthal. 20 (1966) 857–859. Ref. Zbl. ges. Ophthal. 97 (1967) 302.

HAUSER, W.: Lupus erythematodes. In: Dermatologie und Venerologie, Bd. II/1, edited by H. A. Gottron and W. Schönfeld. Thieme, Stuttgart, 1958.

HOTZ, G.: Augenhintergrundveränderungen bei "Kollagenkrankheiten." Ophthalmologica (Basel) 133 (1952) 354.

KLAUDER, J. V., and DELONG, P.: Lupus erythematosus of the conjunctiva, eyelids and lid margins. Arch. Ophthal. 7 (1932) 856.

KORTING, G. W.: Über einige Wesensunterschiede von Sklerodermie, Dermatomyositis und Lupus erythematodes acutus und die darauf basierende differente Therapie. Dtsch. med. Wschr. 92 (1967) 281–288.

KORTING, G. W., and GOTTRON, E.: Sklerosen-Atrophien. Zbl. Haut- u. Geschl.-Kr. 84 (1953) 1–24, 113–139.

KORTING, G. W., and HOLZMANN, H.: Entwicklungslinien der Sklerodermieforschung in der Gegenwart. Ergebn. inn. Med. Kinderheilk. 24 (1966) 1–38.

KORTING, G. W., and HOLZMANN, H.: Die Sklerodermie und ihr nahestehende Bindegewebsprobleme. Thieme, Stuttgart, 1967.

LEINWAND, I., DURYEE, A. W., and RICHTER, M. N.: Scleroderma (based on a study of over 150 cases). Ann. Intern. Med. 41 (1954) 1003–1028.

LISMAN, J. V.: Dermatomyositis with retinopathy. Arch Ophthal. 37 (1947) 155.

MANSCHOT, W. A.: The eye in collagen diseases. Fortschr. Augenheilk. 2 (1961).

MANSCHOT, W. A.: Generalized scleroderma with ocular symptoms. Ophthalmologica (Basel) 149 (1965) 131–137.

MERENLENDER, J., and ZAND, N.: Un cas de sclérémie (scleremia benigna) avec des troubles nerveux. Acta derm.-venereol. (Stockh.) 16 (1935) 352–360.

MEUNIER, A., and TOUSSAINT: Sclérodermie en "coup de sabre" avec lésion du fond d'œil. Bull. Soc. belge Ophthal. 118 (1958) 369.

MIESCHER, P. A., McCLUSKEY, R. T., ROTHFIELD, N. F., and MIESCHER, A.: Der viscerale Lupus erythematodes. In: Hdb. der Haut- und Geschlechtskrankheiten, Bd. II/2, edited by J. Jadassohn. Springer, Berlin, 1965.

MYERSON, R. M., and STOUT, R. E.: Cryoglobulinemia associated with gangrene of the digits. Amer. J. Med. Sci. 230 (1955) 499.

POLLAK, I. P., and BECKER, B.: Cytoid bodies of the retina in a patient with scleroderma. Amer. J. Ophthal. 54 (1962) 655–660.

ROHRSCHNEIDER, W., and EHRING, F. J.: Die Bewertung der Augenbefunde für die Erkennung der Ätiologie beim Erythematodes. Hautarzt 4 (1953) 451–455.

SACKNER, M. A.: Scleroderma. Modern medical monographs. Grune & Stratton, New York, 1966.

SAUTTER, H.: Die Trübungsformen der menschlichen Linse. Thieme, Stuttgart, 1951.

SCHRECK, E.: Veränderungen des Sehorgans bei Haut- und Geschlechtskrankheiten. In: Dermatologie und Venerologie, Bd. IV, edited by H. A. Gottron and W. Schönfeld. Thieme, Stuttgart, 1960.

SCHUBERT, M.: Zbl. Haut- u. Geschl.-Kr. 60 (1938) 298.

SPAETH, G. L.: Corneal staining in systemic lupus erythematosus. New Engl. J. Med. 276 (1967) 1168–1171.

THOMAS, C., CORDIER, J., and DUPRÉZ, A.: Manifestations ophthalmoscopiques des dermatomyosites. Bull. Soc. franç. Derm. Syph. 66 (1959) 349–350.

VILANOVA X. CARDENAL, C., and CAPDEVILA, J. M.: Chronischer Lupus erythematodes der Conjunctiva. Dermatologica (Basel) 113 (1956) 226–231.

DE VRIES, S.: Retinopathy in dermatomyositis. Arch. Ophthal. 46 (1951) 432.

WALSH, F. B.: Clinical neuro-ophthalmology, 2. Aufl. Williams & Wilkins, Baltimore, 1957.

WALTON, J. N., and ADAMS, R. D.: Polymyositis, E. & S. Livingstone, Edinburgh, 1958.

WATRIN, J., MICHON, P., and MICHON, C.: Un cas de scléroedème de Buschke. Bull. Soc. franç. Derm. Syph. 60 (1953) 349–352.

WITTELS: Circumscripte Sklerodermie. Zbl. Haut- u. Geschl.-Kr. 85 (1953) 239.

Atrophies of the Skin

Senile-degenerative atrophy. The regressive process of atrophy represents aging of the tissue and is ordinarily observed in later stages of life. It appears in older persons as colloid degeneration or as senile elastosis of the face, especially on the temples, forehead, and cheeks, and, not least important, also periorbicularly without special ocular changes except blepharochalasis (relaxation of the skin of the eyelid—Translators), arcus senilis, and so on. A part of such farmer's (J. JADASSOHN) or sailor's (UNNA) skin is represented by "Cutis nuchae rhomboidalis" (JADASSOHN-NIKOLSKI).

Facial hemiatrophy. ROMBERG described in 1846 atrophy of one-half of the face. This is hardly a nosologic entity because not rarely

encephalitis or some other trauma to the sympathetic nervous system may cause this effect. In contrast to Romberg's "idiopathic" facial hemiatrophy the atrophic stage in this location is observed in scleroderma only secondarily after the edematous or indurative phases are over. Additional cases appear as heredodegenerative afflictions with relatively weak penetration of the gene (see KORTING and HOLZMANN, 1967). HAUSER emphasizes as an ophthalmological concomitant sign in Romberg's hemiatrophy the narrowing of the palpebral fissure. WOLFE and WEBER noticed in such a case shrinkage of the globe with a corneal ulcer, and GONDESEN noticed amblyopia. A case of KORTING and RUTHER presented in addition to the hemiatrophy, homolateral ptosis with the eye in the adductor position.

Atrophoderma vermiculatum (DARIER) is a cutaneous atrophy of nevoid character. It commonly starts in prepuberty. Peculiar reticular lesions with dilated follicles and the formation of pseudocomedones give the skin a worm-eaten appearance. It is almost exclusively limited to the zygomatic region.

Macular atrophy, or dermatitis maculosa atrophicans, is primarily a disease of the young. It occurs mostly in females. An inflammatory picture suggests anetodermia erythematosa (THIBIÉRGE, JADASSOHN). Anetoderma of the Pellizzari type is diagnosed when macular atrophy develops at the site of a preceding urticarial eruption. The Schweninger-Buzzi type presents saccular or hernia-like protruding atrophy above the level of the skin. (For details see KORTING, CABRÉ, and HOLZMANN.) BLEGVAD and HAXTHAUSEN reported the association of anetodermia with *blue sclera*, osteogenesis imperfecta, and *zonular cataracts*. Follow-up observations by GRIMALT and KORTING, and GRUPPER and BONPARIS, however, failed to find such cataracts in similar cases.

Acrodermatitis chronica atrophicans (BUCHWALD; HERXHEIMER and HARTMANN) favors women in the fifth or sixth decade of life. This slowly developing atrophy of one or more extremities starts with lymphohistiocytic or edematous-infiltrative (PICK: "erythromelia") changes, which later develop into flaccid, atrophic, and rarely into scleroderma-like, hard, atrophic changes with visualization of the large deeper veins. Occasionally, "fibroid" juxta-articular nodes develop. Acrodermatitis atrophicans is probably not a treponematosis, although there have been isolated observations of spirochetes in the tissue, isolated positive results of the pallida serum reaction (LOHEL), and a response of the disease in its preatrophic stage to penicillin (SVARTZ). Changes in regional lymph nodes and bone marrow, an increased sedimentation rate (HAUSER) and also frequently associated macrocryoglobulinemia (BREHM) speak in favor of a systemic disease, originally caused by an infectious agent (transmitted by ticks?). Newer transplantation experiments (GÖTZ) seem to support this view.

In rare cases, acrodermatitis chronica atrophicans involves the face, but marked blepharochalasis (tendency of the skin of the lids to hang in folds—Translators) (FUCHS) occurs only rarely (see ASCHER syndrome, page 32) (case of THIEME). GUGGENHEIM'S observation of scar formation of the conjunctiva has not been confirmed. LEBAS reports recurrent bouts of iridocyclitis.

References

BLEGVAD, O., and HAXTHAUSEN, H.: Blaue Sclerae und Tendenz zu Knochenbruch mit fleckförmiger Hautatrophie und zonulärer Katarakt. Hospitalstidende 64 (1921) 609–616.

GONDESEN: Hemiatrophia et hypertrophia faciei sin. Osteochondrom. Arch. Derm. Syph. (Berlin) 189 (1949) 472. Derm. Wschr. 122 (1950) 848.

GRIMALT, Fr., and KORTING, G. W.: Anetodermie und Osteopsathyrose (Syndrom von Blegvad-Haxthausen). Z. Haut- u. Geschl.-Kr. 22 (1957) 361–365.

GRUPPER, Ch., and BONPARIS, A.: Anétodermie chez un enfant. Troubles osseux associés. Bull. Soc. franç. Derm. Syph. 66 (1959) 270–272.

GUGGENHEIM, I.: Narben in der Conjunctiva bei Acrodermatitis chronica atrophicans. Klin. Mbl. Augenheilk. 74 (1929) 213–216.

HAUSER, W.: Atrophien. In: Dermatologie und Venerologie. Bd. II/2, edited by H. A. Gottron and W. Schönfeld. Thieme, Stuttgart, 1958; especially p. 851.

KORTING, G. W., CABRE, J., and HOLZMANN, H.: Zur Kenntnis der Kollagenveränderungen bei der Anetodermie vom Typus Schweninger-Buzzi. Arch. klin. exp. Derm. 218 (1964) 274–297.

KORTING, G. W., and HOLZMANN, H.: Die Sklerodermie und ihr nahestehende Bindegewebsprobleme. Thieme, Stuttgart, 1967.

KORTING, G. W., and RUTHER, H.: Zur nervalen Genese von Hemihyper- und Arch. Derm. Syph. (Berl.) 198 (1954) 384–395.

LEBAS, P.: Les Syndromes oculo-cutanés, Bruxelles, Imprimerie Médicale et Scientifique, 1959, 708.

THIEME: Fall von Acrodermatitis chronica atrophicans bei 69jährigem Taglöhner. Zbl. Haut- u. Geschl.-Kr. 5 (1922) 214.

WOLFE, M. O., and WEBER, M. L.: Zu Hemiatrophia faciei. J. Nerv. Dis. 91 (1940) 595; 68 (1942) 431. Ref. Zbl. Haut- u. Geschl.-Kr. 84 (1953) 124.

Avitaminoses

At the present time diseases caused by a relative lack of one or several vitamins (poly-hypovitaminoses) are rarely seen in Central Europe. Even complex avitaminoses are rarely the result of exogenous, but more likely of enteric, deficiencies.

Vitamin A

The best-known function of vitamin A is its role in *vision*. The retina of most vertebrates contains for light perception two types of receptors, the cones (they are responsible for vision at high light intensities and color discrimination) and the rods, which are responsible for vision at low degrees of light (night vision). Light perception of both types of receptors is provided by proteinlike pigments whose prosthetic group consists of a carotinoid. The pigment of the rods, the visual purple or rhodopsin, contains, as the carotinoid component, a derivative of vitamin A. The visual pigments of the rods and cones receive light waves falling upon the retina and convert this energy via photochemical reactions into nerve impulses. They are carried by the optic nerves and the optic tracts to the central nervous system (for further details see LANG). Besides, even the lower invertebrates without eyes also have light reception via carotinoids. The utilization of the carotins (less so of vitamin A) is bound to the fat contents of the food, and the resorption of the carotin, less so of vitamin A, is aided by simultaneous consumption of much fat. In the liver (especially in the liver of the polar bear) large amounts of vitamin A can be stored.

While vitamin C's point of action is the mesenchyma, important signs of vitamin A deficiency occur in the *epithelium*. In the eye lack of vitamin A results in keratomalacia due to necrosis of the cells of the cornea with formation of ulcers and perforation. Additional signs are increased cornification of the cells of the conjunctiva and of the epithelial elements of the tear glands, thereby ending the production of tears and causing dryness of the conjunctivae *(xerophthalmia)*. Later, brownishly pigmented or yellowish white, somewhat triangularly shaped, thickened areas on the bulbar conjunctiva, especially within the palpebral fissure, appear as the so-called Bitot spots (1863). LANG also emphasizes the frequency of chalazia due to specific curtailment of the meibomian glands. An early sign of avitaminosis A is nyctalopia, which represents a slowing down process of adaptation.

Besides the additional phenomenon of cornification in other organs with changing borders between skin and mucous membranes (e.g., the respiratory or the intestinal tracts) or where alterations of the epithelium lead to desquamation and formation of concrements (for instance, in the urogenital tract) there are special *cutaneous changes* present. A clinical sign of vitamin A deficiency of the integument is a peculiar slate gray or reddish, pinhead sized, follicular, sometimes slightly pruritic, papule, which is more conspicuous at sites of pressure, sometimes only on a limited area such as the zygoma, and which occasionally shows perifollicular depigmentation—such a papule, especially in Negroes, becomes conspicuous because of its contrast with the surrounding skin. This *multiple* papule resembling pityriasis rubra pilaris or in other cases looking rather acneiform has been named by NICHOLLS "phrynoderma" or toadskin. This author observed such lesions on prisoners on the island of Ceylon; however, they suffered also from neuritis and diarrhea. The more marked expression of this papular eruption occurs as a rule in adults, while children frequently present a flourlike dryness of the skin. Once the follicular horny plugs drop out of the opening of the follicle one sees craterlike depressions. Within this framework of vitamin A deficiency there is *dryness of the skin* accompanied by loss of elasticity, wrinkles, diminished sweat and oil secretion and intensive pigmentation. PILLAT observed in Chinese, not only on the eye (internal canthus) but also over the entire body, "*a peculiar ashy grey discoloration*," which assumed a dark gray brown shade resembling argyrosis. This pigmentation was especially prominent at the site of old scars or scabietic lesions. Additional cutaneous signs of vitamin A deficiency are loss of glossiness, graying of hair, alopecia areata-like loss of hair and longitudinal furrowing of the nails.

On the other hand, intake of very large doses of vitamin A causes pronounced alterations in rats and man. Strangely enough, it causes disturbances of cornification in addition to spontaneous fractures; there are great differences in sensitivity to high doses of vitamin A (in infants hydrocephalus has been observed in addition to other signs). (For further details and references see KORTING, 1959.)

Vitamin B₂ (Riboflavin)

Riboflavin (lactoflavin) exerts its effect on the organism through co-enzymes; it is of fundamental importance for the metabolism of the cornea and lens. Its lack results in vascularization of the cornea; later sometimes interstitial keratitis similar to rosacea-keratitis and clouding of the refractive media follow (see LOPEZ GONZÁLES, and MORENO). In addition, LANG emphasizes the strikingly high vitamin B₂ level of the epithelium of the cornea and lens (180 and 360 gamma per cent, respectively); similarly high is the vitamin B₂ level of the lacrimal fluid (13.5 gamma per cent) as compared to the riboflavin concentration in blood plasma (3.5 gamma per cent). Riboflavin also takes part in the transformation of rhodopsin in the retina.

DERMATOLOGICALLY, in man a deficiency in vitamin B₂ has more significance than a lack of aneurin (thiamine—Translators). SEBRELL and BUTLER experimentally produced with a lactoflavin (riboflavin)-free diet a dermatological condition showing as its most conspicuous feature a condition they called *cheilosis*. Thirteen of 18 women on such a deficient diet presented at the angles of the mouth loss of color, maceration, and transverse fissuring; the latter may occur also in the middle of the lip (CASTELLANI: fissura labialis mediana). Such fissures are also seen on the fuchsin red or purple bluish tongue in connection with coarsening of the papillae. But the essential part of cheilosis is "perlèche" or "stomatitis angularis."

Perlèche of the diabetic is essentially an expression of disturbed growth of the epithelium due to insufficient tissue oxidation as a result of diminished respiratory cell ferment essential for this process. Generally speaking, the etiology of a case of perlèche cannot always be ascertained correctly without considering other clinical findings. Yet it is possible to diagnose the perlèche of ariboflavinosis to some extent since it occurs together with relatively characteristic other manifestations of a functional (dimness of vision) as well as an organic character. Patients with vitamin B₂ deficiency present, besides cheilosis, fine, thinly lamellar, seborrheidal erythemas with formation of rhagades or filiform excrescences, especially in the nasolabial folds, the corners of the eye, on the pinnae of the ears, and the external auditory canals. Besides the signs on the cornea and conjunctiva mentioned above and the nasolabial change just described, there are disturbances of nail growth (koilonychia) and cutaneous changes of the genitals, especially in inhabitants of the temperate zone. All are of great importance and are indicative of ariboflavinosis (vulvitis "pellagrosa," "scrotal dermatitis"; for clinical references see KORTING).

Nicotinic acid (Niacin)

Nicotinic acid and nicotinamide become biochemically active with the help of two coferments, diphosphopyridin-nucleotide (DPN) and triphosphopyridin-nucleotide (TPN), which are prosthetic groups of dehydrogenases. They carry, therefore, hydrogen of substrates capable of dehydrogenation to hydrogen acceptors, especially the flavine enzymes. The animal organism is in a position to furnish, in part, his own supply of niacin; the amino acid tryptophane is the precursor substrate (for details see LANG). The protein of maize (corn) has an unfavorable composition because it is poor in tryptophane and lysine. Clinically, the antivitamin to niacin, isonicotinic-acid-hydrazide (INH) is important; several reports of the last few years have confirmed this. Disturbances due to INH may become clinically manifest as a deficiency of pantothenic acid (sign of "the burning feet"), or as a lack of vitamin B$_6$ (deficiency of pyridoxine). Clinically, the lack of niacin (*pellagra*) is characterized by the four d's ("diarrhea," "dermatitis," "dementia," and "death"). The cutaneous changes of pellagra, pellagrous dermatitis, occasionally show edematous and seldom, rather exudative, symmetrical, sharply outlined erythemas, which after some time assume a mahogony-brown color, after which there follows a thin, small, lamellated desquamation, in the beginning more flourlike, primarily affecting the center of the erythematous areas. Later, the periphery of these erythematous areas turns into dirty hyperkeratoses.

On the arms, these changes are present mostly on the medial aspect. Sites of predilection of this pellagrous dermatitis are the regions of hands, forearms, dorsa of the feet, the neck, and the sternum (Casal's necklace). The palmar aspect of the hands may present marked yellow discoloration simultaneously with thick keratinization. The presence of filiform lesions or folliculopapular keratoses on the nasolabial folds, similar to the seborrhoid-erythematous changes of the genital region, is more likely a manifestation of coincidental ariboflavinosis. The vermilion border of the lips of pellagrins is sore or chapped as if covered with varnish; the tongue is highly red or bluish purple and resembles MÖLLER's glossitis or a beefsteak tongue. Very characteristic is the profuse salivation of the pellagrins. Sun exposure is unquestionably the most important precipitating factor for the appearance of the pellagrous dermatitis, but in contrast to scurvy, heavy bodily exercise or infections are less important contributing factors. Pellagra typically starts in the "stomach" (diarrhea) and occurs seasonally with some regularity between Easter and Whitsuntide (50 days after Easter). In the past few years we have repeatedly seen adults suffering from pellagra because of alcohol abuse (prosperity pellagra).

The older literature (J. JADASSOHN) and a newer textbook (KORTING) do not describe characteristic pellagrous *ocular* changes, although there are some reports on conjunctivitis, keratitis, and retrobulbar neuritis; however, in an individual case, it is hardly possible to make lack of niacin alone responsible for these ocular changes. My own extensive observations of pellagra, especially in prisoner-of-war camps, showed that, among clinical signs, ocular changes receded into the background, but during the last few years several case reports (DJACOS, ROSSETTI) have been published. BIERENT describes keratomalacia.

Vitamin B$_6$ (Pyridoxine), Pantothenic acid

Spontaneous or isolated manifestations of deficiencies caused by lack of pyridoxine (apyridoxinosis) have hardly been seen in man; they simulate lack of vitamin B or iron (Plummer-Vinson syndrome: perlèche, koilonychia, and spastic disturbances of the swallowing reflex). The main symptoms are anorexia, fatigue, nausea and, especially, rhagades of the corners of the mouth, dryness of the skin, and conjunctivitis. Pantothenic acid is essential for man but its lack does not give rise to definite cutaneous signs; it is apparently important for melanogenesis and copper metabolism of the hair. Feeding of corresponding antivitamins gave rise, after some time, to certain neurological disturbances (acroparesthesias).

Vitamin C (Scurvy)

Avitaminosis C is based on a disturbance of the mesenchymal system with severe restriction of the production of collagen. However, the signs and symptoms of this deficiency have at present undergone some revision since our increased knowledge of other vitamins, enzymes, and their deficiencies has made it questionable whether to ascribe certain clinical signs to lack of vitamin C alone. A-avitaminosis of the skin as observed in cachectic individuals shows a tendency toward widely distributed follicular keratosis with a nutmeg-grater or gooseflesh-like appearance. In scurvy, papules may have a harder consistency or an acneiform or hemorrhagic tendency. Looking at a patient with "phrynoderma," which by definition is due to lack of vitamin A (but with absence of subsequent petechiae as seen in scurvy), it is hardly possible to distinguish with certainty this eruption from "lichen scorbuticus" (Fig. 120), "folliculitis scorbutica," "scorbutic gooseflesh," or prescorbutic keratosis pilaris. When the question of the etiology of an epidemic of follicular keratosis arises, internal and external causes such as those seen in times of general food shortage have to be considered (keratosis follicularis "contagiosa," melanodermatitis toxica lichenoides et bullosa, "Basel disease" (an epidemic form of keratosis follicularis occurring in Switzerland—Translators), and others).

"Lichen scorbuticus" prepares the ground for follicular scorbutic bleeding arranged in a chessboard pattern. With further exacerbation cutaneous petechiae, hemorrhages, or internal bleeding into body cavities follow. The suffusions of scurvy prefer areas of special muscular stress, such as the popliteal space and the subcutis of the shins. Palpation of such hematomas of muscle results in reflexlike contractions of the affected muscular areas, called by HEUBNER the "puppet string phenomenon." Dentulous patients develop redness and spongy, dusky red swelling of the interdental papillae, especially of the incisors. In infants, in addition to pseudoparalysis and painful swellings of the extremities, subperiosteal hematomas (cephalhematomas), hematurias (reddish discoloration of diapers soaked with urine), or melena are observed. This infantile scurvy of the type of Möller-Barlow disease (1859, 1885) was in former times observed during the second half of the first year of life, particularly during the winter months; today it is diagnosed primarily radiologically (SCHERZER; KAUER-MAYER et al.).

Ophthalmologically, patients with scurvy may show bleeding of the conjunctiva, but according to my own observations of this occurrence, for instance in prisoner-of-war camps, it was of relatively minor importance compared with the general clinical signs of scurvy.

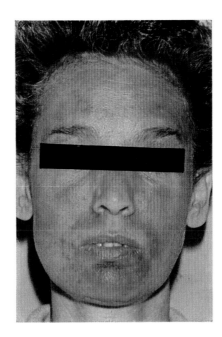

119 Pellagra

120 Lichen scorbuticus.

References

BIERENT, P.: Note sur un syndrome de pellagre avec kératomalacie centrale. Bull. Mem. Soc. Ophthal. 64 (1951) 304.

BITOT, P. A.: Mémoire sur une lésion conjunctivale non encore décrite coincident avec l'héméralopie. Gaz. hebd. méd. chir. 10 (1863) 284–288.

DJACOS, C.: Les altérations oculaires dans la pellagre. Ann. Oculist. (Paris) 182 (1949) 279.

JADASSOHN, J.: Der gegenwärtige Stand der Pellagralehre. In: Hdb. der Haut- und Geschlechtskrankheiten, Bd. IV/2, edited by J. Jadassohn, Springer, Berlin, 1933.

KORTING, G. W.: Hautveränderungen bei Erkrankungen des Magen-Darm-Traktes, Mangelkrankheiten und Avitaminosen (einschl. Pellagra). In: Dermatologie und Venerologie, Bd. III/2, edited by H. A. Gottron and W. Schönfeld. Thieme, Stuttgart, 1959.

KRAUER-MAYER, B., OLAFSON, A., KNÜSEL, E. and KAUFMANN, H. J.: Skorbut im Kindesalter. Schweiz. med. Wschr. 98 (1960) 789–795.

LANG, K.: Die Physiologie der Vitamine. In: Hdb. der allgemeinen Pathologie, Bd. XI/1, edited by E. Büchner, E. Letterer, and F. Roulet. Springer, Berlin, 1962.

LOPEZ Gonzáles, G.: Oftalmopatia ariboflavinósica y Pellagra. Arch. argent. Derm. 10 (1960) 119. Ref. Zbl. Hautu. Geschl.-Kr. 109 (1961) 310.

ROSSETTI, D.: Alterazioni oculari, in corso di eritema pellagroide. Ann. Ophthal. 83 (1957) 376.

SCHERZER, E.: Noch immer Säuglingsskorbut. Münch. med. Wschr. 110 (1968) 535–540.

SEBRELL, W. H., and BUTLER, R. E.: Riboflavin deficiency in man (Ariboflavinosis). Publ. Hlth. Rep. (Wash.) 53 (1938) 2282; 54 (1939) 2121.

Benign Tumors of the Skin

Nevi and nevoid syndromes including phakomatoses

Nevi are cutaneous dysplasias manifest at birth or later in life. They are characterized by an excess or, less often, by an absence of one or more tissue components. Sometimes nevoid or organoid tumors are defined as formations of too much immature tissue.

As examples of pigmented nevi—with the exception of pericellular nevi containing nevus cells—we mention only nevi spili or the flat lentigines, which represent accumulations of pigment *without* nevus cells. There is a slight difference between ephelides (freckles) and lentigines: the latter show a longer rete crest. Syndromes with lentigines are Darier's lentiginosis profusa, Touraine's lentiginosis centrofacialis, inverse ephelides of Siemens, the Peutz-Jeghers syndrome (pigmented spots, polyposis) (PEUTZ, KLOSTERMANN), and the lentigo type Leschke of von Recklinghausen's disease (without neurinoma) (Fig. 121), which brings us to the phakomatoses.

Phakomatoses, a term introduced by the Dutch ophthalmologist van der Hoeve, refers to a group of syndromes presenting anomalies, which are due to disturbed tissue specialization in early embryonic life and continue to develop even after the embryonal period is over. VAN BOGART suggested a synonym for these syndromes by the term "neuroectodermal dysplasias." These comprise the following:

Von Recklinghausen's disease, neurofibromatosis (1882). This polymorphous phakomatosis favors the male sex. Frequently it is present at birth; it originates in the nervous system and is caused by an autosomal dominant gene, whose expression varies greatly. Environmental factors such as pregnancies, trauma, and infectious diseases have a great influence on the course of the disease. Manifestations such as tumors (neurinomas, neurofibromas) and macular pigmentations are present side by side. Small tumors can be pressed down but snap back like bell-buttons. Occasionally, they are distributed along the course of a peripheral nerve or we find neuromas branching out, giving the impression of "dermatolysis" or lobulated elephantiasis (Figs. 122 and 123). Some of the circumscribed pigmentations are large and have the color of milk coffee (café-au-lait spots—Translators); the small ones simulate ephelides and are a dark brown color. If there is a group of more than six rather closely aggregated nevi spili and especially if these happen to be located in the axilla (CROWE: axillary freckling), von Recklinghausen's disease should be suspected. Isolated manifestations of von Recklinghausen's disease favor the acoustic nerve, the vulva, the gastrointestinal tract, and the skeleton (kyphoscoliosis). Vascular neurofibromatosis is, according to FEYRTER, neurofibromatosis of the vessels, and according to REUBI, a unique vascular disease occurring in neurofibromatosis. As a general principle a neurofibroma can degenerate into a sarcoma (QUÉRÉ, RICHIR, and DAVENNE: schwannoma of the limbus of the cornea with malignant degeneration). In general, patients suffering from von Recklinghausen's disease have a dreamy, tired, and melancholic facial expression.

From an *ophthalmological* point of view one can, according to FRANÇOIS, distinguish a central, peripheral, and hydrophthalmos form. The latter is due to a schwannoma of the ciliary nerve with hyperplasia of the connective tissue of the chorioid. As a general principle, every part of the eye or its appendages can be affected (HAGER). Relatively frequent findings are: a possible semi-lateral hypertrophy of the face; elephantiasis-like cutaneous lobule formations, principally of the eyelids; neurofibromas of the lids (H.V.

ALLINGTON and J.H. ALLINGTON), of the conjunctiva as well as of the inner sheaths of the eye; also (one-sided and, under certain circumstances, also pulsating) exophthalmos; and, as mentioned in the beginning, buphthalmos (see SCHRADER). Patients suffering from von Recklinghausen's disease may have involvement of the cornea with hypertrophy of the nerve fibers. It shows gray lines with minute nodular endings (for references see KLEIN and FRANCESCHETTI). SCHRECK emphasizes that the ocular changes of this disease may be confined to the orbit; however, defects of the bony structure of the orbit have also been described. According to JAEGER children may present only a doughy swelling of the eyelids, while formation of nodes appears later in the deeper parts of the tissue of the lids; the tumors, if located superficially, need not interfere with the function of the lid. The conjunctiva shows chemosis or "elephantiasis"; the conjunctiva bulbi may present café-au-lait-like spots, although such pigmentations are perhaps more frequent on the iris or chorioid. Nodules in the iris were described by UNGER and PIETRUSCHKA. Not infrequently, a glaucoma (BÖCK) and, as mentioned before, hydrophthalmos were observed in von Recklinghausen's disease; SCHRECK explains this as happening with vascular changes of neurofibromatosis. The eyeground shows atrophy of the optic nerve or neurofibromatous tumors of the optic nerve; the tumors have more vascular changes than localized tumors of the Pringle type (SCHWAB). Involvement of other orbital nerves (oculomotor and abducens) apparently occurs more rarely.

Bourneville-Pringle Phakomatosis. In 1880, BOURNEVILLE described *tuberous cerebral sclerosis* (its main location being the area of the basal ganglia), which was its significant central finding, as a disease transmitted chiefly as a regular autosomal dominant trait. Additional hamartias are located in the kidneys, heart (v. RECKLINGHAUSEN, 1862), and brain. The chief cutaneous sign is *"adenoma sebaceum"*

(PRINGLE, 1890) (Figs. 124 and 125). One distinguishes, according to BALZER and MENETRIER, the "white variety" with the multiple skin-colored, symmetrically facial nevi located chiefly in the nasolabial area. If the group of papules show telangiectases, the "red variety" (PRINGLE) is present; if extensive proliferation of the connective tissue predominates, the "hard variety" (HALLOPEAU and LEREDDE) is in evidence. The same tendency to proliferation is seen in perigingival nodules, periungual, and subungual fibromas (Koenen's tumors), and lumbosacral *shagreen patches* or connective tissue nevi, which usually develop as early as the second to sixth year of life.

Ocular signs, which according to SCHRECK are present in 3 to 4 per cent of patients with Pringle's disease, but according to SCHWAB are found in much higher percentages, are seen as retinal tumors, which were first described in 1921 by van der Hoeve. They are mostly multiple, indistinct, or sharply circumscribed, yellow white or grayish white, sometimes mulberry- or cauliflowerlike nodules without vascular connections. Additional findings are white sheaths of the retinal vessels, bleeding of the retina, opacities (FRANÇOIS) of the central cornea, lens, or vitreous. SCHRECK reports, presumably as an early stage of the subsequently occurring nodules, thin whitish deposits within the retina similar to retinopathia punctata; according to SCHWAB these deposits are important for an early diagnosis (Fig. 126). JAEGER emphasizes further that the retinal tumors hardly grow in the course of life and thus do not represent any danger for the eye. Subjective impairment of vision of patients with Pringle's disease seems to be rather unknown. On the other hand, such retinal changes already present in children with this disease may suggest a diagnosis of adenoma sebaceum even before the cutaneous changes are fully developed or are present only in an abortive stage. Finally, edema of the papilla may occur because of an increase of pressure of the cerebrospinal fluid in cases

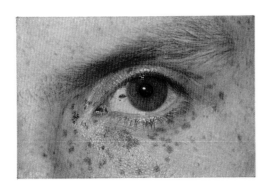

121   Lentigo.

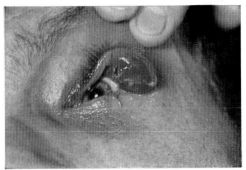

123   Neurofibromatosis *(von Recklinghausen's disease)*.

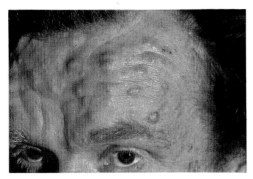

122   Neurofibromatosis *(von Recklinghausen's disease)*.

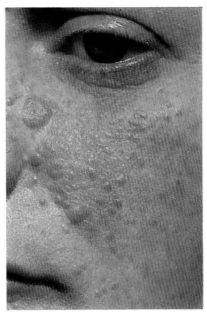

124   Adenoma sebaceum *(Pringle's disease)*.

with simultaneous involvement of the brain (SCHWAB, 1956).

Pigmented spots and intestinal polyposis (The Peutz-Jeghers syndrome) (PEUTZ-KLOSTERMANN). This phakomatosis is most likely transmitted as an autosomal dominant. The gene shows full penetrance. The leading dermatologic signs in dark-complexioned persons are frecklelike melanin spots grouped radially around the facial openings (Fig. 127); fewer spots occur on fingers and toes. Another sign is the intestinal polyposis with the predilection of the small intestines (jejunum), but patients with such an "anlage" may in rare instances show polyposis also of other organs. The polyps cause intussusception, loss of blood, and secondary anemia. They become evident chiefly in the second decade, whereas the melanin spots on the oral mucosa and the lips may be present as early as birth. The tendency of the intestinal hamartomas to become malignant is, at the present time, not considered very high.

Ophthalmologically, the pigment spots have been reported to spread to the palpebral conjunctiva (ANDREW), or the conjunctiva of the bulbus; slit lamp examinations have shown melanotic rods parallel to the limbus (KLOSTERMANN). However, JEGHERS, McKUSICK, and KATZ, as well as DORMANDY and BRADFORD and DANZIG, had observed similar changes mainly adjacent to the limbus. KLOSTERMANN saw, in addition, involvement of the sclera and iris with fine granular spreading of the pigment; STÜTZER confirmed these findings later. Finally, Klostermann saw in one case, as a fundus change, above the papilla, a brown spot of the size of the papilla with somewhat heterogeneous saturation of color. GUILLARD et al. described in 1955 a pigmented spot of the cornea as part of the complete syndrome.

Melanophakomatoses. MUSGER considered that Touraine's melanoblastose neuro-cutanée, Ota's nevus fusco-caeruleus-ophthalmo-maxillaris, or Fitzpatrick's "oculodermal melanocytosis" were melanophakomatoses; the latter was discussed earlier as a pigmentary disorder. In the type described by TOURAINE, bathing suit nevi of different sizes are located on the skin, which has been somewhat hyperpigmented from the beginning. They are associated with changes in the central nervous system such as hydrocephalus, cerebral convulsions, and others, but the melanin-forming tissues of the eye do not seem to take part in the process. In 1968, however, KOZLOWSKI and ZARADA observed this phakomatosis in a ten-year-old girl, who showed fundus changes of the eye consisting of a distinct pigmented ring around the papilla of the optic nerve. The cerebral convulsions mentioned earlier have an unfavorable prognosis; so has a xanthochromic cerebrospinal fluid (BALTZER).

Angiomatous phakomatoses. MUSGER considers syndromes with telangiectatic anomalies (Klippel-Trénaunay, von Hippel-Lindau, Sturge Weber, and Louis-Bar) as angiomatous phakomatoses. They are discussed in connection with the angiomas (see page 149).

Epitheliomatous phakomatoses. This concept, suggested by MUSGER, comprises early embryonal developmental disturbances such as the Hermans-Herzberg phakomatosis, described as nevus epitheliomatosus multiplex and also called basal cell nevus (with extremely numerous basaliomatous nodules on the skin of the trunk, face, and neck).

KNOTH and EHLERS consider also Brooke's epitheliomas (pinhead- to pea-sized semiglobular, skin-colored papules on seborrheic predilection sites with onset in childhood) and the characteristic turban tumors (Spiegler's tumors with manifestations in early adulthood) as phakomatoses.

In individual cases basal cell nevus, epithelioma basocellulare and epithelioma adenoides cysticum cannot always be differentiated histologically; this means that the whole clinical picture remains important (HERZBERG and WISKEMANN). The usual forms of basaliomas occur about 250 times more frequently than the basal cell nevus form and,

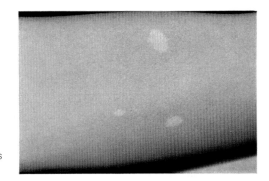

125 Leaf-like depigmentation in Pringle's disease *(Fitzpatrick)*.

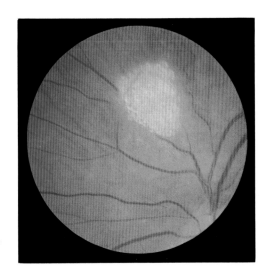

126 Pringle's disease. Eyeground (Collection of Prof. *Jaeger*, Heidelberg).

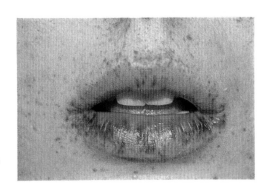

127 Peutz-Jeghers disease (pigment spots on the lip associated with intestinal polyposis).

furthermore, almost always without a familial basis, but the basal cell nevus-jaw cyst syndrome (Ward syndrome) represents a familial trait with an autosomal-dominant mode of inheritance (GERBER). This syndrome has been reported in the literature about 100 times; BINKLEY and JOHNSON (1951) described it as syntropy of basal cell epitheliomas starting in adolescence, with multiple ectodermal defects of bones and canine teeth (jaw cysts) and anomalies of the gonads and the CNS (see review by TOBIAS).

Ophthalmologically, hypertelorism with a roundish skull formation has been observed. CLENDENNING, BLOCK, and RADDE add to these signs dystopia canthorum, congenital cataracts or, especially, congenital blindness (case of OLIVER). GORLIN and GOLTZ observed congenital opacities of the cornea. The two cases of HERMANS, GROSFELD, and SPAAS presented severe ocular anomalies, which required enucleation soon after birth. Colobomas were seen by GORLIN, YUNIS, and TUNA. Basal cell nevi are observed most frequently on the eyelids, where they may cause destruction of the lids unless the basal cell nevi are removed at an early stage.

Hard nevi

On the eyelids such verrucous or hyperkeratotic lesions are rather rare; "striated" lichen planus or lichen simplex is hardly ever seen in the periorbital region. An "anonymous" linear dermatosis may resemble a "systematized nevus." The differential diagnosis has to consider a *verrucous nevus* (histology: elongated papillae, acanthosis, and hyperkeratosis).

Organic nevi

The main representative of these genetically determined organic or glandular nevi is the *nevus sebaceus*; it is frequently present at birth and is located on the scalp near the sutures of the skull or behind the ears, or in a systematic linear arrangement consisting of finger-phalanx-sized groups of closely aggregated, reddish yellow, relatively soft and sometimes more firm, papillomatous nodules (Figs. 128, 129). According to my own observation such patients may show a nevus sebaceus (J. JADASSOHN), especially when multiple lesions with wide distribution occur, and, in addition, epileptic seizures and mental retardation are noted. This establishes a special neurocutaneous syndrome (FEUERSTEIN and MIMS; LANTIS et al.). Besides, at the site of the nevus sebaceus or in close proximity to it, basal cell epitheliomas or cancers of the sebaceous glands may develop (PARKIN and MICHALOWSKI).

The main representative of the sweat gland nevus is the syringoma (JAQUET and DARIER, 1887) or hidradenoma (Figs. 130 and 131). These hamartias become evident late, presenting on chest and eyelids multiple, pinhead- to lentil-sized, moderately firm, transparent yellow to reddish brown, smooth nodules, which may exist throughout life. On the eyelids syringomas are also observed as solitary growths. DAICKER saw them more frequently after the fourth decade; he examined 56 syringomas of the eyelids histologically and found definite interrelationships between the outer root sheaths of the "anlagen" of the hairs, the hairs themselves, and the tumor. Daicker established a connection between the histogenesis of the syringoma and the anlagen of the apocrine sweat glands; he, therefore, considered the syringoma an atavistic apocrine glandular nevus. However, based upon histochemical and electron-microscopical studies, HASHIMOTO, LEVER et al. are of the opinion that the differentiation of the syringoma points toward the eccrine sweat glands (see WORINGER and EICHLER). Nosologically, it is of note that BUTTERWORTH et al. found syringomas of the eyelids among 37 of 200 mongoloids; however, examinations of a

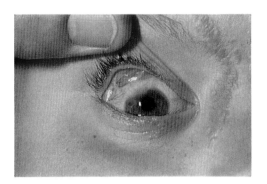

128   Nevus sebaceus.

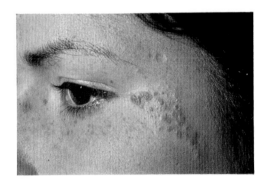

129   Nevus sebaceus.

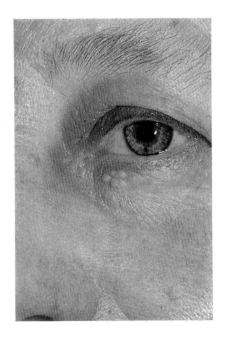

130   Syringoma.

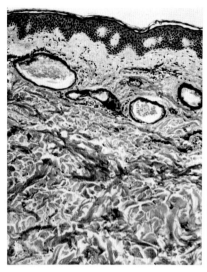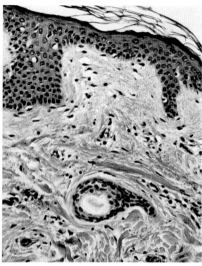

131 Syrin-
goma.
Histologic
appearance.

series of patients by KORTING and HOLZMANN (1966) failed to confirm these findings.

Among the 100 published reports of unilateral comedo nevus two instances of cataract development can be found (POPOV and BOIANOV; and WHYTE).

Fibromas

Soft fibromas are solitary or multiple tumors, usually of hazelnut size and on a broad base. Skin-colored, pedunculated, filiform fibromas *(tags)* develop chiefly on the neck, nasal introitus, and, especially, on the thin wrinkles of the eyelids of older persons; moreover, their development is connected with certain phases of proliferation—for instance, pregnancy.

Hard fibromas can also appear, rarely, in the medial corners of the eyelids and there at first glance look like a meningocele lying in the fissure. Most of the cutaneous nodules (ARNING and LEWANDOWSKI), level with the surrounding skin, characteristically show sinking in a flat fashion on palpation with two fingers; they are found, on the other hand, chiefly in the region of the lower half of the body. But

they also appear in the region of the eyelid (see, for example, HENKIND and SCHULTZ).

Histologically, the following extensive synonymous concepts are in use with certain qualifications: histiocytoma (LEVY-COBLENTZ), fibrome en pastille (CIVATTE), dermatofibroma lenticulare SCHREUS), nodular subepidermal fibrosis (MICHELSON), and sclerosing hemangioma (GROSS and WOLBACH) (Figs. 132 and 133). Cutaneous nodules and histiocytomas, which often owe their formation to a band traumatization, as for example a scratch, and which, perhaps incidentally, show the cell-rich connective tissue's final condition of an inflammatory stroma reaction, are basically a few assorted states of development, rather than probably a priori established differentiations.

The nevoxanthoendothelioma, more than likely another form of histiocytoma, from which it is differentiated by its ability to revert back, is discussed with the lipid storage diseases.

The so-called *Bindegewebsnaevi* (GUTMANN), *Pflastersteinnaevi* (LIPSCHÜTZ), or *Naevi elastici* (LEWANDOWSKI) are circumscript shagreenlike lesions located almost exclusively in the lumbosacral area and primarily occurring in Bourneville-Pringle disease.

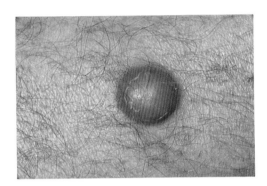

132   Histiocytoma (sclerosing hemangioma).

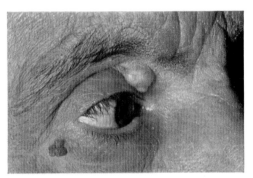

135   Large keratin cyst.

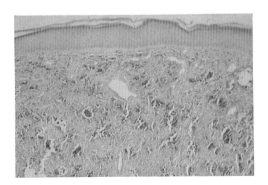

133   Histiocytoma. Histologic appearance
(Berlin blue reaction).

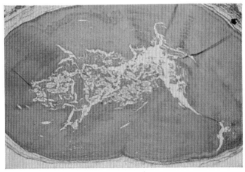

136   Keratin cyst. Histologic appearance.

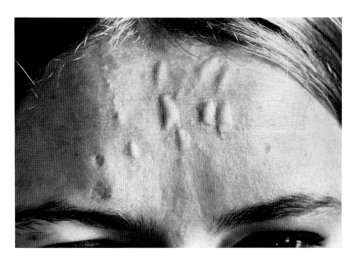

134   Ceremonial keloidal scars.

Keloids

In 1810, ALIBERT gave this name to sharply circumscribed, firm, brownish red, sometimes also lighter colored, plaquelike, elevated tumors, which increase in size like a lobster's claw and have a finely wrinkled and telangiectatic surface (Fig. 134). The peculiar disposition to keloid formation may be due to familial, racial, or endocrine (thyroid, hypercalcemia) factors as well as to metabolic disturbances (porphyria) (review by NIKOLOWSKI). In contrast to scar keloids the rapidly growing hypertrophic scar remains limited to the original wound area. The so-called spontaneous keloid, located typically over the sternum, develops frequently on the basis of folliculitis sclerotisans. Otherwise, the keloid occurs most commonly on face, chest, and ears.

Lipomas

Lipomas occur on the skin as *solitary* growths or *multiple* crops, and frequently in a distinctly familial pattern, often at the time of hormonal crises and finally, not infrequently, in adolescents suffering from peripheral-hypotonic circulatory disorders. They are hardly ever seen on the eyelids, and only if lipomas are already present on other parts of the body.

Multiple lipomas may occur also in greater numbers in Pringle-Bourneville phakomatosis (see page 140), or with the dominantly inherited special form of palmo-plantar keratoses of type Hanhart. Subconjunctival lipomas are part of the oculoauricular Goldenhar syndrome (auricular appendages, fistulas of the ear, malformation of the pinna, dysplasia of one-half of the face) and the Wildervanck syndrome observed in female patients (labyrinth deafness, congenital paresis of the abducens, hypoplasia of one-half of the face, and auricular appendages).

Leiomyomas

Like lipomas, leiomyomas occur as solitary, but more frequently, as multiple aggregated lesions. They arise from the muscles of the arrectores pilorum, the blood vessels, or the tunica dartos. The diagnosis is evident when characteristic sensitivity to contact is noted, such as exists in eccrine spiradenomas, glomus tumors, or some neurinomas, and is observed also in muscle tumors after the effect of caffeine or cold.

Cysts

These nodular, tight, or fluctuating tumors (Figs. 135 and 136) arise either from retention of secretory material of the cutaneous glands or as genuine neoplasms from dispersed epidermal particles such as epidermoid or dermoid cysts. Dermoids occur not uncommonly on the lateral or medial sides of the upper eyelid. Cysts of Moll's gland appear as skin- or rose-colored, transparent-vesicle-like growths on the ciliary border. They may attain a size of 1.0 cm. or more in diameter. If the overlying skin is relatively thick the transparency of the tumor is not evident, so that an imitation of a solid tumor such as a nodular basal cell epithelioma results. Cysts of Krause's glands, which according to H.V. ALLINGTON and J.H. ALLINGTON develop primarily along the conjunctival fornix, are observed as inflammatory retention cysts, for instance, in pemphigus of the conjunctiva.

Hydrocystomas (ROBINSON) are millet-sized, firm vesicles equidistant from the eyes, and are observed mainly in persons exposed to severe heat. The walls of the tumors consist of smooth to cubical stratified epithelium.

The most frequent cyst within the area of the eye is the pinhead-sized milium or grutum, which is found chiefly in groups on the upper two-thirds of the face (Fig. 137). The retention cysts originating from Zeiss-sebaceous glands (on the free edges of the eyelids — Translators) are considered milia by SCHRECK. According to dermatological experience secondary milia develop especially after bullous cutaneous processes such as burns, epidermolysis bullosa, or porphyria cutanea tarda, but also at the point of contact with the skin of the cone of the x-ray tube. EPSTEIN and KLIGMAN consider the milium a benign keratinizing tumor.

Angiomas and angiophakomatoses

If one also considers their location in inner organs such as the liver, then hemangiomas belong without doubt to the most common tumors in man. Oncologically, they occupy a position between hyperplasias and hamartias. They are not infrequently associated with developmental anomalies in other tissues; they have a peculiar tendency to spontaneous regression (VIRCHOW), which is most often seen in capillary hemangiomas. Nevi flammei are at present classified with the nevi telangiectatici or telangiectases. SCHNYDER draws a sharp line between *medial-symmetric* (main example: Unna's nevus of the nape, 1894) *and lateral-telangiectatic nevi flammei* (see, however, H.O. Curth and Goldensohn's facial midline-nevus, a variant of the Sturge-Weber syndrome — Translators). These plain angiomas are frequently present at birth; they may later show nodular transformation. On the face, they frequently involve the conjunctiva and are often associated with dilatation of the lymph vessels. Their initially functional character, evident in the beginning, is discernible in their various colors, for example, under thermic influences, when the child is crying or sleeping, or in their complete pallor after death.

The strawberry type of cavernous hemangiomas appears—genetic factors apparently exert no special influence on their development—as a *flat tuberous or nodular tuberous* growth (Figs. 138 and 139). One of its main locations on the face is the *eyelids*. Clinical experience has taught us that what appear to be small angiomas of eyelids extend far into the orbit and derive their blood supply from several large, deeply located vessels; this is of therapeutic importance (OFFRET and ROSA). HOLLAND stresses the upper eyelid as the site of predilection, the preponderance of girls, and the tendency to spontaneous involution. Sometimes, however, periorbicular angiomas (in a narrow sense) are not present and are only simulated by phlebectasias (DELACRETAZ and CHRISTELER).

Lymphangiomas develop usually before the fourteenth year of life, without definite sex preference and mostly on the head and neck; in contrast to the angiomas they rarely regress spontaneously. One distinguishes clinically lymphangioma simplex (small, frogspawn-like, transparent, grapelike vesicles); lymphangioma cavernosum (as soft as a pillow filled with water); and lymphangioma cysticum, chiefly in submaxillary location.

MASSON'S (1923) *glomus tumor* represents one of the few painful cutaneous tumors and corresponds to a neuromyoarterial angioma. As single lesions they appear in adults, but multiple and less painful manifestations can occasionally occur in children. Glomus tumors or glomus angiomas *on the eyelids* have been observed only occasionally (MORTADA; JENSEN). OTTO and ALNOR report the simultaneous occurrence of glomus tumors and Horner's syndrome.

Pyogenic granuloma (Granuloma telangiectaticum benignum)

Pyogenic granuloma, also known as botryomycosis, sits on a small or wide basis and is a dark, fleshy-red, mostly solitary tumor, which reaches a certain size and remains stationary without tendency to involution. Favorite sites are the lips or other parts of the face in general, and the hands. The tumor corresponds to a vascular granuloma; it really is an eruptive angioma, whose development and growth are caused by trauma and secondary infection. Neither sex is particularly favored; it can appear at any age, although it is most common in adolescents. NALIK, SOOD, and AURORA described a pyogenic granuloma on the lower eyelid measuring 3.25 by 3.75 cm. in diameter.

There are *other* forms of angiomas, for instance, Darier's "multiple progressive angiomas," Bean's viscero-cutaneous hemangiomatosis ["blue rubber bleb nevus with hematemesis, seepage of blood (due to gastrointestinal hemangiomas — Translators), melena, and secondary anemia"], the Mafucci syndrome (asymmetrical chondromatosis, also known as dyschondroplasia with hemangiomas in which the angioma is occasionally a

"hydroangioma" [KORTING and BREHM], or the Kasabach-Merritt syndrome (partly viscerally located giant hemangiomas probably with resulting thrombocytopenic purpura). The following angiomatous phakomatoses require special oculodermatological consideration.

Angiomatous phakomatoses

Angiomatosis retinae (v. HIPPEL-LINDAU)

JAEGER stresses the point that this syndrome rather rarely shows dermatological correlations such as cavernomas or café-au-lait spots. However, in autopsies hemangiomas were regularly found in the cerebellum, cerebrum, or spinal cord and on the retina and papilla; they were observed in the first case by v. HIPPEL and LEBER, and also in other internal organs such as the kidneys, spleen, pancreas, and ovaries. LINDAU (1926) reported this disorder's striking predilection for the cerebellum. *Ophthalmological* signs, according to SCHWAB, are enormous dilation and tortuosity of the arteries and veins of the retina, sometimes with tumorlike elevation, later also exudative changes and detachment of the retina, and iridocyclitis with secondary glaucoma.

Closely allied nosologically to Hippel-Lindau's disease is the syndrome of Bonnet-Dechaume-Blanc; it shows neurooculo-facial arteriovenous aneurysms with extensive facial angiomas and rather innocuous vascular convolutes of the fundus of the eye.

Angiomatosis trigemino-cerebralis or encephalo-oculo-cutanea

Sturge-Weber syndrome (encephalo-trigeminal angiomatosis)

The (Schirmer-) Sturge-Weber (Krabbe) syndrome (Fig. 140) belongs to the nevus flammeus syndromes; it comprises "angiome

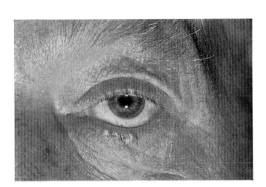

137 Milia.

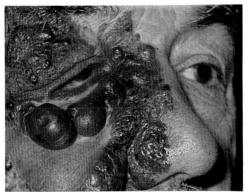

139 Systematized nodular angioma.

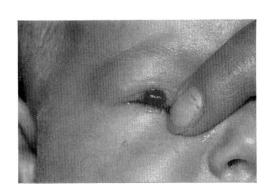

138 Cavernous hemangioma

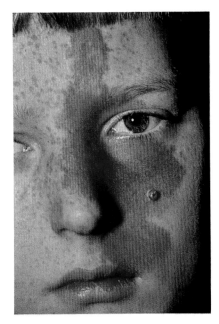

140 Sturge-Weber syndrome.

trigéminé" manifestations primarily along the course of the superior branch of the trigeminal nerve, rarely along its third branch; in addition, occurrence of glaucoma, jacksonian epilepsy, or its migraine equivalents is noted. This systemic angiomatosis of the skin, deep ocular membrane, and the leptomeninges is probably transmitted as an irregularly dominant trait. Important for the diagnosis are the typical changes in the x-ray picture of the skull, first described by WEBER: sinuous double-contoured, tortuous lines of calcium deposits (called "tram lines" — Translators), which correspond to calcifications of cerebral gyri (KRABBE).

According to JAEGER the origin of glaucoma occurring in this syndrome is not uniform (see also KREIBIG). Causative factors which have to be considered are: aplasia of Schlemm's canal, adenoma of the chorioid, or obstruction of outflow of the aqueous chamber by conjunctival angiomas. The question remains unanswered whether a patient who at first does not present glaucoma associated with a nevus flammeus in this location may develop such a condition later. On the other hand, this oculocutaneous or encephalotrigeminal angiomatosis also presents an *oculocephalic* type without *telangiectatic skin changes*.

FRANÇOIS thoroughly discussed the so-called Lawford syndrome (facial nevus and late glaucoma).

Klippel-Trénaunay-Weber Phakomatosis

These osteohypertrophic nevi may be present in metameric distribution but are widely distributed, systematized, telangiectatic nevi associated with hypertrophy of nearby connective tissue and bones. Widespread varicosities, representing rather *phleb-(arterio-)ectasies*, do not occur in every case. The latter develop sometimes rather early in life in the area of the radial or ulnar artery. Besides, functional impairment, pigmentary disturbances, color differences, atrophies, and osseous

changes, as well as protruding vascular convolutions, are present in these cutaneous areas. In a related disorder, angioma racemosum, characterized by its peculiar grapelike aspect, arteriovenous shunts may become (very rarely) a dangerous load for the right side of the heart. Recent investigations by LINDEMAYR, LOFFERER, MOSTBECK and PARTSCH showed that the lymphatic system can also be involved in the Klippel-Trénaunay syndrome. *Ophthalmologically*, reports of an atypical coloboma of the iris (SCHNYDER, LANDOLT, and MARTZ) and cataracts (CALMETTES, DÉODATI, BEC, and BÉCHAC) are noteworthy.

Ataxia telangiectatica (Louis-Bar syndrome)

This cerebello-oculocutaneous telangiectatic syndrome, now observed in about 100 cases, was first described in 1941 by the Belgian neurologist Madame Louis-Bar in a nine-year-old Caucasian boy; he presented *cerebellar ataxia* with inability to walk and stand and with imperfect articulation in speech and *dilatations of the terminal blood vessels, especially of the conjunctivae and pinnae* (Fig. 141). This syndrome probably is related to angiomatosis of v. Hippel-Lindau (GRÜTZNER). Presumably, the atrophying cerebellar changes are caused by analogous telangiectases. Sometimes the butterfly type, grouped analogous netlike telangiectases of the face

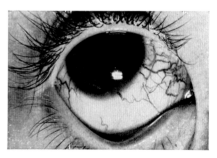

141 Ataxia telangiectatica (Louis-Bar syndrome).

are replaced by café-au-lait pigmentations. There are other accompanying signs which reduce the life expectancy such as bronchitis or bronchiectases, otitis, sinusitis, hypoplasia of the thymus gland, and immunoelectrophoretic disturbances (diminution of the beta 2μ fraction, and absence of the gamma 1μ globulin (IgE) fraction). Finally, *peculiar ocular movements* were observed in these patients manifesting the oculocutaneous telangiectases; the patients die at an early age (SMITH and COGAN; WILLIAMS, DENIS and HIGDON).

References

ALLINGTON, H. V., and ALLINGTON, J. H.: Eyelid tumors. Arch. Derm. Syph. Chic. 97 (1968) 50–65.

ANDERSON, T. E., and BEST, P. V.: Linear basal-cell naevus. Brit. J. Derm. 74 (1962) 20–23.

ANDREW, R.: Generalized intestinal polyposis with melanosis. Gastroenterology 23 (1953) 495–499.

BALTZER, J.: Ein Beitrag zu den Melanophakomatosen. Beitr. path. Anat. 137, 99–119, 1968.

BISHOP, D. W.: Trichoepithelioma. Arch. Ophthal. 74 (1956) 4.

BÖCK, J.: Über eigenartige Gefäßveränderungen im Auge bei v. Recklinghausenscher Neurofibromatose. Ber. Dtsch. ophthal. Ges. 53 (1948) 275–282.

VAN BOGAERT, L.: Les dysplasies neuro-ecto-dermiques congénitales. Rev. neurol. 63 (1935) 802–808.

BONNET, P., DECHAUME, J., and BLANC, E.: L'anévrysme cirsoide de la rétine (anévrysme racémeux): ses rélations avec l'anévrysme cirsoide de la face et avec l'anévrysme cirsoide du cerveau. J. méd. Lyon 18 (1937) 165–178.

BRADFORD, M. L., and DANZIG, L. E.: Gastrointestinal adenomatosis with oral pigmentation. Boston med. Quart. 1 (1950) 21–24.

BUTTERWORTH, Th., STREAN, L. P., BEERMAN, H., and WOOD M. GRAY: Syringoma and Mongolism. Arch. Derm. Syph. (Chic.) 90 (1964) 483–487

CALMETTES, L., DÉODATI, F., BEC, P., and BÉCHAC, G.: Evolution de la maladie de Sturge-Weber-Krabbe. Bull. Soc. Ophthal. Fr. 65 (1965) 82–83.

CLENDENNING, W. E., BLOCK, J. B., and RADDE, I. C.: Basal cell nevus syndrome. Arch. Derm. Syph. (Chic.) 90 (1964) 38–53.

CROWE, F. W., and SCHULL, W. J.: Diagnostic importance of café-au-lait spot in neurofibromatosis. Arch. intern. Med. 91 (1953) 758–766.

CURTH, H. O. and GOLDENSOHN, E.: Die Assoziation von Gefäßmälern in der Mittellinie des Gesichts mit Konvulsionen und intrakraniellen Gefäßmißbildungen. Der Hautarzt 10 (1959) 366–370.

DAICKER, B.: Das Lidsyringom. Dermatologica (Basel) 128 (1964) 417–463.

DELACRÉTAZ, J., and CHRISTELER, A.: Phlébectasies progressives pseudoangiomateuses. Dermatologica 137 (1968) 242–249.

EPSTEIN, W., and KLIGMAN, A. M.: The pathogenesis of milia and benign tumors of the skin. J. Invest. Derm. 26 (1956) 1–11.

FEYRTER, F.: Über die vasculäre Neurofibromatose nach Untersuchungen am menschlichen Magen-Darmschlauch. Virchows Arch. path. Anat. 317 (1949) 221.

FRANÇOIS, J.: Les manifestations oculaires de la maladie de Recklinghausen. Ann. Oculist (Paris) 181 (1948) 753.

FRANÇOIS, J.: Angiomatose oculo-cutanée de Lawford. Ophthalmologica (Basel) 122 (1951) 215.

FRANÇOIS, J., GILDEMYN, H., and RABAEY, M.: Mélanose cancéreuse de la cornée. Bull. Soc. belge Ophthal. 110 (1955) 298–306.

GERBER, N. J.: Zur Pathologie und Genetik des Basalzell-Naevus-Syndroms. Humangenetik 1 (1965) 354–373.

GOLDENHAR, M.: Associations malformatives de l' œil et de l'oreille. J. Génét. hum. 1 (1952) 243–282.

GORLIN, R. J., and GOLTZ, R. W.: Multiple nevoid basalcell epithelioma, jaw cysts and bifid rib. New Engl. J. Med. 262 (1960) 908–912.

GORLIN, R. J., YUNIS, J. J., and TUNA NAIP: Multiple nevoid basal cell carcinoma, odontogenic keratocysts and skeletal anomalies. Acta derm.-venerol. (Stockh.) 43 (1963) 39–55.

GRÜTZNER, P.: Augensymptome bei Ataxia teleangiectatica. Klin. Mbl. Augenheilk. 135 (1959) 712–717.

HAGER, G.: Augenärztliche Beobachtungen und Probleme bei der v. Recklinghausenschen Erkrankung. Klin. Mbl. Augenheilk. 152 (1958) 350–363.

HASHIMITO, K., GROSS, B. O., and LEVER, W. F.: Syringoma. J. Invest. Derm. 46 (1966) 150–166.

HASHIMOTO, K., DIBELLA, R. J., BORSUK, G. M., and LEVER, W. F.: Eruptive hidradenoma and syringoma. Arch. Derm. Syph. (Chic.) 96 (1967) 499–519.

HENKIND, P., and SCHULTZ, G.: Dermatofibroma of the eyelid. Am. J. Ophthal., Ser. 3, 65 (1968) 420–425.

HERMANS, E. H., GROSFELD, J. C. M. and VALK, L. E. M.: Eine fünfte Phakomatose, Naevus epitheliomatodes multiplex. Hautarzt 11 (1960) 160–164.

HERMANS, E. H., GROSFELD, J. C. M., and SPAAS, J. A. J.: The fifth phacomatosis. Dermatologica (Basel) 130 (1965) 446–476.

HERZBERG, J. J., and WISKEMANN, A.: Die fünfte Phakomatose. Dermatologica 126 (1963) 106–123.

VAN DER HOEVE, J.: Eye symptoms in phakomatoses. Trans. ophthal. Soc. U. K. 52 (1932) 380–401.

VAN DER HOEVE, J.: Les phakomatoses de Bourneville, de Recklinghausen et de v. Hippel-Lindau. J. belge Neurol. Psychiat. 33 (1933) 752–762.

HOLLAND, G.: Hämangiome der Lider. Klin. Mbl. Augenheilk. 152 (1968) 365–373.

HOWELL, J. B., and CARO, M. R.: The basal-cell naevus, its relationship to multiple cutaneous cancers and associated anomalies of development. Arch. Derm. Syph. (Chic.) 79 (1959) 67–80.

JAEGER, W.: Facomatosis. In: Dermato-Oftalmologia, edited by J. Casanovas and X. Vilanova. Alhacen, Barcelona, 1967.

JAQUET, L., and DARIER, J.: Hydradénomes éruptifs. Ann. Derm. Syph. 8 (1887) 317.

JENSEN, O. A.: Glomus tumor (glomangioma) of eyelid. Arch. Ophthal. (Chic.) 73 (1965) 511–513.

KLEIN, D., and FRANCESCHETTI, A.: Mißbildungen und Krankheiten des Auges. In: Humangenetik, Bd. IV, edited by P. E. Becker. Thieme, Stuttgart, 1964.

KLOSTERMANN, G. F.: Pigmentfleckenpolypose, klinische, histologische und erbbiologische Studien am sogenannten Peutz-Syndrom. Thieme, Stuttgart, 1960; especially pp. 30, 32, and 35.

KNOTH, W., MEYHÖFER, W.: Zur Nosologie des Adenoma sebaceum Typ Balzer, der Koenenschen Tumoren und des Morbus Bourneville-Pringle. Hautarzt 8 (1957) 359–366.

KORTING, G. W.: Lentiginosis profusa perigenito-axillaris. Z. Haut- u. Geschl.-Kr. 42 (1967) (19–22).

KORTING, G. W., and BREHM, G.: Hidroangioma cutis. Z. Haut- u. Geschl.-Kr. 38 (1965) 5–8.

KORTING, G. W., and HOLZMANN, H.: Hautveränderungen bei mongoloider Abartung. Med. Welt (Stuttg.) 17 (1966) 2801.

KOZLOWSKI, J., and ZARADA, A.: A case of neuro-cutaneous melanosis, supravital diagnosis. Dermatologica (Basel) 136 (1968) 129–140.

KREIBIG, W.: Aderhautveränderungen bei Neurofibromatose (Recklinghausen). Klin. Mbl. Augenheilk. 105 (1940) 369–370.

KREIBIG, W.: Feuermal des Gesichtes, Glaukom und Skeletveränderungen. Klin. Mbl. Augenheilk. 110 (1944) 208–216.

LANTIS, S., LEYDEN, J., THEW, M., and HEATON, Ch.: Nevus sebaceus of Jadassohn. Arch. Derm. 98 (1968) 117.

LINDEMAYÉR, W., LOFFERER, O., MOSTBECK, A., and PARTSCH, H.: Das Lymphgefäßsystem bei der Klippel-Trenaunay-Weber'schen Phakamatose. Z. Haut- u. Geschl.-Kr. 43 (1968) 183–191.

LOUIS-BAR, D.: Sur un syndrome progressif comprenant des téléangiectasies capillaires cutanées et conjunctivales symmétriques, à disposition naevoide et des troubles cérébelleux. Confin neurol. (Basel) 4 (1941) 32–42.

MALIK, S. R., SOOD, G. S., and AURORA, A. L.: Granuloma pyogenicum. Brit. J. Ophthal. 48 (1964) 502.

MICHALOWSKI, R.: Naevus sébacé de Jadassohn – un état précancéreux. Dermatologica (Basel) 124 (1962) 326–340.

MORTADA, A.: Glomangioma on the eyelid. Brit. J. Ophthal. 47 (1963) 697–699.

MUSGER, A.: Melano-Phakomatosen, I. Melano-Phakomatose vom Typus der sog. Mélanoblastose neuro-cutanée Touraine. Hautarzt 14 (1963) 106–110.

MUSGER, A.: Was sind Phakomatosen? Hautarzt 15 (1964) 151–156.

NIKOLOWSKI, W.: Pathogenese, Klinik und Therapie des Keloids. Arch. klin. exp. Derm. 212 (1961) 550–569.

OFFRET, G., and ROSA, D. A.: Ne touchez pas aux angiomes des paupières sans les avoir bien explorés. Bull. Mem. Soc. franç. Ophthal. 76 (1963) 418–427.

OLIVER, R.: Basal-cell nevus. Arch. Derm. Syph. (Chic.) 81 (1960) 284.

OTTO, K., and ALNOR, P. C.: Glomustumoren. Med. Klin. 55 (1960) 1673–1678.

PARKIN, Th.: Naevus sebaceus (Jadassohn) with squamous cell epithelioma. Brit. J. Derm. 62 (1950) 167–170.

PIETRUSCHKA, G.: Zur Frage der Irisveränderungen bei der Recklinghausenschen Krankheit. Klin. Mbl. Augenheilk. 121 (1952) 663–672.

POPOV, L., and BOIANOV, L.: Undescribed congenital cutaneo-ocular syndrome (congenital cataract-comedo syndrome). Süvr. Med. 13 (1962) 49–50; zit. n. Whyte.

PRINGLE, J.: Über einen Fall von kongenitalem Adenoma sebaceum. Mh. prakt. Derm. 10 (1890) 197.

QUÉRÉ, M. A, and DAVENNE, C.: Schwannome limbique à évolution maligne. Arch. Ophthal. (Paris) 24 (1964) 285–290.

REUBI, F.: Les vaisseaux et les glandes endocrines dans la neurofibromatose. Schweiz. Z. Path. 7 (1944) 168.

SCHNYDER, U. W., LANDOLT, E., and MARTZ, G.: Syndrome de Klippel-Trénaunay avec colobome irien atypique. J. Génèt. hum. 5 (1956) 1–8.

SCHRADER, K. E.: Auge und Allgemeinleiden. Schattauer, Stuttgart, 1966.

SCHRECK, E.: Veränderungen des Sehorgans bei Haut- und Geschlechtskrankheiten. In: Dermatologie und Venerologie, Bd. IV, edited by H. A. Gottron and W. Schönfeld. Thieme, Stuttgart, 1960.

SCHREIBER, M. M., and MCGREGOR, J. G.: Pseudolymphoma syndrome. Arch. Derm. Syph. (Chic.) 97 (1968) 297–300.

SCHWAB, F.: Die Augenhintergrundsveränderungen bei tuberöser Hirnsklerose. Klin. Mbl. Augenheilk. 128 (1956) 257–296.

SCHWAB, F.: Über die Mitbeteiligung des Auges bei den sogenannten Phakomatosen. Wien. klin. Wschr. 70 (1958) 253–258.

SMITH, J. L., and COGAN, D. G.: Ataxia-teleangiectasia. Arch. Ophthal. 62 (1959) 364–369.

STÜTZER, H.: Das Peutz'sche Syndrom, eine präventivmedizinische Aufgabe des HNO-Arztes. HNO (Berl.) 8 (1959) 23–24.

TEMPLETON, H. J.: Cutaneous tags of the neck. Arch. Derm. Syph. (Chic.) 33 (1936) 485–505.

TOBIAS, Cs.: Zum Basalzellnaevus-Kiefercysten-Syndrom (Ward-Syndrom) mit familiärem Auftreten. Schweiz. med. Wschr. 97 (1967) 949–953.

UNGER, K.: Neurofibromatosis iridis (Recklinghausen's disease). Arch. Ophthal. 38 (1947) 754–759.

WHYTE, H. J.: Unilateral comedo nevus and cataract. Arch. Derm. Syph. ((Chic.) 97 (1968) 533–535.

WILDERVANCK, L. S.: Een cervico-oculo-acusticus-syndroom. Ned. T. Geneesk. 104 (1960) 2600–2605.

WILLIAMS, H. E., DEMIS, D. J., and HIGDON, R. S.: Ataxia-teleangiectasia. Arch. Derm. Syph. (Chic.) 82 (1960) 937–942.

WINKELMANN, R. K., and MULLER, S. A.: Sweat gland tumors, I. Histochemical studies. Arch. Derm. Syph. (Chic.) 89 (1964) 827–831.

WITTEN, V. H., and LAZAR, M. P.: Multiple superficial benign basal cell epithelioma of the skin. Brit. J. Derm. 64 (1952) 97–103.

WORINGER, Fr., and EICHLER, A.: Constatations et réflexions au sujet d'un cas d'hidradénomes éruptifs. Ann. Derm. Syph. (Paris) 78 (1951) 152–164.

Malignant and Possibly Malignant Tumors of the Skin, Pigmented Tumors

Basal cell and squamous cell epitheliomas, (metastatic) cutaneous cancers

The problem of the benign or malignant character of a tumor, which is so important for clinical medicine, is not free of slogans, e.g., anaplasia; HANSEMANN understood this to mean all morphological or functional differences from tumor cells to normal cells; opposed to it is the concept of "maturity of tissue" ("Gewebsreife") created by BORST. In any case, STRÄULI is of the opinion that, considering our present knowledge of cell division and chromosomes, a general differentiation between benign and malignant tumors according to karyological principles is as impossible as a definition of the malignant cell by morphological methods. The dermatologist faces the problem of malignancy or semi-malignancy among the epithelial tumors in the outstanding example of the *basal cell epithelioma*; the Dublin ophthalmologist ARTHUR JACOB established this concept in 1827 as a special characteristic group ("observations respecting an ulcer of peculiar character which attacks the eyelids and other parts of the face") (Fig. 142). The present view toward this favorite of dermatologic oncology can be stated as follows: the squamous cell cancer (prickle cell carcinoma, keratinizing squamous cell epithelioma, cancroid, or spinalioma) represents the typical malignant neoplasm of the prickle cell layer of the skin; KROMPECHER's concept of the basalioma as a basal cell *carcinoma* (BORRMANN: Corium carcinoma), which he established in many publications between 1900 and 1913, is now considered obsolete; this tumor has now been classified as *epithelioma* basocellulare or basalioma (NÉKAM).

HANNOVER in 1852 established the concept of epithelioma as a tumor derived from the "epithelium" (RUYSCH); nowadays the epithelial position of the cellular elements is the important factor. Similarly, the origin of the basal cell epithelioma from normal basal cells is not stressed any longer, but its derivation from special basal cell elements is emphasized. Some authors, especially formerly E. HOFFMANN and nowadays LEVER, consider the basal cell epithelioma a tardive, hamartoid or nevoid growth starting from the primary epithelial germ cell (MARKS and RÖMER), which is the matrix for hairs, sebaceous glands, and apocrine sweat glands. In a similar way, RICKER and SCHWALBE consider the derivation from the epithelium as the determining factor in the genesis of the basal cell epithelioma. OBERSTE-LEHN arrives at a different conclusion from his studies of epidermocutaneous borderline reliefs; he considers this epithelioma a "vicarious hyperplasia of residual epithelial crests based on epithelial atrophy." Other authors call the basal cell epithelioma an "endophyte" (a tumor growing inward — Translators) (FEYRTER), or a "dependent tumor" (FURTH), or a tumorlike adaptation hyperplasia, in the sense of BÜNGELER (for further details and literature see KORTING, 1964).

Finally, certain changes have resulted in the classical concept of the basal cell epithelioma due to Gertler's recent investigations into the fundamental affinity of this tumor to the epithelium; this has led to the exclusion of the special type of "pagetoid" basalioma (glued like a swallow's nest to the epithelium) and to the rejection of Darier's concept of the so-called metatypic epithelioma; today this is considered an atypical, deformed, or uncontrolled (MIESCHER) tumor, which means a basal cell carcinoma (as in the concept of GOTTRON and NIKOLOWSKI) (Fig. 143).

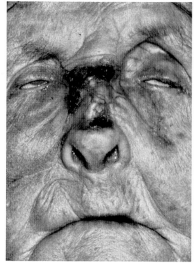

other findings subpleural pulmonary metastases. Finally, a new essential point of the concept of the basal cell epithelioma is the basal cell nevus syndrome which has been discussed above under the heading of Epitheliomatous Phakomatoses. A discussion of a maximal variant (the Pinkus tumor) and its mesenchymal-parenchymal reciprocal effect, so striking in a basal cell epithelioma (and its site of predilection in the lumbosacral region), will not be discussed here.

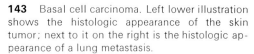

143   Basal cell carcinoma. Left lower illustration shows the histologic appearance of the skin tumor; next to it on the right is the histologic appearance of a lung metastasis.

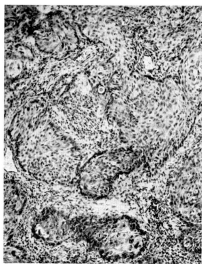
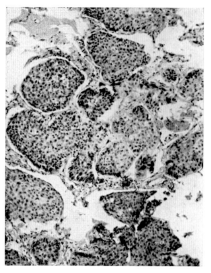

These are the cases which again and again are described as metastasizing basal cell epitheliomas. In spite of many case reports, to which a few have lately been added, one cannot say that a growing basal cell epithelioma of long duration may finally metastasize. A thorough histologic study of such a primary tumor, its recurrences, and its so-called metastases will always show several areas which, in a strict sense, have to be considered basal cell carcinomas. My own case (Fig. 143) bears this out; it shows among

It seems unnecessary to discuss here the various clinical types of basal cell epithelioma (mainly nodular, ulcerating, destructive, superficially growing, scarring, erythematoid and morphealike variants) (Figs. 144 to 147), or to mention that the chronic, slowly growing rodent ulcer of the cornea (Moorens ulcer) has nothing to do with the cutaneous rodent ulcer, which is the chief representative of the basal cell epithelioma.

Among the numerous *eyelid tumors*, which according to HEYDENREICH present 70 single

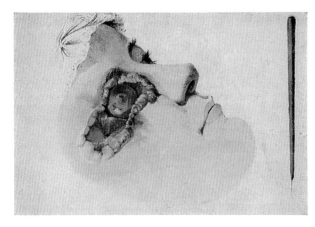

142 First picture of a basal cell epithelioma in the publication of *Arthur Jacolt* (1827).

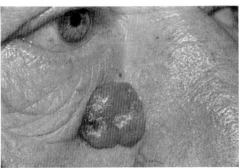

144 Nodular basaloma.

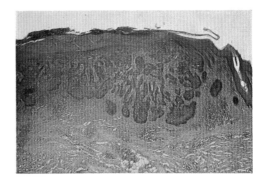

146 Epithelioma basocellulare planum cicatrix.

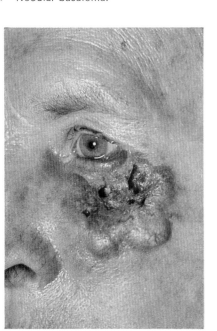

147 Histologic picture of a basal cell epithelioma.

145 Nodulo-ulcerative basal cell epithelioma (nodular basaloma) developing into an ulcus rodens.

types, epithelial tumors are five times more prevalent than those of the conjunctiva (BARRON). According to HOLLAND and BELL-MANN basal cell and squamous cell epithelio-mas are the most common tumors; they were observed in 0.5 per cent of all patients in their Eye Hospital.

There are, however, considerable variations within the statistics, e.g., H.V. ALLINGTON and J.H. ALLINGTON found only seven basal cell epitheliomas among 2156 eyelid tumors. According to MARTIN the lower eyelid showed epitheliomas in 54 per cent and the upper eyelid in 13 per cent. In both locations most of these were basal cell epitheliomas (HEYDENREICH; VANCEA and CHISLEAG: 73.1 per cent; HOL-LAND and BELLMAN: 78 per cent); 30 percent within the periorbital region are of the cystic or adenoid type; on the temporal area, how-ever, they show styloid structure and are macroscopically sclerodermalike, or appear as cicatrizing and flat epitheliomas. EHLERS collected 1785 basal cell epitheliomas; 88.8 percent were in the facial region, most of them on the cheeks (271 cases), on the alae nasi (210 cases), on the temples (210 cases), and on the forehead (191 cases). In fifth place appeared the naso-orbital angle with 162 cases.

Malignant tumors of the eyelids—for in-stance, keratinizing squamous cell carcinomas (EPSTEIN et al.: 2 per cent in 7000 cases)—and uncontrolledly growing basal cell epithelio-mas may metastasize; this is mentioned only in respect to comparative pathology—for in-stance, in dogs and cattle such metastases have not been found (BARRON). The average age of patients with a basal cell or squamous cell epithelioma of the eyelid is 65 years (see HOL-LAND and BELLMANN), although exceptions occur. The development of such lid tumors may extend over many years. Almost two-thirds of patients with an epithelioma of the lids are men.

In contrast to the conditions of the eyelids just described in which the primary site of basal cell epitheliomas is near the inner canthus

of the eye, the conjunctiva presents squamous cell cancers five times more frequently than basal cell epitheliomas; the latter remain flatter than the ones of the "squamous" type (REESE). As far as further details of tumors of the con-junctiva, cornea, and so on, are concerned, I have to refer the reader to special ophthal-mological literature. Special mention, how-ever, has to be made of the so-called *calcifying epithelioma* (MALHERBE and CHENANTAIS, 1881) which today, according to recent histo-genetic studies (see KORTING and HASSEN-PFLUG), is described as "trichomatrioma" (HULETT) or as "pilomatrixoma" (FORBIS and HELWIG) (Figs. 148 and 149). (According to PINKUS and MEHREGAN the correct etymo-logical name should be "pilomatricoma" — Translators.) Besides the scalp, neck, and the upper extremities, *the eyebrow region* is the site of predilection of this type of tumor (BONIUK and ZIMMERMANN: 10 per cent). The Malherbe epithelioma presents mostly as a hard, red-dish, small, pea-sized node, frequently firmly attached to the skin, freely movable on its base, and atheromalike; histologically, one sees shadow cells in solid masses with an osteoclast-cell-like border (see NIKOLOWSKI and SEITZ; GASTEIGER and MASSHOFF).

With reference to histological details of several other cutaneous tumors near the eye (BONIUK and HALPERT: myoepithelioma of the eyelids; GASTEIGER and MASSHOFF: myo-blastmyoma, oncocytoma), one has to consult textbooks of dermatohistology (GANS and STEIGLEDER; LEVER; CIVATTE; ALLEN; MONT-GOMERY, and others). Mention, however, is made of tumors of the eyelids derived from ocular appendages such as the meibomian glands (GOVEY and YERN), frequently simulat-ing a chalazion (WOLKOWITZ). Not infre-quently, *sebaceous gland carcinomas* are observed, especially on the eyelids; otherwise the skin of the orbit with regard to malignant adnexal tumors presents a preponderance of *sweat gland carcinomas* (ALLINGTON and ALLINGTON). Ac-cording to GINSBERG 1 per cent of eyelid carcinomas are derived from the meibomian

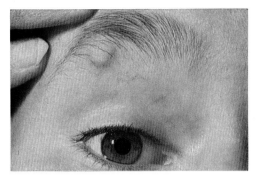

148 Epithelioma of Malherbe in the area of the upper eyelid.

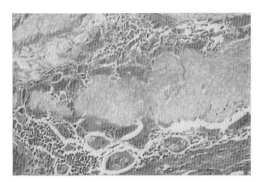

149 Histologic appearance of the epithelioma of Malherbe.

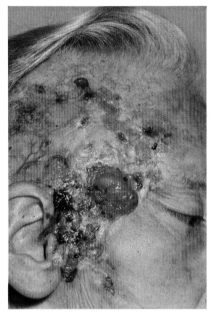

151 Squamous cell carcinoma appearing in a radiodermatitis.

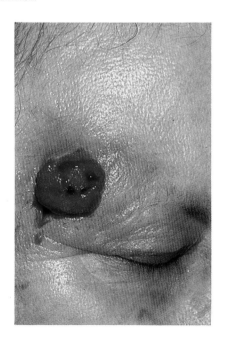

150 Squamous cell carcinoma.

152 Histologic appearance of squamous cell carcinoma grades II or III.

glands; they are twice as frequent on the upper as on the lower lid and show a tendency to recurrence after removal and to regional lymph node metastasis (Figs. 150 to 152). Extramammary Paget's disease derived from Moll's glands of the lid was observed (KNAUER and WHARTON); this tumor is not any longer considered a precancer in the strict sense but biologically a developing carcinoma d'emblée. HAGEDOORN described a case of extramammary Paget's disease on the eyelid.

Finally, secondary cutaneous cancers or cutaneous metastases of cancers of internal organs may appear on the eyelids; this seems important because such secondary cutaneous tumors occur in only 1 percent of all diagnosed internal neoplasms (review by KORTING, 1964). The cutaneous metastases of breast cancers take first place in the statistics of metastases; on the eyelids metastases have in most instances (MUENZLER, OLSON, and EUBANK) their origin in a breast cancer. There are additional reports of metastases of the lids from, for instance, a bronchial cancer (WRIGHT and MEYER), a cancer of the parotid gland (CASANOVAS), the urinary bladder (HOLLANDER and GROTS), or a neuroblastoma sympathicum (SIVARAMASUBRAMANIAM).

As regards the occurrence of secondary or metastatic tumors of other parts of the eye, see REESE. It may be mentioned, however, that ALBERT, RUBINSTEIN, and SCHEIE found metastases, among 213 adults with metastatic malignancies, five times in the chorioid and five additional cases in the orbit. In 41 children with metastatic neuroblastoma or Ewing's sarcoma only orbital metastases were observed, and in no case intraocular metastases.

Precancers

DUBREUILH was the first to use this expression, at the Third International Congress of Dermatology in London in 1896; there has been much discussion about its significance (MIESCHER) and also about its clinical value (GOTTRON). Many heterogeneous cutaneous diseases were brought together which in a larger sense, or facultatively, or in a stricter sense, or as a circumscribed epithelial proliferation of "preinvasive character" (MONACELLI), lead or may lead to carcinoma. From a dermato-ophthalmological point of view Paget's disease has been mentioned briefly above; among the numerous precanceroses grouped together by the dermatologists and of importance topographically is the cutaneous horn, which histologically may deviate from case to case (see CRAMER and KAHLERT) (Fig. 153) and the senile keratosis (DUBREUILH) (Figs. 154 and 155), which can be distinctly separated histologically from the senile verruca (NEUMANN; FREUDENTHAL) (Figs. 156–158), and above all the tumors of the Bowen and erythroplasia type (Figs. 159 and 160). It should be mentioned that the psoriasiform lesion, commonly single, of Bowen's dermatosis is equated today with erythroplasia (QUEYRAT, 1911) and occurs on the mucous membrane or near it (the glans penis and prepuce — Translators); it is characterized by sharply marginated, bright red, velvety, finely granular lesions. In addition, it should be strongly emphasized that in each case of Bowen's disease (as was already mentioned in the original description), in view of the histological similarity, arsenical changes (history of arsenical intake in the past) and the association of Bowen's disease with a visceral cancer (1 out of 15 of my own cases, BEUDT) should be considered. In such instances one should perhaps consider the possibility that the skin manifestations suggestive of Bowen's disease and progressing toward Bowen's carcinoma may be metastases with involvement of the epidermis resembling Bowen's disease (KLOSTERMANN). (The regular and causal association of Bowen's disease with internal malignant tumors seems questionable — Translators.) In any case, the occurrence of this precancer on the eyelid is well known (ALLINGTON and ALLINGTON).

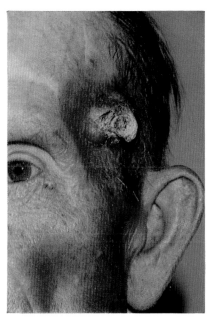

153 Cutaneous horn.

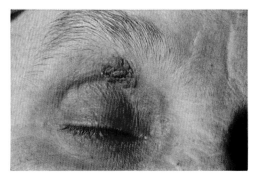

157 Isolated seborrheic keratosis of the upper eyelids.

155 Development of a squamous cell carcinoma in a senile keratosis.

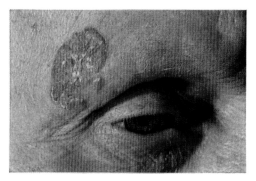

159 Bowen's disease.

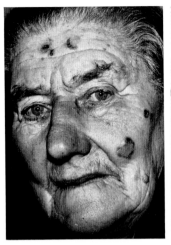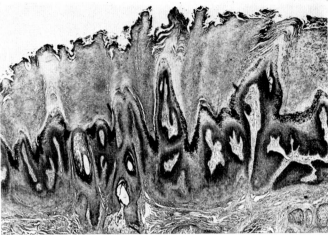

154   Senile keratosis.

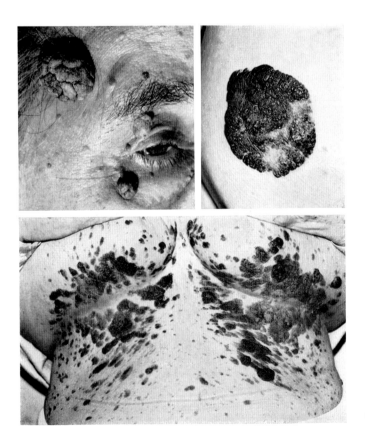

156   Seborrheic keratosis.

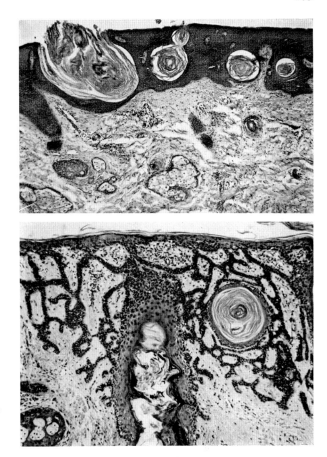

158 Typical histologic appearance of seborrheic keratosis (adenoid and keratotic growth).

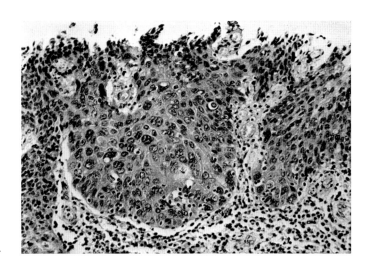

160 Bowen's disease. Typical histologic appearance.

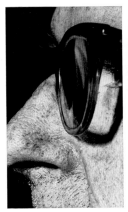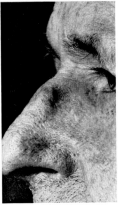

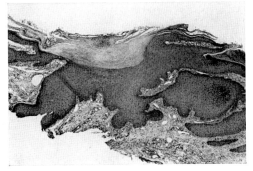

161   Acanthoma secondary to irritation by eyeglass frames.

Intraepidermal epitheliomas can ordinarily be diagnosed only by microscopic examination. Since the first description of the spinalioma by BORST (1904), on the one hand, and the basal cell epithelioma by J. JADASSOHN (1926), on the other hand, the enormous degree of variation of these neoplasms, which in the final analysis present rather heterogeneous growths, has been increasingly recognized. CABRÉ and HOLZMANN made a comprehensive report on our own cases. It will simplify matters if many of such intraepidermal tumors are diagnosed as "intraepidermal acanthoma" (HABER).

Pseudocancers

Pseudocarcinomas (v. ALBERTINI) or pseudocancers (GOTTRON) present macroscop-

ically for the most part as warty papillomas, which histologically often show the characteristic signs, including horny pearl formation, of highly differentiated spinaliomas; sometimes, however, they present nothing else but "acanthomas due to external irritation," for instance from the pressure of an eyeglass frame (Fig. 161). The most important (topographically) and, in the differential diagnosis, greatly significant neoplastic acanthoma is the keratoacanthoma, which fails to develop final differentiation (Figs. 162 and 163). It was already known in 1888 to Sir JONATHAN HUTCHINSON as "crateriform ulcer of the face," but has become well recognized after the reports of DUPONT, 1930, McCOMAC and SCARFF *(molluscum sebaceum)* and ROCK and WHIMSTER. (Following the suggestion of Freudenthal the name keratoacanthoma was generally accepted.) We observe on glabrous skin a rapidly growing, hemispheric, flesh-colored to reddish-violet node, with either a central crater or a basketlike depression; its histologic separation from a highly differentiated squamous cell epithelioma can be difficult. Because of the pseudocancerous nature of this growth, a complete excision need not be performed routinely; instead, a number of diagnostic criteria are available (among others: short duration, definite tendency to involution, molluscumlike structure, eosinophilic infiltration, a high degree of PAS-positivity, and almost always absence of enlarged lymph nodes). Histologically, this tumor derives mainly from the outer root sheath of the upper third of the hair follicle (KALKOFF and MACHER). Its etiology remains unsolved in spite of occasional findings of supposedly viruslike particles by electron microscopy. In any case, there is no doubt today that this is a neoplasm with a pleonasm of nosological individuality. In our opinion it should be removed surgically and examined histologically in spite of its tendency to spontaneous involution (FERGUSON-SMITH), as already mentioned above. An unusual variant of the keratoacanthoma is represented

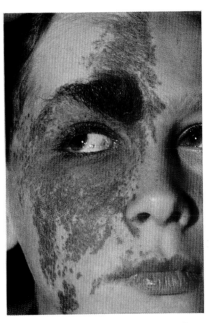

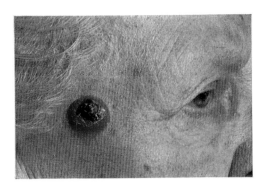

162 Keratoacanthoma.

164 Dermal nevus cell nevus partially appears as hairy nevus.

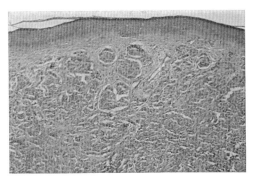

163 Keratoacanthoma. Dish- or saucerlike appearance on low magnification.

165 Dermal nevus cell nevus with typical alveolar structure.

by the basosquamous acanthoma, or acro-
trichoma, which corresponds to the "inverted
follicular keratosis", originally reported by
HELWIG (for details see H. PINKUS).

Ophthalmologically, the occurrence of keratoacanthomas on the eyelids has not only been
repeatedly reported (HAGER; RODENHÄUSER;
PERDRIEL, LOUBIÈRE, and PFISTER), but this
localization represents a significant place
(BAER and KOPF: 6 per cent) in a large series
of cases; although the monograph of POIARES
BAPTISTA states that the keratoacanthoma
occurs on the face in 82.4 per cent and among
these most frequently on the cheeks (30.3 per
cent) and nasal area (23.04 per cent). Other
statistics, however, describe a frequent location on the lids; for example, of 44 keratoacanthomas reported by BONIUK and ZIMMER-
MANN, 25 were on the lower lid and 13 on the
canthus of the eyelid.

Seborrheic wart

In 1869 NEUMANN described verruca seborrhoica or senilis as a basalomalike hyperplasia of the epidermis of not uncommon
occurrence; LEVER called it a "basal cell
papilloma." Such protruding and crumbling
lesions are most numerous on the back, the
face, and not rarely around the orbit; they are
sometimes excessively dark (melanin); histopathologically, they present a rather surprising
wealth of differentiation. More important than
the subdivision of these senile warts into
keratotic or adenoid forms is the knowledge
of those coarsely structured, acanthotic,
seborrheic warts with whorls or centers of
keratinization; they may occasionally give rise
to rather epitheliomatous formations (WEID-
MANN: pseudoepitheliomatous hyperplasia);
in 1924, FRIBOES already indicated the same
idea and, lately, CRAMER discussed this aspect
thoroughly. According to classical teaching
the malignant transformation of the seborrheic

wart is denied, but lately cases have been observed which showed the opposite (see CRA-
MER and DELACRÉTAZ).

Sarcomas

The malignant connective tissue tumors
lately reviewed by GOTTRON and NIKOLOWSKI
are characterized by a structural appearance
which differs from case to case, and by differentiation into fibroplastic, angioblastic,
myxoblastic (see case of KORTING and NÜRN-
BERGER), and other types; clinically, they are
characterized by rapid growth of a more
nodular than lamellated pattern with a tendency to ulceration. The tumorous growth of
dermatofibrosarcoma protuberans is mostly
present on the back; it looks like a keloid, is
resistant to radiation, and may recur after
removal, sometimes locally but may (only
rarely) metastasize (lungs). Kaposi's sarcoma
(multiple idiopathic hemorrhagic sarcoma) is
a form of sarcoma with multipotential vascular
new formations, pronounced participation of
spindle cells, and characteristic deposits of
hemosiderin. It favors the extremities; according to GROENOUW the eyelids are rarely involved in the beginning, but in some instances
they become invol ed later. There are more
recent reports about Kaposi's sarcoma on the
lids by ALEXANDER and SEZER and ERCIKAN.
Sarcomatous changes of the uvea in spite of
its endowment with abundant vessels and
reticuloendothelial elements are conspicu-
ously rare (ROSSELET and DELLER). The retothelial sarcoma (ROULET and RÖSSLE) has its
origin not infrequently in lymphatic tissue
such as the tonsils; it shows mycosislike
tumors on the skin. RECUPERO and PETRO-
SILLO report a case of reticulum cell sarcoma
on the bulbar conjunctiva. In general, facial
sarcomas may occur in lupus vulgaris treated
with x-rays (SCHNEIDER) or in xeroderma
pigmentosum. Primary polymorphic cell sarcomas affecting the lids or conjunctiva (BER-

NARDI and PALMIERI) or fibromyxosarcoma (SHUKLA and AGRAWAL) were reported, however.

Pigmented tumors

Nevus cell nevi

According to v. ALBERTINI the nevus cell nevi are dysontogenetic tumors on the order of hamartomas with a highly differentiated organoid structure. They are subject to surprisingly evolutional and involutional forces, which manifest themselves in the third decade by the highest frequency of these lesions with twenty nevus cell nevi, on an average; after the 50th year there are only about three, whereas a person of 90 years is practically free of nevus cell nevi. In advanced age they project themselves distally. In any case, these nevi can disappear within a few years; their involution may take place either with formation of pedunculated lesions, which might drop off, with fibrosis, or by fatty infiltration of the resting atrophic nevus.

The classification proposed by TRAUB and KEIL, and ALLEN and SPITZ of differentiating nevi according to their position within the layers of the skin has been accepted:

1. *Nevus cell nevi* located on the border of the epidermis and cutis (called borderline nevi by UNNA, now known as *junction nevi*)

2. *Nevi of the dermis* or the *corium* (Figs. 164 and 165)

3. Intermediate or *compound nevi*, or so-called combination nevi (in several layers)

Junction nevi are seen almost entirely in childhood, and compound nevi until puberty; intradermal nevus cell nevi are almost entirely limited to adults. However, one can also argue that these three varieties, differing in their position, represent several phases of development of the single nevus cell nevus. According to the investigations of GREIFELT and EKBLAD, the chances of *malignant degeneration of a nevus cell*

nevus are about 1:1 million, which means that such a nevus cannot rate without qualifications as a preblastomatous condition. The malignant melanoma, on the other hand, probably develops mainly from a junction nevus, but this contradicts the experience in children, whose nevus cell nevi are almost always showing junctional activity, and who develop malignant melanomas much more rarely than adults. The hypothesis of junctional activity as a necessary presupposition for the development of a melanoma can hardly be brought in accord with the fact, borne out by experience, that an intradermal nevus cell nevus appearing as "clinically resting," such as a papillary or hairy nevus cell nevus or even a classic hairy bathing suit nevus ("nevus pellitus"), can show transformation into a melanoma, contrary to accepted teaching (see case KORTING).

Ophthalmologically, ALLINGTON and ALLINGTON, in their statistics of eyelid tumors, put nevus cell nevi in third place of frequency; they observed all three above-named types on the eyelids as well as on the conjunctiva. These authors call a "divided nevus" a nevus cell nevus occurring on the upper and lower eyelids of the same eye; EHLERS found 19 such cases in the literature. As to junctional activity, REESE arrives at a percentage of 90 to 95 per cent of nevus cell nevi in this location.

Blue nevus (nevus caeruleus)
JADASSOHN-TIÉCHE

This nevus is usually a solitary, up to lentilsize, dark bluish tumor and is slightly lighter in the center because the follicles are spared; the lesion sometimes presents a nodular formation because of a three-dimensional heaping of spindle-shaped melanocytes in the cutis. Malignant degeneration of the blue nevus is an extraordinary occurrence. If histopathologically there are additional fibromatous, neurinomatous, or sarcomalike structures, we are dealing with the special type of the cellular blue nevus (neuronaevus bleu Masson, or

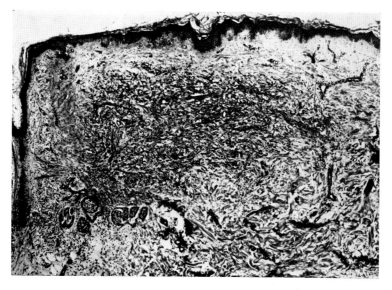

167   Blue nevus. Three dimensional collection of spindle-shaped melanocytes in the middle of the cutis.

cellular blue nevus — Allen, for details see GARTMANN, 1961 and 1965, and CRAMER, 1966; Figs. 166 and 167). OTA's naevus caeruleus as well as the mongolian spot were discussed above with the pigmentations.

So-called juvenile melanoma

This special type of compound nevus was described by SOPHIE SPITZ in 1948 as a distinct entity. A better name, which at the same time differentiates this benign tumor from malignant melanoma with its completely different prognosis, is simply fascicular nevus (MIESCHER) or "spindle cell nevus" (GERTLER) (Fig. 168).

It occurs chiefly between the third and eighth year of life, but exceptionally is already present at birth or occurs in an adult. A preferred location is the face, but it is present also on other parts of the body as a pinhead-sized, flat conelike, elevated reddish-brown or red-brown nodule, which under glass pressure appears either homogeneously light brownish or somewhat slate gray. Its surface is, in the majority of cases, smooth, so that neither oozing, erosions, nor other changes belong to the classic picture; neither does multiplicity.

Histologically, we observe distinct subepidermal edema, formation of large spindle- or starlike-shaped cells, also characterized as "tennis-racket-like," "myoblastoid" cells with a round or lobulated nucleus and one to two nucleoli (Fig. 169). Of diagnostic importance are dilatations of vessels and, above all, giant cells frequently close to the Tuton type (STEIGLEDER and WELLMER). Mitoses, but not typical mitotic figures, can be present. The great importance of recognizing this special form of nevus lies in the avoidance of drastic, radical therapeutic measures, because conservative excision (before puberty!) with subsequent histologic verification is sufficient. In one of my cases recurrence in loco was observed.

Ophthalmologically, a juvenile melanoma occurred on the eyelids (JONES and DUKES: DHERMY and DUGELEY-MAGNAUD), also on the conjunctiva (GARTMANN and THURM; TIMM), in the uveal tract (ELLSWORTH) and, finally, on the iris (SAMUELS).

Melanosis circumscripta preblastomatosa

HUTCHINSON was the first to mention in 1890 "melanotic freckles," and in 1894 and 1912 DUBREUILH called this disease "melanosis

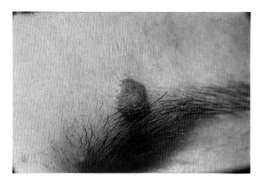

166 Blue nevus.

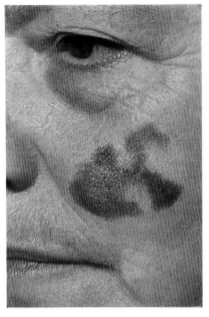

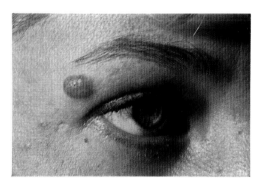

168 Juvenile melanoma.

170 Malignant freckle (melanosis circumscripta preblastomatosa-*Dubreuilh*).

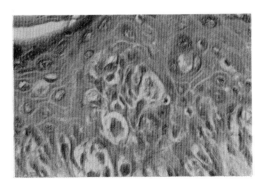

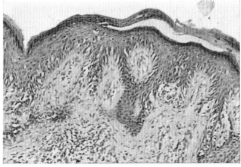

169 Juvenile melanoma. Formation of large spindle-shaped tennis racketlike or "myoblastoid" cells.

171 Melanosis circumscripta-*Dubreuilh*. Histologic appearance showing the beginning segregation of larger groups of cells.

circumscripta precancerosa" or, as we call it today, "preblastomatosa"; it does not always represent the starting point of a biologically malignant melanoma spreading without limitation, but the removal of such spots without doubt can be considered a genuine prophylaxis against melanoma because three-quarters of such cases (if not excised — Translators) will in the long run develop into melanomas. In the beginning such a melanotic freckle appears washed out and light brown; then slowly it enlarges and runs into a spotted, dark brown to brownish-black pigmented, or in places, depigmented lesion, occasionally looking like spotted leopard skin. The condition becomes dangerous if within the lesion or close to it, solitary or multiple, hemispheric or nipplelike protruding, smooth, black, shiny elevations ("ink spots" — Translators) develop. In two-thirds of all cases the face is the chief site of "Dubreuilh's melanosis"; the upper two-thirds of the face are mainly affected. However, such developments may occur also on other parts of the body, for instance, as a speckled black "balanitis circumscripta preblastomatosa" or on the mucous membrane as "melanoplakia."

Histologically, the main feature of this preblastomatous, circumscribed melanosis is the "abtropfung" (proliferative extension of cells from the dermoepithelial junction into the corium — Translators), with segregation of clusters of larger round cells accompanied by an unrest of cells; this cellular unrest also occurs along the follicular ducts. MISHIMA recently made a statement, based upon electron-microscopic and histochemical examinations, that Dubreuilh's melanosis does not depend upon proliferation of pathologically changed nevus cells but on a conglomeration of atypical dendritic melanocytes (Figs. 170 and 171).

As far as the *eye* is concerned, Dubreuilh's melanosis is usually found in the 40- to 50-year-old patient, either on the eyelids (ALLINGTON and ALLINGTON) or on the conjunctiva; the latter shows partly granular pigmentations at the level of the surface on the palpebral as well as the bulbar portion (see FRANÇOIS et al.). REESE emphasizes that metastasis may develop from these pigmented areas without local exophytic development in such a melanosis. In any case, besides rapid enlargement and increased pigmentation, a certain degree of pain and uncharacteristic signs of inflammation are indicative of malignant transition; but REESE (1966) believes this change in Dubreuilh's melanosis of the conjunctiva is not too high (17 per cent).

(Malignant) melanoma

Malignant melanoma (CARSWELL, 1838) is a rare tumor not only when compared with the frequency of the nevus cell nevi but also when its incidence is considered in the population at large. Of all the malignant tumors of patients in departments of dermatology, melanomas amount to only about 3 per cent. One of the chief characteristics of the melanoma, the end product of malignantly deteriorated melanocytes, is, besides its affinity to internal organs, its tendency to peripheral projection, which in a typical case manifests itself by definite affinity to the epidermis; not only is the pigment deposited throughout the horny layer, but isolated cellular elements pushed to the skin surface or formation of subcorneal colonies of cells are noteworthy.

This tumor, highly malignant in man, exists also in animals; for instance, in fishes (PRIEBE), dogs, and so on. Eighty per cent of all white horses more than twenty years old die of melanoma (literature VIRCHOW). The melanoma of man can be transplanted into the *anterior chamber of the eye of the guinea pig* in a certain percentage (see COTTINI) and can be grown in isolated cell cultures.

As far as the *age* is concerned, the average is 52.3 years in about 160 cases of melanoma observed by myself; this corresponds to most of the reports in the literature. The proportion between the sexes is 45 per cent for males and

54 per cent for females (1 per cent among children).

The *location of the melanoma* presents, on the one hand, a preponderance of head and face and, on the other hand, predilection for the lower extremity. Besides, there are reports of primary melanomas in the anorectal area, larynx, bronchi, urethra, stomach, and even hypophysis.

Familial occurrence of the melanoma is certainly rare but has been reported without doubt (personal observations of mother and daughter and of brother and sister).

Customarily, the origin of malignant melanoma is considered a nevus cell nevus degenerated toward malignancy. In addition (see GARTMANN), Dubreuilh's melanosis circumscripta preblastomatosa should also be considered a point of origin. But in each series of cases there remain those of "melanom d'emblée" pointing to a tumor, which is a priori a melanoma.

As far as the skin is concerned, *differential diagnostic* possibilities are histiocytomas with storage of hemosiderin, telangiectatic granulomas, small eruptive angiomas, flat mushroomlike, greasy-friable, basal cell papillomas, acanthotic senile warts, and others.

The *typical melanoma* is round or kidney-shaped with a suggestion of a polycyclic outline; when not colorless, it shows dark hues, which vary between brown and blue black and, more important, may appear with different intensity of color saturation (Figs. 172 and 173). If the tumor is not too dark, we find not infrequently in the periphery very thin, darker, sometimes spiral or branchlike lines. The typical melanoma further shows, especially when hemispherically elevated, a smooth, frogspawn-like, glassy shiny, easily traumatized surface, which, however, may in later stages assume an uneven, papillomatous or ulcerated aspect. But hard, solid types of tumors also exist. The melanoma's well-known propensity to bleed when subject to the least amount of surface pressure is shared with small angiomas and pyogenic granulomas,

while carcinomas, spindle cell- and reticulosarcomas do not bleed. The diagnosis is definite when there are perifocally one or more pinpoint-sized satellites; just as important is the presence of an erythematous halo. Further danger signs are itching, rapid increase in size of the lesion, increasing pigmentation, either homogeneous or speckled, increase of moisturization, or even oozing, bleeding, or ulceration.

The melanoma shows *histologically*, sometimes quite early, but not in all cases, a loss of organoid structure, at least in peripheral areas. In contrast to other malignant tumors, the inflammatory reaction of the base, i.e., the reaction of the stroma or the cellular attempt of defense against the imminent invasion, may be completely absent (for details see KORTING, HOEDE, and HOLZMANN). Of special importance is the presence of mitoses and, above all, atypical mitoses. Another stigma of malignancy is the active junctional proliferation, which is absent in metastases. Cellular anaplasia, cell distortion, and, especially, emigration of melanoma cells into the upper epidermal layers are of further importance; so also is the projection of very fine quicksand-like particles of melanin toward the surface.

The melanomas developing from melanosis preblastomatosa as well as those occurring in later life and the (preponderantly spindle cell) melanomas of the eye (JÄNISCH and SCHULZE) have (perhaps) a better prognosis clinically. Considering the histologic picture, the fascicular or fusocellular melanomas are (perhaps) less malignant, whereas alveolar and globocellular melanomas probably possess a higher degree of malignancy; thus, the former differentiation between "nevocarcinomas" and "melanosarcomas" would regain its importance.

Laboratory examinations such as Thormälen's or Jaksch's tests for melanuria have hardly any importance, except that, when positive, they point to the presence of hepatic metastases containing melanin.

Rare reports refer to multiple-autochthonous melanomas (1 among 160 of my own observations) and melanomas co-existing with other multiple growths, as especially in xeroderma pigmentosum. Dermatomyositis may occasionally have a melanoma as the underlying malignant tumor (for further details see KORTING).

Specific *involvement of the lymph nodes* in melanoma takes place early but escapes palpation frequently so that it remains subclinical. The impregnation of a group of lymph nodes with ink-black foci can be found only when an en-bloc extirpation and subsequent pathologic-anatomic examination of all nodes take place. Therefore, in practice the diagnosis of stage one, i.e., a primary melanoma without lymph node metastasis, is actually made too frequently.

As far as the involvement of internal organs by a *metastasizing melanoma* is concerned, I frequently observed in twenty autopsies, besides the well-known sites of metastasis such as lung, liver, and brain, the heart (in third place in our series).

As mentioned several times before, the course of an individual case may be so erratic that one cannot make definite statements about the actual degree of malignancy and, therefore, about the further course of the disease. In this connection it has been emphasized before that although the validity of such criteria is, generally speaking, uncertain, spindle-celled melanotic tumors, the circumscript preblastomatous melanosis of Dubreuilh, and perhaps the melanomas of older people on the whole and those of the eye especially (JÄNISCH and SCHULZE) have a better prognosis. The reverse is true of melanotic tumors of the mucous membranes, the anogenital region, and those of the periphery in a craniocaudal direction. They have an a priori worse prognosis. Furthermore, patients with a primary tumor of up to 2 cm in diameter have (perhaps) a better chance of survival than those with a primary tumor measuring 3 to 7 cm. However, not every case follows this rule.

Spindle cell tumor types are common in melanomas of the eye, a fact which could explain the generally better prognosis of these growths compared with those of the skin (JÄNISCH and SCHULZE) (Figs. 174 and 175). SAUTTER and HAGER could prove experimentally the influence of the vegetative system on the melanoma of the eye of the rabbit (acceleration through adrenalin and retardation through acetylcholine). From a racial point of view there are, according to ZIMMERMAN, 150 reports of uveal melanomas of whites compared to one of a Negro; histopathologic examinations by PAUL have shown that strong pigmentation of the iris seems rather to exclude a melanoma. KROLL and KUWABARA report on electron-microscopic findings in the melanoma of the chorioid. In a histological specimen it is possible to differentiate between the types of malignant melanoma of the chorioid, but in tissue cultures this differentiation is lost (LUND). This leads to the assumption that these tumors are not variants of various tumor forms but that extratumorous factors cause a change in an a priori uniform type of tumor cell. This results in different degrees of malignancy. Furthermore, the histologic picture of melanoblastomas treated by light-coagulation is similar to spontaneous necrosis of these tumors (LUND). Besides the location on the chorioid malignant melanomas may occur infrequently on the iris or the ciliary body; this location affects patients 10 to 20 years younger than those with a melanoma of other parts of the uvea; perhaps the explanation of these findings is that a tumor on this site can be more easily discovered (REESE). Malignant melanomas of the *cornea*, which are similar to epithelial tumors, are very rare, while malignant mesenchymal tumors of the cornea are practically unknown. According to NOESKE, NOTTER, and SANDRITTER, melanomas of the cornea and conjunctiva are derived from "Schwann cells or special basal cells of the corneal epithelium

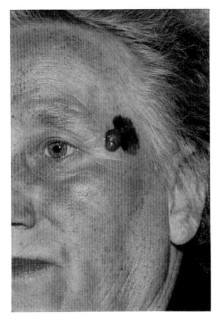

172 Malignant melanoma.

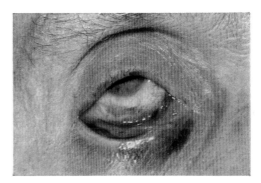

174 Malignant melanoma of the conjunctiva.

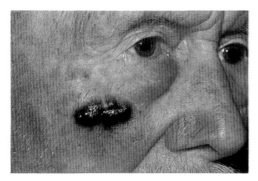

173 Malignant melanoma.

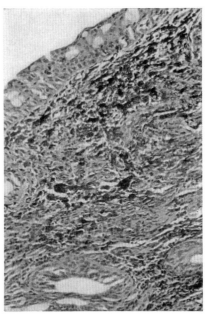

175 Histologic appearance of the malignant melanoma shown in Fig. 174.

or corresponding cells of the limbus or out-side of it." According to Reese, reports on bilateral melanomas of the eyes can be con-sidered proved only in very few instances. Finally, the primary melanoma of the skin may metastasize into the chorioid or the orbit (for references see SCHRECK); I witnessed,

however, twenty autopsies, presenting wide-spread metastasizing primary cutaneous mela-nomas and there was not one instance with involvement of the eye. (However, metastases to the eye have been reported by others — Translators.)

References

ALBERT, M. D., RUBENSTEIN, R. A., and SCHEIE, H. G.: Tumor metastasis to the eye. I. Incidence in 213 adult patients with generalized malignancy. Amer. J. Ophthal. Ser. 3, 63 (1967) 723—726.

ALBERT, D. M., RUBENSTEIN, R. A., and SCHEIE, H. G.: Tumor metastasis to the eye. II. Clinical study in infants and children. Amer. J. Ophthal. 63 (1967) 727—732.

ALEXANDER, C. M.: Kaposi's sarcoma with ocular mani-festations. Amer. J. Ophthal. 55 (1963) 625.

ALLEN, A. C.: Juvenile melanomas of children and adults and melanocarcinomas of children. Arch. Derm. Syph. (Chic.) 82 (1960) 325—335.

ALLEN, A. C.: Juvenile melanomas. Ann. N. Y. Acad. Sci. 100 (1963) 29—48.

ALLEN, A. C.: The Skin. 2d Ed. Grune & Stratton, New York, 1967.

ALLINGTON, H. V., and ALLINGTON, J. H.: Eyelid tumors. Arch. Derm. Syph. (Chic.) 97 (1968) 50—61.

ANDRADE, R.: Die präanceröse und canceröse Wucherung von Epidermis und Anhangsgebilden. In: Hdb. der Haut-und Geschlechtskrankheiten, Bd. I/2, edited by J. Jadas-sohn. Springer, Berlin, 1964.

ANDREW, R.: Gastroenterology 23 (1953) 495—499; see also Steigleder.

BAER, R. L., and KOPF, A. W.: Yearbook of Dermatology. Year Book Med. Publ., Chicago, 1962—1963.

BARRON, Ch. N.: The comparative pathology of neoplasms of the eyelids and conjunctiva with special reference to those of epithelial origin. Acta derm.-venereol. (Stockh.) 42, Suppl. 51 (1962) 1—100.

BEUDT, M.: Der jetzige Stand der Lehre vom Morbus Bowen. Inaug.-Diss., Mainz, 1965.

BISHOP, D. W.: Trichoepithelioma. Arch. Ophthal. 74 (1965) 4.

BONIUK, M., and HALPERT, B.: Clear cell hidradenoma or myoepithelioma of the eyelid. Arch. Ophthal. 72 (1964) 59—63.

BONIUK, M., and ZIMMERMANN, L. E.: Pilomatrixoma (benign calcifying epithelioma of the eyelids and eyebrow). Arch. Ophthal. 70 (1963) 399.

BONIUK, M., and ZIMMERMAN, L. E.: Eyelid tumors with reference to lesions confused with squamous cell carci-noma: III. Keratoacanthoma. Arch. Ophthal. 77 (1967) 29—40.

BRADFORD, M. L., and DANZIG, L. E.: Boston. med. Quart. 1 (1950) 21—24; quoted by Steigleder.

CASANOVAS, J.: Epitelioma metastásico. Arch. Soc. oftal. hisp.-amer. 22 (1962) 1103—1108. Ref. Zbl. ges. Ophthal. 90 (1964) 215.

CARSWELL, R.: Pathological anatomy: illustrations of the elementary forms of disease. Longman, Orme, Brown, Green and Longman, London, 1838.

CHRISTELER, A., and DELACRÉTAZ, J.: Verrues séborrheiques et transformation maligne. Dermatologica (Basel) 133 (1966) 33—39.

CIVATTE, J.: Histopathologie cutanée. Flammarion, Paris, 1967.

CIVATTE, J.: Verrue séborrhéique. In: Histopathologie cuta-née. Flammarion, Paris, 1967.

CRAMER, H. J.: Verruca seborrhoica und sog. "Basosqua-mous Cell Acanthoma." Arch. klin exp. Derm. 212 (1960) 49—63.

CRAMER, H. J.: Über den "Neuro-Nevus blue" (Masson). Hautarzt 17 (1966) 16—21.

CRAMER, H. J., and KAHLERT, G.: Das Cornu cutaneum. Selbständiges Krankheitsbild oder klinisches Symptom? Derm. Wschr. 150 (1964) 521—531.

CRAMER, R., and CRAMER, H. J.: Über die pseudobasalio-matöse Epithel-Hyperplasie der Haut. Arch. klin. exp. Derm. 216 (1963) 231—245.

DHERMY, P., and DUGELAY-MAGNAUD, J.: Mélanome juvénile de la paupière. Mélanome de Sophie Spitz. Bull. Soc. Ophthal. Fr. 64 (1964) 503—512.

DORMANDY, T. L.: "Peutz-Jeghers Syndrome." In: Modern Trends in Gastroenterology, Ser. 2 Jones, London, 1958.

DUBREUILH, W.: Kératose sénile. In: Des hyperkératoses circonscrites. Ann. Derm. Syph. (Paris) 7 (1896) 1158—1204.

EHLERS, G.: Zur Klinik der Basalzellepitheliome unter Be-rücksichtigung statistischer Untersuchungen. Z. Haut- u. Geschl.-Kr. 41 (1966) 226—238.

EHLERS, N.: A case of divided nevus. Arch. Ophthal. 73 (1965) 664—666.

ELLSWORTH, R. M.: Juvenile melanoma of the uvea. Trans. Amer. Acad. Ophthal. 64 (1960) 148—149.

EPSTEIN, E., EPSTEIN, N. N., BRAGG, K., and LINDEN, G.: Metastases from squamous cell carcinomas of the skin. Arch. Derm. Syph. (Chicago) 97 (1968) 245—249.

EPSTEIN, W., and KLIGMAN, A. M.: The pathogenesis of milia and benign tumors of the skin. J. Invest. Derm. 26 (1956) 1—11.

FRANÇOIS, J., GILDEMYN, H., and RABACY, M.: Melanose cancereuse de la cornée. Bull. Soc. belge ophthal. 110 (1955) 298—306.

FRANÇOIS, J., et al.: Tumours of the eye, edited by A. B. Reese. Harper & Brothers, New York, 1953.

FREUDENTHAL, W.: Verruca senilis und Keratoma senile. Arch. Derm. Syph. (Berl.) 152 (1926) 505—528.

GANS, O., and STEIGLEDER, G. K.: Histologie der Haut-krankheiten, Bd. I and II, 2d Ed. Springer, Berlin, 1955, 1957.

GARTMANN, H.: Über den zellreichen blauen Naevus. Derm. Wschr. 143 (1961) 297—307.

GARTMANN, H.: Das sog. juvenile Melanom. Münch. med. Wschr. 104 (1962) 587—592; 633—635.

GARTMANN, H.: Neuronaevus bleu Masson-cellular blue nevus Allen. Arch. klin. exp. Derm. 221 (1965) 109—121.

GARTMANN, H., and THURM, K.: Juveniles Melanom der Augenbindehaut. Derm. Wschr. 142 (1960) 805—811.

GASTEIGER, H., and MASSHOFF, W.: Über seltene Neubil-dungen im Bereich der Lider. Wien. klin. Wschr. 76 (1964) 503—509.

GERTLER, W.: Zur Epithelverbundenheit der Basaliome. Derm. Wschr. 151 (1965) 673—677.

GERTLER, W.: Faszikulärer Spindelzellnaevus. Derm. Wschr. 133 (1966) 110–111.

GINSBURG, J.: Present status of meibomian gland carcinoma. Arch. Ophthal. 73 (1965) 271.

GOTTRON, H. A.: Präkanzerosen und Pseudokanzerosen. Derm. Wschr. 147 (1963) 338–344.

GOTTRON, H. A., and NIKOLOWSKI, W.: Sarkom der Haut. In: Dermatologie und Venerologie, Bd. IV, edited by H. A. Gottron and W. Schönfeld. Thieme, Stuttgart, 1960.

GOWEY, R. J., and KERN, W. H.: Metastasizing adenocarcinomas of the tarsal glands. Calif. Med. 103 (1965) 126.

GROENOUW, A.: Beziehungen des Auges zu den Hautkrankheiten. In: Hdb. der Haut- und Geschlechtskrankheiten, Bd. XIV/1, edited by J. Jadassohn. Springer, Berlin, 1930.

GUILLARD, J., LAUMONIER, R., SEYER, J., and BOULET, R.: L'association polypose intestinale-lentiginose (syndrome de Peutz-Jeghers). Presse méd. 65 (1958) 1076–2079.

HABER, H.: Intraepidermal Acanthoma. Dermatologica (Basel) 117 (1958) 304–316.

HAGEDOORN, A.: Paget's disease of the eyelid associated with carcinoma. Brit. J. Ophthal. 21 (1937) 234–241.

HAGER, G.: Das Keratoakanthom, seine Abgrenzung von anderen epithelialen Lidtumoren, besonders dem Carcinom. v. Graefes Arch. Ophthal. 158 (1956/57) 393–402.

HAMPERL, H., and KALKOFF, K. W.: Zur Kenntnis des Molluscum pseudocarcinomatosum. Hautarzt 5 (1954) 440–447.

HEYDENREICH, A.: Differentialdiagnostische Erwägungen bei Tumoren der Augenlider. Dtsch. Gesundheitswesen 1964, 2240–2246.

HOLLAND, G., and BELLMANN, O.: Zur Klinik und Therapie der Basaliome und Spinaliome. Ophthalmologica (Basel) 150 (1965) 138–152.

HOLLANDER, A., and GROTS, J. A.: Oculocutaneous metastases from carcinoma of the urinary bladder. Arch. Derm. Syph. (Chicago) 97 (1968) 678–684.

HOLZMANN, H., and CABRÉ, J.: Über intraepidermale Epitheliome der Haut. Frankf. Z. Path. 73 (1964) 659–667.

JÄNISCH, W., and SCHULZE, B.: Vergleichende Untersuchungen über die Malignität der Melanocytoblastome des Auges und der Haut. Arch. Klin. exp. Derm. 217 (1963) 60–70.

JEGHERS, H., McKUSICK, V. A., and KATZ, K. H.: Generalized intestinal polyps and melanin spots of the oral mucosa, lips and digits. New Engl. Med. J. 241 (1949) 993–1005; 1031–1036.

JONES, S. T., and DUKES, T. E.: Juvenile melanoma of the eyelid. Amer. J. Ophthal. 56 (1963) 816.

KALKOFF, K. W., and MACHER, E.: Zur Histogenese des Keratoakanthoms. Hautarzt 12 (1961) 8–15.

KLOSTERMANN, G. F.: Über Epidermisbeteiligung bei metastatischem Hautkrebs, zugleich eine Studie über die Ausbreitung des Bowen-Carcinoms. Arch. klin. exp. Derm. 216 (1963) 18–35.

KNAUER, W. C., Jr., and WHARTON, C. M.: Extramammary Paget's disease originating in Moll's glands of the lids. Trans. Amer. Acad. Ophthal. Otolaryng. 67 (1963) 829, see also Allington and Allington.

KOPF, A. W., and ANDRADE, R.: Year Book of Dermatology. Year Book Med. Publ., Chicago, 1965–1966.

KOPF, A. W., MORRILL, S. D., and SILBERBERG, I.: Broad spectrum of leucoderma acquisitum centrifugum. Arch. Derm. Syph. (Chic.) 92 (1965) 14–35.

KORTING, G. W.: Ausschnitte aus der dermatologischen Onkologie der Gegenwart. Ärztl. Fortbildg. 14 (1964) 39–48.

KORTING, G. W.: Über Klinik und Therapie der Melanome. Dtsch. Ärztebl. 61 (1964) 367–380; 455–461.

KORTING, G. W.: Therapie der Melanome. Dtsch. med. Wschr. 91 (1966) 501–505.

KORTING, G. W.: Drei Demonstrationen zum Thema Melanom und Pseudo-Melanom. Med. Welt (Stuttg.) 18 (1967) 917–918.

KORTING, G. W., BREHM, G., and NÜRNBERGER, F.: Zur klinischen Variationsbreite des sog. juvenilen Melanoms. Z. Haut- u. Geschl.-Kr. 43 (1968) 233–238.

KORTING, G. W., and HASSENPFLUG, K.-H.: Zur Histogenese des Epithelioma Malherbe und von seinen topischen Beziehungen zum Keratoakanthom. Arch. klin. exp. Derm. 222 (1965) 11–22.

KORTING, G. W., HOLZMANN, H., and HOEDE, N.: Bemerkungen zur Stromareaktion beim Melanom. Med. Welt (Stuttg.) 18 (1967) 1786–1794.

KORTING, G. W., and NÜRNBREGER, F.: Zur Frage des Lipidgehaltes von Myxosarkomen. Arch. klin. exp. Derm. 230 (1967) 172–182.

KREIBIG, W.: Das Auge und sein Hilfsapparat. In: Lehrbuch der speziellen pathologischen Anatomie, Bd. III/2, edited by E. Kaufmann and M. Staemmler. De Gruyter, Berlin, 1961.

KROLL, A. J., and KUWABARA, T.: Electron microscopy of uveal melanoma. A comparison of spindle and epitheloid cells. Arch. Ophthal. 73 (1965) 378–386.

LEVER, W. F.: Pathogenesis of the benign tumors of cutaneous appendages and of basal cell epithelioma. II. Basal cell epithelioma. Arch. Derm. Syph. (Chic.) 57 (1948) 709–724.

LEVER, W. F.: Histopathologie of the skin. 4th Ed. J. B. Lippincott Co. Philadelphia, 1967.

LUND, O.-E.: Histologische Untersuchungen an lichtcoagulierten Melanoblastomen der Aderhaut. v. Graefes Arch. Ophthal. 164 (1962) 433–456.

MALIK, S. R., SOOD, G. S., and AURORA, A. L.: Granuloma pyogenicum. Brit. J. Ophthal. 48 (1964) 502.

MARTIN, H. E.: Cancer of the eyelids. Arch. Ophthal. 22 (1939) 1–20.

MASSON, P.: Neuronaevus bleu. Ann. Anat. Path. 3 (1926) 417, 657.

MISHIMA, Y.: Melanosis circumscripta praecancerosa (Dubreuilh). J. Invest. Derm. 34 (1960) 361–375.

MISHIMA, Y.: Macromolecular changes in pigmentary disorders. Arch. Derm. Syph. (Chic.) 91 (1965) 519–557.

MONACELLI, M.: Fibroepitheliale Hauttumoren und Präkanzerosen. Med. Klin. 62 (1967) 1154–1156.

MONTGOMERY, H.: Dermatopathology, Bd. I and II. Harper & Row, New York, 1967.

MORTADA, A.: Glomangioma of the eyelid. Brit. J. Ophthal. 47 (1963) 697.

MUENZLER, W. S., OLSON, J. R., and EUBANK, M. D.: Metastatic tumors of the eyelid. A report of two cases and review of the literature. Amer. J. Ophthal. 55 (1963) 791–794.

NIKOLOWSKI, W., and SEITZ, R.: Epithelioma Malherbe im augapfelnahen Hautgewebe. Klin. Mbl. Augenheilk. 136 (1960) 825–836.

NOESKE, K., NOTTER, H., and SANDRITTER, W.: Maligne Tumoren der Kornea. Med. Welt (Stuttg.) 17 (1966) 1671–1673.

NOTTER, H.: Bericht über ein malignes Melanom der Hornhaut. Klin. Mbl. f. Augenheilk. 147, 50–57, 1965.

PAUL, S. D.: Iris pigment in ocular melanomata. J. All. India Ophthal. Soc. 13 (1965) 62–64.

PERDRIEL, G., LOUBIÈRE, R., and PFISTER, A.: Un cas de kératoacanthome palpebral. Bull. Soc. Ophthal. Fr. 65 (1965) 191–193.

PIERARD, J., and KINT, A.: Kératose sénile. Arch. belge Derm. Syph. 20 (1964) 137–157.

PINKUS, H.: Differenzierung bösartiger und relativ gutartiger Spinaliome und Melanome. Med. Klin. 62 (1967) 1160–1162.

POIARES BAPTISTA, A. V. B.: Querato-acantoma. Coimbra, Coimbra/Portugal, 1964.

PRIEBE, K.: Über Pigmentierungsanomalien der Haut bei Nutzfischen des Nordatlantik. Lebensmittelhygiene 18 (1967) 97–99.

RECUPERO, C., and PETROSILLO, O.: Su di un caso di reti-colosarcoma della congiuntiva. Contributo clinico ed istologico. Arch. Ottal. 67 (1963) 219—228.

REESE, A. B.: Tumors of the eye. Harper & Brothers, New York, 1953; especially p. 509 ff.

REESE, A. B.: Precancerous and cancerous melanosis. Amer. J. Ophthal. 61 (1966) 1272; especially pp. 294, 222.

RODÉ, I.: Die klinischen und strahlenbiologischen Eigenschaften des Melanoblastoms. Akadémiai kiadó, Budapest, 1962.

RODENHÄUSER, J. H.: Zur Klinik und Histologie des Molluscum pseudocarcinomatosum (Molluscum sebaceum, Kerato-Akanthoma). v. Graefes Arch. Ophthal. 158 (1956/1957) 468—484.

ROSSELET and DELLER: Lésions choriorétiniennes de l'angiosarcomatose de Kaposi, Ophthalmologica 133, 361—364, 1957.

SAMUELS, S. L.: Juvenile melanoma of the iris. Trans. Amer. Acad Sci. Ophthal. Otolaryng. 67 (1963) 718—722.

SAUTTER, H., and HAGER, H.: Vom Einfluß des vegetativen Systems auf das Wachstum der Geschwülste, speziell der Melanoblastome. v. Graefes Arch. Ophthal. 151 (1950) 156—166.

SCHNEIDER, W.: Sarkome und Carcinome in ihren Wechselbeziehungen auf röntgenbestrahltem Lupus vulgaris. Strahlentherapie 80 (1949) 335—366.

SCHRECK, E.: Veränderungen des Sehorgans bei Haut- und Geschlechtskrankheiten. In: Dermatologie und Venerologie, Bd. IV, edited by H. A. Gottron and W. Schönfeld. Thieme, Stuttgart, 1960.

SEZER, N., and ERICKAN, C.: Kaposi's sarcoma with ocular manifestations. Brit. J. Ophthal 48 (1964) 223.

SHULER, J. M., and AGRAWAL, S.: Fibromyxosarcoma of the eyelid. Brit. J. Ophthal. 51 (1967) 403.

SIVARMASUBRAMANIAM, P.: Solitary metastatic orbital tumour due to sympathetic neuroblastoma. Brit. J. Ophthal. 47 (1963) 312—313.

STEIGLEDER, G. K., and WELLMER, K.: Zur Abtrennung des sog. juvenilen Melanoms. Arch. klin. exp. Derm. 202 (1955/1956) 556—566.

STRÄULI, P.: Gut- und Bösartigkeit von Tumoren. In: Hdb. der Medizinischen Radiologie, Bd. XVIII, edited by L. Diethelm et al. Springer, Berlin, 1967.

STÜTZER, H.: HNO (Berl.) 8 (1959) 23, quoted by Steigleder.

TIMM, G.: Zur Mikromorphologie des Spitzschen Melanoms der Conjunctiva bulbi. Klin. Mbl. Augenheilk. 143 (1963) 391—400.

VANCEA, P., and CHISLEAG, Gh.: Beiträge zum klinischen und therapeutischen Studium des Lidkrebses. Klin. Mbl. Augenheilk. 142 (1963) 1064—1070.

VIRCHOW, R.: Die krankhaften Geschwülste. Bd. II. Hirschwald, Berlin, 1862—1865.

WINKELMANN, R. K., and MULLER, S. A.: Sweat gland tumors: I. Histochemical studies. Arch. Derm. Syph. (Chic.) 89 (1964) 827—831.

WOLKOWITZ, M. J.: Chalazion-like neoplasms of the lids. Amer. J. Ophthal. 54 (1962) 249.

WRIGHT, J. C., and MEYER, G. E.: Metastatic bronchogenic carcinoma of the eyelid. Amer. J. Ophthal. 54 (1962) 135.

ZIMMERMAN, L. E.: Melanocytes, melanocytic nevi, and melanocytomas. Invest. Ophthal. 4 (1965) 11—41.

Leukoses, Reticuloses, and Reticuloendothelioses of the Skin

Leukoses

Leukoses are today regarded as irreversible proliferations of elements of one of the three blood cell lines. They are either genuine tumors or hyperplasias, or they may represent the result of viral infections. The cutaneous phenomenon of lymphadenosis (BISIADECKI, 1876) (lymphatic leukemia—Translators), or myelosis (BRUUSGARD, 1911) (myeloid leukemia—Translators), either specific or nonspecific (PINKUS and AUDREY), is a concomitant manifestation occurring as a rule at an advanced age. In childhood we are frequently dealing with acute leukemias with immature cells and, as a main sign, a severe hemorrhagic diathesis. In the leukosis of the third blood cell line, monocytic leukemia (RESCHAD and SCHILLING; NAEGELI; HITTMAIR), necrotic, ulcerative lesions of the gums, or similar changes on the palatal arch or on the tonsils are predominant; however, in lymphadenoses and myeloses polymorphic exanthematic, chiefly pruriginous, reactions prevail (prurigo lymphatica). Chronic lymphatic leukemia favoring males of an advanced age shows the complete picture of acrodermatosis, with a leonine face and bluish red to reddish blue, nodular, band- or plaquelike infiltrations of the cheeks, forehead, lips, and ear lobes; the arches of the eyebrows are transformed into bulging tumors. Lymphatic leukemic infiltrations of the eyelids may become so extreme that they prevent opening and closing of the eyes; such tumorous infiltration of the orbit may also cause severe exophthalmos. SCHRECK gives references on iridocyclitis, retinopathia, chorioiditis, and edema of the papilla of the optic nerve; in some of these cases histologic verification made definite the diagnosis of their leukemic nature. This is the place to emphasize that the thick horizontal bulging masses sometimes seen on all four eyelids may be due to "lymphomas" of different origins (HEYDEN-REICH, MORTADA, and others). I am referring especially to "plasmomas" and the cutaneous changes observed in macroglobulinemia. The "sludged blood phenomenon" observed in vessels of the fundus of the eye in cases of leukemia may possibly be due to (perhaps coincidental) dys- or paraproteinemias (see NOVER, 1962). Associated signs in (chiefly lymphatic) leukemia are generalized herpes zoster and priapism, as well as possible changes of the conjunctiva (KREIBIG), secondary damage to the cornea, shrinkage of the eyelids, and ectropion, chiefly, in leukodermic erythrodermias (LUTZ).

Reticuloses in their narrow sense

In 1924, LETTERER observed a disorder related to the leukoses showing irreversible-autonomous proliferation of the reticuloendothelial system (RES). These disorders begin chiefly as multicentric proliferations bound to the RES; they are neoplasias not at all rarely, and more often they start to develop autochthonously in the skin. Compared to such cutaneous reticuloses in the narrow sense, "chronic inflammatory aggregates of reticulum cells" (KÖLLIKER, 1889), which are capable of being stimulated and are also reversible, should more aptly be called "reticular hyperplasia" (GOTTRON), or "reticulocytosis" (LENNERT and ELSCHNER). The malignant, chiefly sarcomatous, retothelial tumors and the lipid storage disorders should not be classified with the reticuloendothelioses; such disorders are not based on primary reticulum cell elements. Anglo-American publications frequently use the term reticulosis and subordinate it to a vaguely defined "lymphoma," lymphoblastoma, or other groups (LEINBROCK).

The above-defined "genuine" cutaneous reticuloses with their multiple-autochthonous involvement of the skin (GOTTRON: so-called reticulosarcomatosis cutis, rarely also of an erythrodermatic aspect; MUSGER) are located initially around the skin appendages and, to a lesser degree, perivascularly. At least initially, the infiltrates consist of uniform cells.

Sézary's syndrome (see FLEISCHMAJER and EISENBERG) is an entity with the cutaneous aspect of an erythroderma and the hematological aspect of leukemia, while reticuloses, in general, involve systemically the viscera, including the lymph nodes, and are associated relatively late (before death) with an abundance of circulating monocytes.

No definite sites of predilection exist for the several variants—in one case, they are large-tuberous and plaquelike; in another, small-tuberous and exanthematous. They have been described as exceptional occurrences in an infant (SCHULZ, JÄNNER, and WEX) or even at birth (AMBS, BIRREN, and KLINGMÜLLER). Frequently, involvement of the *periorbital region* was observed; the erythrodermic reticuloses of the skin may also present a stubborn conjunctivitis, at times together with ectropion or "dark red discolorations" of the conjunctiva, even before generalization (LEINBROCK) takes place (MUSGER, 1966).

Lipoid storage reticuloses of the skin

Letterer-Siwe disease, which was the prototype of reticulosis, Hand-Schüller-Christian lipoid granulomatosis and the eosinophilic granuloma are now grouped together as a disease entity, called histiocytosis X (LICHTENSTEIN). Each of these three syndromes starts out as a reticulosis. Letterer-Siwe disease has a rapid febrile course with anemia and hepatosplenomegaly. Children up to the second year present seborrhoid erythemas, scaling-crusted, rice-corn sized, sometimes (rather rarely) purpuric papules resembling Darier's disease. Among the bones the mastoid is affected, but in contrast to Hand-Schüller-

Christian disease the *orbital bone* is only rarely involved. *Ophthalmologically*, FRANÇOIS and BACSKULIN described tapetoretinal degeneration in this disorder.

Hand-Schüller-Christian disease begins slowly in children up to about the fifteenth year of life; their skin is dry and faintly brownish in color, with disseminated even tuberous xanthomas; a characteristic hoarseness is caused by xanthomas located on the vocal cords. Sites of preference of these *xanthomas* are the neck, axillae, lower sides of the trunk and, especially, the *eyelids*. *Ophthalmologically*, another characteristic feature of this disease is the unilateral or bilateral *exophthalmos* caused by constriction of the orbital space; the specific granulations, which cause the globe to be pushed out, originate in the bones of the roof of the orbit (HERZAU and PINKUS). The multiple defects of the cranial bones of the skull (maplike skull) and diabetes insipidus are other characteristic signs of the disease.

Exophthalmos, so characteristic of Hand-Schüller-Christian disease, is absent in *eosinophilic granuloma* (FINZI; LICHTENSTEIN and JAFFÉ). This granuloma is nosologically and phenomenologically very close to Hand-Schüller-Christian disease but is recognizable only by a few large, eroded, or granulomatous, vegetating lesions in the axillae or genitoanal region.

MYSKA et al. report a case of eosinophilic granuloma with involvement of the eyelids and xanthomatous infiltrations of both corneae. This case seems to belong more likely to Hand-Schüller-Christian disease.

Urticaria pigmentosa

Urticaria pigmentosa in a larger sense is now considered a mast cell reticulosis (SÉZARY); this disease was originally observed by NETTLESHIP in 1869; the present name goes back to SANGSTER, 1878. As an expression of its more specific reticular hyperplastic character, urticaria pigmentosa presents, although less frequently, single tumorlike lesions and more often an exanthematic eruption of pale yellowish to brownish yellow, small, coin-sized lesions (Figs. 176 and 177). In children (75 percent) and young adults, face, palms, soles, and oral mucosa ordinarily are spared; if one rubs the lesions they swell up characteristically and this capability of erection due to friction has been explained by the release of histamine and heparinlike substances in the mast cell granules, which are discharged when the lesion is rubbed.

Other manifestations of the disease are mast cell infiltrates in the sternal marrow in the gastrointestinal tract in the liver and spleen, and in the osseous system (SAGHER), easily diagnosed by roentgenologic examination. In spite of the occurrence of urticaria pigmentosa in many locations (see the monograph of SAGHER and EVEN-PAZ), reports about functional or structural involvement of the eye have remained surprisingly rare or at least cannot be included without further qualifications (CHATARD: coincidental with congenital syphilis; BOWDLER and TULLETT: coincidental with polycythemia vera; FINDLAY, SCHULZ, and PEPLER: ocular changes not until 22 years after appearance of the mastocytosis).

Cutaneous and ocular manifestations in plasmocytoma and macroglobulinemia (Waldenström's disease)

BENCE-JONES in 1848 was the first to discover a special kind of protein in the urine; RUSTIZKY in 1873 established the entity of "myeloma"; KAHLER in 1889 reported the clinical findings in a physician; and APITZ in 1940, in continuation of a train of thought started by WALLGREN, coined the term "paraprotein" as well as *plasmocytoma*—the latter occurs in 10 to 15 percent of cases together with amyloidosis.

Laboratory findings of importance are: protein-chemical evidence of paraproteinemia (frequently, extremely high sedimentation rate; tent-shaped small alpha, beta, and gamma peaks in the electrophoretic pattern; ultracentrifugation of whole serum showing increased macroglobulins; urinary paraproteins: Bence-Jones proteins), corresponding cellular findings (plasmocytes in the bone marrow and in peripheral blood), and, last but not least, osseous changes (sharply outlined transparent zones or diffuse osteoporosis).

From a general clinical point of view, the plasmocytomatous kidney disease ("paraproteinemic nephrosis") is the leading sign (for further details and additional references see KLEMM). *Cutaneous changes* are uncharacteristic. They present reddish-blue, indolent nodes and infiltrates reminiscent of tumors of mycosis fungoides or reticuloendothelial sarcomas or unspecific erythemas, pigmentations, alopecias, ichthyosiform changes resulting in atrophy, or a seborrhoidlike dermatitis (BLUEFARB; LEINBROCK and others).

Ophthalmologically, the changes of the fundus in plasmocytoma are chiefly known as fundus paraproteinemicus (BERNEAUD-KÖTZ); the irregular alterations of the walls of the vessels, sometimes together with papillary edema, bleeding, and similar changes, result in only moderate reduction of vision, and this will happen only exceptionally bilaterally (NOVER).

Of importance to pediatricians is the moderate macroglobulinemia sometimes appearing in the serum of healthy children during the first year of life and symptomatically among the complex bleeding tendencies of syphilitic infants. In later life, especially in old age, symptomatic macroglobulinemias accompany systemic disorders (reticuloses, leukemias, and plasmocytomas) or other underlying neoplastic diseases. The cutaneous lesions of "essential" Waldenström's disease (1944), accompanied by hepatosplenomegaly and enlarged lymph nodes, are secondary and uncharacteristic, especially when associated with cryoglobulinemia. They appear as purpura, Raynaud's syndrome, ulcerative lesions of the acra (Nödl) or acrodermatitis chronica atrophicans (Brehm), and are accompanied by bleeding from the nose and oral cavity. The tumorous, cutaneous type of Waldenström's disease, which Gottron, Korting, and Nikilowski described for the first time in 1960 (Fig. 178), has been observed several times since (Bureau et al., Röckl et al., Orfanos and Steigleder).

Two observations reported at the same time present almost identical histologic findings:
1. Development of multicentric infiltrates located deep cutaneously/subcutaneously,
2. Sarcomalike architecture,
3. Alveolar, gridlike structure of the infiltrate,
4. Oligomorphic cytology with plasmocytoid transformations,
5. Mixed color changes, especially changes in intensity of PAS-reactivity of the connective tissue within the infiltrate,
6. PAS-positive nuclear inclusions.

Case 1 of Gottron, Korting, and Nikolowski presented macroscopically on the skin club- or plaquelike infiltrates with erythema in places. The *periorbicular* region showed partly lacrimal saclike, partly sausage- or cordlike, freely movable infiltrates together with a jellylike infiltration of the conjunctivae. This condition resembled the case of macroglobulinemia of Nover and Glees caused by plum-sized swelling of the tear glands. In addition, *changes of the fundus* were observed several times (see the review of Klemm), similar to fundus paraproteinemicus due to plasmacytoma and mentioned above; in macroglobulinemia additional changes of the fundus are punctate, streak- or plaquelike bleedings, especially in the central part of the fundus. Patients with extreme dys- and paraproteinemias often show a grainlike appearance or fragmentation of the blood column within the blood vessels of the conjunctiva

bulbi (sludged-blood phenomenon); Nover and Nover-Berneaud-Kötz ascribe this phenomenon on the one side to hemodynamic obstruction, and on the other, to sensitization. This hypothesis is based on appropriate animal experiments.

Reticulogranulomatoses

Systemic Reticulogranulomatoses

Lymphogranulomatosis of the skin (Paltauf-Sternberg's or Hodgkin's disease) (Fig. 179) shows a peak incidence around the second to fourth decade of life, and slightly favors the male sex. The etiology of the disorder is not clearly determined. Are we dealing with an inflammation (granuloma), a possibly virus-induced tumor with questionable possibilities of development into a genuine sarcoma, or a tumor with a benign course ("paragranuloma"—Jackson and Parker)?

The classification of clinical signs into cutaneous-lymphonodular, cervical, or cervico-axillary, mediastinal-plurinodular subtypes is easier. Initial abdominal manifestations may not be correctly diagnosed for a long time. Signs and symptoms of importance are high continuous or undulating fever (Pel-Ebstein), leukopenia, acute dysproteinemia, diazoreaction in the urine, and (perhaps less so) pain at the site of the disease after the ingestion of alcohol ("alcoholic pain"). In this "malignant granuloma" specific infiltrations, nonspecific cutaneous manifestations (Grosz, 1906) or a mixture of both are present. Paroxysms of intractable pruritus are frequently associated with varying signs of urticarial, ichthyotic-squamous and pruriginous eruptions (see also Knoth et al.). Lichenified, eczematoid, partly erythrodermic or partly poikilodermic eruptions are less frequently seen. Still rarer are exudative, nodular, or hemorrhagic eruptions. The specific cutaneous changes of lymphogranulomatosis are

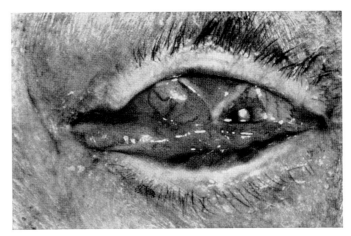

178 Gelatinous infiltration of the conjunctiva in Waldenström's macroglobulinemia (Case 1 of *Gottron, Korting* and *Nikolowski*).

preponderantly tumorlike, papular, or plaque-like infiltrations which often break down, forming long-lasting "ulcera lymphogranulo-matosa" (ARZT and RANDAK); these may become rather widespread. Allochthonous lymphogranulomatosis develops when infiltrations of contiguous specific granulomatous tissue from other organs have extended into the skin.

Ophthalmologically (see reviews of SCHRECK or NOVER and also SALM-SALM), initially specific tumors or those clinically appearing as Mikulicz syndrome, conjunctiva, within the orbit, "corneal infiltrates" (see case of REINSBERG-KADLICKY), and retinal changes with hemorrhages between papilla and macula were repeatedly reported. RIDLEY'S report of swelling of both tear glands, caused by infiltration of small lymphocytes, resembles the observation of NOVER mentioned earlier on macroglobulinemia as the underlying disease of such a swelling.

Mycosis fungoides (granuloma fungoides) (Figs. 180 and 181) affects adults in the fourth to sixth decades, without definite preference of sex. We observe most frequently the classic progressive type (ALIBERT-BAZIN) with a polymorphous initial phase (pruritus, urticaria, prurigo, eczema- or parapsoriasis-like erythema with a tendency to lichenification). After a period of years the stage of mycotic

infiltration begins, and the final stage of mycotic tumors follows. In exceptional cases mycosis fungoides begins with erythroderma (type Hallopeau-Besnier or Leredde) with uninvolved areas ("nappes clairs"), whereas "mycose d'emblée" (VIDAL-BROCQ) usually represents reticuloendotheliomatous sarcoma. In contrast to Hodgkin's disease, pruritus or other sensations present no certain diagnostic criteria.

The basis of *reticulogranulomatosis* is probably a chronic inflammatory, pathological condition, which may result in neoplasia, i.e., reticuloendotheliomatous sarcoma as shown by HOLZMANN and HOLDE in one of my own cases or lately in another observation by HUNDEIKER, BERGER, and PETRES. Latent blastomatous or preblastomatous changes of the reticulohistiocytic system, however, may cause an inflammatory reaction appearing as the first sign (HORNSTEIN).

Ophthalmologically, the older literature (GROENOUW, especially referring to BESNIER and HALLOPEAU, and also ERDMANN) and more recent reviews by SCHRECK mention mycosis fungoides as responsible for lesions of the eyelids, conjunctiva, cornea, and tear glands, in a fashion similar to that stated above for macroglobulinemia and lymphogranulomatosis. Lately GÄRTNER furnished by histologic examination proof of an affection of

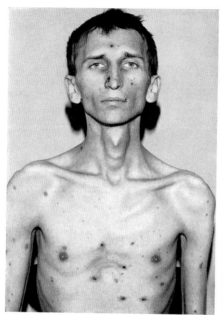 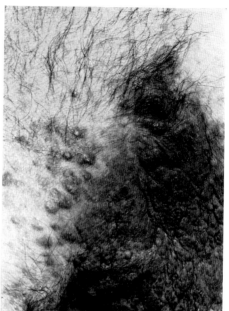

179 Lymphogranulomatosis of the skin (left illustration shows the nonspecific skin manifestation of prurigo, right illustration shows the specific infiltrates).

the chorioid in mycosis fungoides. HERZBERG published a detailed description of metastatic involvement of the uveal tract in an instance of "mycosis d'emblée." However, other recent reports deal with retinal involvement with papilledema (FRANCESCHETTI), therapeutically reversible changes of the fundus (PAILLARD), protrusion of the bulbus caused by retrobulbar infiltration (DEGOS), and additional cases of keratitis (LEKOW and NYQUIST) and parenchymatous keratitis (HERMANN). In view of this and the review of STÜTTGEN and MEISTERERNST, one has to conclude that some sort of involvement of the eye in mycosis fungoides is more frequent than has been supposed before.

Circumscribed lympho- and plasmocytic hyperplasias of the skin

The main representatives of this group of reversible cutaneous hyperplasias, (supposedly) subject to stimuli, are (for the dermatologist) eosinophilic granuloma and lymphadenosis benigna cutis (BÄFVERSTEDT). These types may show various lymphocytic, large follicular, or plasmocellular modulations. The expression "eosinophilic granuloma" was first used by MARTINOTTI (1919, 1921, and 1941); he applied this expression to a group of heterogeneous pathological conditions, which had nothing in common except a striking eosinophilia of the tissue. NANTA and GADRAT in 1937 described gingival and perianal vegetating ulcers, which they considered to be part of eosinophilic granulomas, osseous foci in the jaw, and diabetes insipidus. The osseous lesions had been called eosinophilic granuloma of the bone by FINZI (1929) before and later by LICHTENSTEIN and JAFFÉ (1940). KUSKE (1952) added to the already overdone concept of eosinophilic granuloma by including a tumorous variety; KRESBACH and NÜRNBERGER (1967) of my own Institute reported other ecthymatous and keratoacanthoma-like forms of eosinophilic granuloma. Granuloma eosinophilicum faciale (WOERDEMANN and

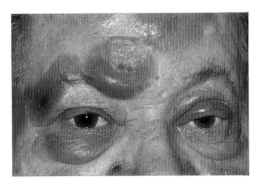

180 Mycosis fungoides.

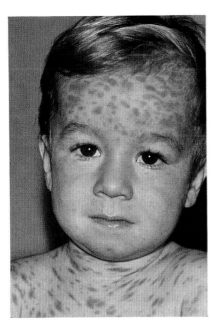

176 Urticaria pigmentosa.

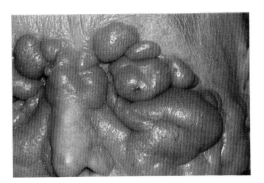

181 Mycosis fungoides.

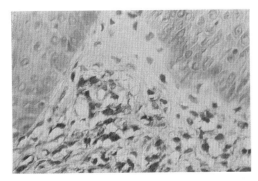

177 Urticaria pigmentosa. Mast cell infiltrates (Toluidine blue).

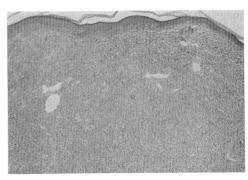

182 Mycosis fungoides. Histologic appearance (tumor stage).

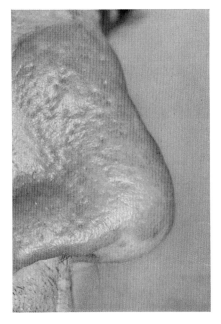

183   Eosinophilic granuloma faciale.

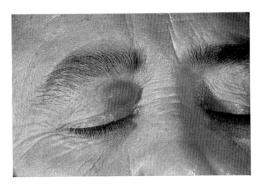

186   Lymphadenosis benigna cutis *(Bäfverstedt)*.

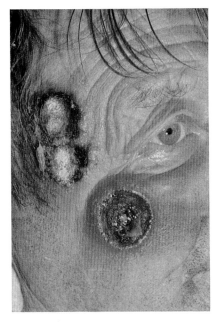

185   Keratoacanthoma-like eosinophilic granulomas in the facial area.

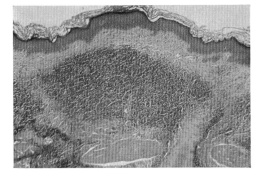

187   Lymphadenosis benigna cutis. Histologic appearance.

184 Eosinophilic facial granuloma. Histologic appearance.

PRAKKEN; PINKUS; Figs. 183, 184, and 185), has to be separated from the rather unmethodical group of cases. This is a circumscribed reticulogranulomatosis of the skin, mysteriously favoring the face (see also case of KORTING). Subjective signs are absent; therefore, the physician is usually consulted by the patient fearing a malignant tumor or for cosmetic reasons.

Under the name of lymphadenosis benigna cutis (Figs. 186 and 187), BÄFVERSTEDT grouped the following disorders: Spiegler-Fendt's (1894 and 1900) sarcoid and lymphocytoma (BIBERSTEIN, 1923), which was described earlier (1921) by MARIE KAUFMANN-

WOLF as "benign, lymphocytic neoplasia of the scrotal skin of the child" (Figs. 186 and 187). In detail, *single or multiple* lesions develop relatively slowly in the cutis or subcutis. Multiple lesions are either disseminated or generalized, gray brown, small or large, plaquelike or flat atrophic, tuberous, or miliary, but rarely ulcerating. They cause little discomfort and tend to regress spontaneously. The lesions may simulate lupus vulgaris nodules, but a probe will not sink in. Characteristic are (especially in children) light or dusky red-blue infiltrations of the ear lobe, and, less frequently, of the scrotum or the areola of the breast. Development of such

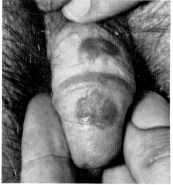

188 Circumorificial plasmacytosis: plasmacellular conjunctivitis and balanoposthitis in the same patient.

solitary lymphocytomas is sometimes secondary to a tick bite, as in erythema migrans and acrodermatitis chronica atrophicans. The association with an underlying malignant disease may be present only exceptionally (see BÄFVERSTEDT, 1960). Transmission of lymphadenosis benigna cutis was successfully done in clinical experiments (PASCHOUD).

A few cases of lymphadenosis benigna cutis of the face present periorbicular localization, as shown on pictures by Bäfverstedt (1968) and Bluefarb. GODTFREDSEN and LINDGREN described identical localizations within the orbital region; Bäfverstedt (1956) calls this condition *"lymphadenosis benigna orbitae."*

A clinical variant of such benign "lymphoplasias" (MACH) is lymphocytic infiltration (JESSNER and KANOF, 1953) with brownish-red, flat discoid lesions, which in contrast to chronic lupus erythematosus are smooth and without follicular hyperkeratosis. They are located predominantly on the face and may be provoked by exposure to light.

The lymphoreticular cutaneous hyperplasias belong in our opinion to the same category as certain "tumorlike degenerations of the conjunctiva" (OPPEL), however, with emphasis on the plasmocytes. The ophthalmologists, on the other hand, reported these conditions as "plasmomas" (Fig. 188) (PASCHEFF, SCHWARZKOPF, KREIBIG, OPPEL, HEYDENREICH, ALLINGTON and ALLINGTON). Pascheff (1908) described conjunctivitis plasmocellularis: "All these hypertrophies and tumors consist of plasma cells and produce a special kind of tumor, which I called plasmoma." However, the presence of plasma cells in the conjunctivae is not at all unusual; Schwarzkopf remarks in this connection: "The conjunctiva of the eye is normally as well as pathologically a chief depository for plasma cells (see granulosis and vernal catarrh)." According to Kreibig such conjunctival plasmomas have to be considered "inflammatory formations." In addition, as Oppel states, these formations are not "strictly localized" but the expression of circumorifical plasmocytosis as shown by my own observation of simultaneous occurrence of circumscribed plasmocellular conjunctivitis *and* plasmocellular balanoposthitis in the same person (KORTING and THEISEN).

References

ALLINGTON, H. V., and ALLINGTON, J. H.: Eyelid tumors. Arch. Derm. Syph. (Chic.) 97 (1968) 50–65.
AMBS, E., BIREN, P., and KLINGMÜLLER, G.: Über eine angeborene Retikulose. Hautarzt 17 (1966) 63–68.
BÄFVERSTEDT, B.: Über Lymphadenosis benigna cutis. Acta derm.-venereol. (Stockh.) 24, Suppl. 11 (1943).
BÄFVERSTEDT, B.: Lymphadenosis benigna cutis as a symptom of malignant tumors. Acta Derm.-Venereol. (Stockholm) 33 (1953) 171–180.
BÄFVERSTEDT, B.: Lymphadenosis benigna cutis (LABC), its nature, course and prognosis. Acta Derm.-Venereol. (Stockholm) 42 (1962) 3–10.
BÄFVERSTEDT, B.: Lymphadenosis benigna cutis. Acta Derm.-Venereol. (Stockh.) 48 (1968) 1–6.
BÄFVERSTEDT, B., LUNDMARK, C., MOSSBERG, M., and STENBECK, A.: Lymphadenosis benigna orbitae. Acta Ophthal. 34 (1956) 367.
BERNEAUD-KÖTZ, G.: Augenbeteiligung bei Dys- und Paraproteinämien (Fundus paraproteinaemicus). Beih. Klin. Mbl. Augenheilk. 32 (1959) 1.
BESNIER, E., and HALLOPEAU, H.: Sur les érythrodermies du mycosis fongoides. Ann. Derm. (1892) 987.
BIBERSTEIN, H.: Lymphocytome. Zbl. Haut- u. Geschl.-Kr. 6 (1923) 70.

BLUEFARB, S. M.: Cutaneous manifestations of multiple myeloma. Arch. Derm. Syph. (Chic.) 72 (1955) 506–522.
BLUEFARB, S. M.: The cutaneous manifestations of the benign inflammatory reticuloses, edited by A. C. Curtis. Thomas, Springfield, Ill., 1960.
BOWDLER, A. J., and TULLETT, G. L.: Urticaria pigmentosa and polycythemia vera. Brit. Med. J. 1960/I, 396.
BUREAU, Y., BARRIÈRE, H., LITOUX, P., and BUREAU, L.: Maladie de Waldenström avec manifestations cutanées à type de lympho-réticulose diffuse du visage. Presse méd. 71 (1963) 2101–2103.
CHATARD, M. H.: Urticaire pigmentaire chez un enfant à antécédents spécifiques. Bull. Soc. franç. Derm. Syph. 61 (1954) 242.
ERDMANN, S.: Über Mycosis fungoides und Auge. Klin. Mbl. Augenheilk. 66 (1921) 296–297.
FENDT, H.: Beiträge zur Kenntnis der sogenannten sarcoiden Geschwülste der Haut. Arch. Derm. Syph. (Wien, Leipzig) 53 (1900) 213–242.
FINDLAY, G. H., SCHULZ, E. J., and PEPLER, W. J.: Diffuse cutaneous mastocytose. S. Afr. med. J. 34 (1960) 353.
FLEISCHMAJER, R., and EISENBERG, S.: Sézary's reticulosis. Arch. Derm. Syph. (Chic.) 89 (1964) 9–19.

FRANCESCHETTI, A.: Mycosis fungoides mit Augenbeteiligung. Ann. Ottalm. 76, 413 zit. bei J. J. Herzberg. Z. Haut- und Geschl.-Kr. 14 (1953) 180–187.

FRANÇOIS, J., and BACSKULIN, J.: Chorio-retinal degeneration of the pigmentary type in the Abt-Letterer-Siwe disease. Ophthalmologica (Basel) 153 (1967) 241–258.

GÄRTNER, J.: Mycosis fungoides mit Beteiligung der Aderhaut. Klin. Mbl. Augenheilk. 131 (1957) 61–69.

GERTLER, W.: Retikulosarkomatöse Umwandlung tumorartiger Lymphozytome. Derm. Wschr. 132 (1955) 1035–1042.

GODTFREDSEN, E., and LINDGREN, S.: Orbital, palpebral and epibulbar lymphomas. Acta Ophthal. 31 (1923) 29.

GOTTRON, A. H.: Dermoleucohaemoblastosen. Regensbg. Jb. ärztl. Fortb. 8 (1959/1960).

GOTTRON, H. A., KORTING, G. W., and NIKOLOWSKI, W.: Die makroglobulinämische Hyperplasie der Haut. Arch. klin. exp. Derm. 210 (1960) 176–201.

GROENOUW, A.: Beziehungen des Auges zu den Hautkrankheiten. In: Hdb. der Haut- und Geschlechtskrankheiten, Bd. XIV/1, edited by J. Jadassohn. Springer, Berlin, 1930.

HERMANN, P.: La kératite du mycosis fungoide. Arch. Ophthal. (Paris) 11 (1951) 39–45. Ref. Zbl. Haut- u. Geschl.-Kr. 79 (1952) 360.

HERZAU, W., and PINKUS, H.: Beitrag zur Schüller-Christianschen Krankheit. Klin. Mbl. Augenheilk. 89 (1932) 721.

HERZBERG, J.: Die Stellung der Mycosis fungoides im System der Hautkrankheiten. Derm. Wschr. 25 (1954) 422.

HERZBERG, J. J.: Mycosis fungoides d'emblée mit metastatischem Befall des Uvealtraktes. Z. Haut- u. Geschl.-Kr. 14 (1953) 180–187.

HEYDENREICH, A.: Differentialdiagnostische Erwägungen bei Tumoren der Augenlider. Dtsch. Gesund. Wes. 19 (1964) 2240–2246.

HOLZMANN, H., and HOEDE, N.: Zur Frage der nosologischen Selbständigkeit der Mykosis fungoides anhand einer Beobachtung eines Retothelsarkoms bei einem Kranken mit Mykosis fungoides. Z. Haut- u. Geschl.-Kr. 39 (1965) 451–457.

HORNSTEIN, O.: Reticulosen der Haut. In: Fortschritte der Dermatologie und Venerologie, Bd. III, edited by A. Marchionini. Springer, Berlin, 1960.

HUNDEIKER, M., BERGER, H., and PETRES, J.: Mycosis fungoides nach Thorotrastanwendung mit Entwicklung zur Reticulumzellsarkomatose. Arch. Klin. exp. Derm. 232 (1968) 56–65.

JESSNER, M., and KANOF, N. B.: Lymphocytic infiltration of the skin. Arch. Derm. Syph. (Chic.) 68 (1953) 447–449.

KAUFMANN-WOLF, M.: Über gutartige lymphocytäre Neubildungen der Scrotalhaut des Kindes. Arch. Derm. Syph. (Berl.) 130 (1921) 425–435.

KLEMM, D.: Die paraproteinämischen Hämoblastosen. Ergebn. inn. Med. Kinderheilk. 26 (1967) 110–192; especially p. 173.

KNOTH, W., BREITWIEDER, Pl., and KLEINHANS, D.: Zur Kenntnis unspezifischer Begleitdermatosen bei Lymphogranulomatose. Med. Welt (Stuttg.) 19 (1968) 170.

KORTING, G. W.: Granuloma eosinophilicum faciale. Med. Welt (Stuttg.) 17 (1966) 397–398.

KORTING, G. W., and THEISEN, H.: Circumscripte plasmacelluläre Balanoposthitis und Conjunctivitis bei derselben Person. Arch. klin. exp. Derm. 217 (1963) 495–504.

KREIBIG, W.: Über die Plasmocytome der Bindehaut. v. Graefes Arch. Augenheilk. 131 (1933) 89–101.

KREIBIG, W.: Über Bindehautveränderungen bei leukämischen Erkrankungen. Z. Augenheilk. 84 (1934) 120–136.

KRESBACH, H., and NÜRNBERGER, F.: Ecthyma- und keratoakanthomähnliche eosinophile Granulome der Haut. Arch. klin. exp. Derm. 230 (1967) 286–303.

KUSKE, L.: Tumorförmige eosinophile Granulome der Haut. Dermatologica (Basel) 104 (1952) 254–259.

LEINBROCK, A.: Das Plasmocytom und seine pathologischen Hautveränderungen. Hautarzt 9 (1958) 249–259.

LEINBROCK, A.: Das Plasmocytom und seine pathologischen Hautveränderungen. Hautarzt 9 (1958) 249–259.

LEINBROCK, A.: Reticulosen der Haut. In: Aktuelle Probleme der Dermatologie, Bd. I, edited by R. Schuppli. Karger, Basel, 1959.

LEKOW, P., and NYQUIST, B.: Keratitis in mycosis fungoides with squamous cell carcinoma of the skin. Acta Ophthal. 39 (1961) 3.

LENNERT, K.: Reticulumzell-Hyperplasie (Reticulocytose). In: Hdb. der speziellen pathologischen Anatomie und Histologie, Bd. I. Edited by E. Mehlinger. 3d Ed. Springer, Berlin, 1961.

LENNERT, K., and ELSCHNER, H.: Zur Kenntnis der lipomelanotischen Reticulocytose. Frankf. Z. Path. 65 (1954) 559–577.

LUTZ, W.: Die auf das Auge übergreifenden Hautkrankheiten. In: Lehrbuch der Augenheilkunde, edited by M. Amsler, A. Brückner, and A. Franceschetti. Karger, Basel, 1948.

MACH, K.: Die gutartigen Lymphoplasien der Haut. Arch. klin. exp. Derm. 222 (1965) 325–349.

MARTINOTTI, L.: Zur Frage des eosinophilen Granuloms. Derm. Wschr. 112 (1941) 25–30.

MATHUR, S. P.: Conjunctival plasmoma. Brit. J. Ophthal. 43 (1959) 499–500.

MORTADA, A.: Nature of lymphoid tumours. Of the orbit, conjunctiva, eyelids and lacrimal gland. Amer. J. Ophthal. 57 (1964) 820–826.

MUSGER, A.: Erythrodermatische Hautretikulosen. Hautarzt 17 (1966) 148–152.

MUSGER, A.: Hautretikulosen. Med. Klin. 62 (1967) 1157–1160.

MYŠKA, V., OTORADOVEC, J., KLOUČEK, F., SOBRA, J., and PROCHÁZKA, B.: Mukokutane Form eines eosinophilen, xanthomatösen Granulomes mit schwerer Beteiligung der Hornhäute bei einem Erwachsenen. Cs. Oftal. 20 (1964) 360–368. Ref. Zbl. ges. Ophthal. (1965/66) 313.

NANTA, A., and GADRAT, J.: Sur un Granulome eosinophilique cutané. Bull. Soc. franç. Derm. Syph. 44 (1937) 1470–1479.

NEXMAND, P. H.: Discussion of diagnosis (mycosis fungoides with exfoliative erythrodermia or psoriasis with exfoliative erythrodermia and roentgen-dermatitis. Acta Derm.-Venereol. (Stockh.) 35 (1955) 248–249.

NOVER, A.: Störungen der Mikrozirkulation und Gefäßpermeabilität des Auges bei Dys- und Paraproteinämien. Wien. klin. Wschr. 74 (1962) 377–380.

NOVER, A., and BERNEAUD-KÖTZ, G.: Experimentelle Untersuchungen über die Permeabilität der Bindehautgefäße. v. Graefes Arch. Ophthal. 159 (1958) 582–595.

NOVER, A., and BERNEAUD-KÖTZ, G.: Beobachtungen an den Bindehautgefäßen bei der Makroglobulinämie Waldenström. Medizinische (1959) 1364.

NOVER, A., and ZU SALM-SALM, E.: Augenveränderungen als Initialsymptom bei Allgemeinerkrankungen. Klin. Mbl. Augenheilk. 134 (1959) 323–329.

NOVER, A., and GLEES, M.: Tränendrüsenbeteiligung bei der Makrogloublinämie Waldenström. Klin. Wschr. 44 (1966) 462–464.

OPPEL, O.: Geschwulstähnliche Entartungen der Bindehaut. Klin. Mbl. Augenheilk. 128 (1956) 145–158.

ORFANOS, C., and STEIGLEDER, G. K.: Die tumorbildende kutane Form des Morbus Waldenström. Dtsch. med. Wschr. 92 (1967) 1449–1454, 1475–1477.

PAILLARD, R.: Deux cas de mycosis fungoide traités par l'urèthane. Ref. Dermatologica (Basel) 98 (1949) 49.

PASCHEFF, C.: Plasmacelluläre Bildungen (Plasmome) der Bindehaut und der Hornhaut. v. Graefes Arch. Ophthal. 68 (1908) 114–125.

PASCHOUD, J. M.: Die Lymphadenosis benigna cutis als übertragbare Infektionskrankheit. Hautarzt 8 (1957) 197–211; 9 (1958) 153–165, 263–269.

PINKUS, H.: Granuloma faciale. Dermatologica (Basel) 105 (1952) 85—99.

REINSBERG, V., and KADLICKY, R.: Lymphogranulomatosis cutis et conjunctivae bulbi. Cs. Derm. 4 (1923) 160—172. Ref. Zbl. ges. Ophthal. 14 (1925) 102.

RIDLEY, N. C.: An unusual complication of Hodgkin's disease. Trans. Ophthal. Soc. U. K. 49 (1929) 399—406. Ref. Zbl. ges. Ophthal. 23 (1930) 261—262.

RIVA, G.: "Makroglobulinaemie Waldenström." Schwabe, Basel, 1958.

RÖCKL, H., BORCHERS, H., and SCHRÖPL, F.: Lymphoretikulose der Haut mit Makroglobulinämie als Sonderform der Makroglobulinämie Waldenström. Hautarzt 13 (1962) 491—499.

SAGHER, F., and EVEN-PAZ, Z.: Mastocytosis and the mast cell. Karger, Basel, 1967.

SCHRECK, E.: Veränderungen des Sehorgans bei Haut- und Geschlechtskrankheiten. In: Dermatologie und Venerologie, Bd. IV, edited by H. A. Gottron and W. Schönfeld. Thieme, Stuttgart, 1960.

SCHULZE, K. H., JÄNNER, M., and WEX, O.: Akute Retikulose bei einem Säugling. Dermatologica (Basel) 135 (1967) 392—402.

SCHWARZKOPF, G.: Ein Fall von symmetrischer Geschwulstbildung aller vier Lider (Plasmome) mit pathologisch-anatomischem Befund. Z. Augenheilk. 45 (1921) 142—158.

SÉZARY, A.: Une nouvelle reticulose cutanée. La réticulose maligne leucémique à histio-monocytes monstrueux et à forme d'erythrodermie œdemateuse et pigmentée. Ann. Derm. Syph. 9 (1949) 5—22.

SPIEGLER, E.: Über die sogenannte Sarkomatosis cutis. Arch. Derm. Syph. (Vienna and Leipzig) 27 (1894) 163—174.

STÜTTGEN, G., and MEISTERERNST, W.: Mycosis fungoides. In: Hdb. der Haut- und Geschlechtskrankheiten, Bd. III/1, edited by J. Jadassohn. Springer, Berlin, 1963.

WALDENSTRÖM, I.: Incipient myelomatosis or essentiel hypergamma-globulinemia with fibrinogenopenia. Acta Med. Scand. 117 (1944) 216.

WALDENSTRÖM, I.: Zwei interessante Syndrome mit Hyperglobulinämie. Schweiz. med. Wschr. 78 (1948) 927.

Diseases of the Sebaceous Glands

Acne vulgaris

Juvenile acne vulgaris is characterized by the simultaneous occurrence of osteofollicular keratoses, comedones, and papulopustular infiltrations and their limited distribution to the face and upper parts of the back (Figs. 189 and 190). This dermatosis of peripuberty usually runs in families, is accompanied by gastrointestinal dysfunction and endocrinopathies (this is questionable—Translators), and is provoked by certain foods. In contrast, acne conglobata (SPITZER and LANG), characterized by grouped or giant comedones, connecting and irregular scars with pressure atrophies and a cutis laxa type of integument, is present on the buttocks, thighs, and so forth. The lesions of occupational acne usually are morphologically as monomorphic as drug-induced acneiform eruptions (following administrations of ACTH, testosterone, isoniazid, vitamin D, and others).

Ophthalmologically, it seems remarkable that in spite of the multiple occurrence of modified sebaceous and sweat glands, formation of comedones at the tarsal glands hardly ever takes place even in extensive acne vulgaris, unless one wants to consider retention of sebum with occasional inflammation of the meibomian glands analogous to acne vulgaris. Only older poeple show a conspicuous number of *palpebral and periorbital* comedones (CSILLAG) (Fig. 191). This may lead to an extroversion of the lower lacrimal punctum with epiphora (tears trickling down the cheeks—Translators) (PICHLER). On page 113 xanthoma sebaceum and nodular elastoidosis of the skin with cysts and comedones (FAVRE and RACOUCHOT) was mentioned. We cannot confirm SPRAFKE's statement that there is a connection between acute eruptions of acne and the presence of iritis.

Rosacea

The chronic vasoneurosis "rosacea" shows a characteristic butterfly configuration (corresponding to the second and third segment of the medial parts of the nucleus terminalis spinalis of the trigeminal nerve in the medulla) and consists of small macular, slightly scaling, erythematous lesions (rosacea erythematosa). Occasionally, these lesions are slightly elevated (rosacea papulosa) and show more or less numerous telangiectases (Figs. 192 and 193). It must be added that by hyperplasia of the sebaceous glands and by periglandular fibrosis, mostly in males, rosacea changes into *rhinophyma*. In the differential diagnosis rosacealike tuberculid (BLASCHKO-LEWANDOWSKI) has been considered rather too often. Pathogenetically, the preceding and later more persisting macular hyperemia of rosacea depends, with individual differences, first of all on labile blood supply in hypertensive patients, blood volume, ovarian dysfunction, and gastrointestinal disturbances, primarily in the areas of origin of the portal vein. But even banal constipation may be a contributing factor.

Blepharitis, conjunctivitis, and keratitis as *ocular* manifestations in patients with rosacea have been known for a long time. But it was TRIEBENSTEIN who made precise studies of rosacea keratitis, known until then only inexactly. He distinguished between three typical clinical pictures: marginal keratitis, keratitis with subepithelial infiltration, and progressive ulcus rodenslike keratitis. According to this author, in almost each instance rosacea of the cornea (which, by the way, affects males in a third of the cases) starts with rather typical changes in the marginal convolutional rete of the cornea. According to Triebenstein typical rosacea keratitis may exist without extensive cutaneous rosacea. Cutaneous and ocular lesions of rosacea need not coincide, so that the "ocular disease is not secondary to the dermatosis." The rare occurrence of a perforation of the sclera in rosacea keratitis was recently described by RICHTER.

References

CSILLAG, J.: Neigung alternder Personen zur Bildung palpebraler und periorbitaler Komedonen. Derm. Wschr. 88 (1929) 609–611.
PICHLER, A.: Comedonen als Ursache von Auswärtsdrehung des unteren Tränenpunkts mit Tränenträufeln. Klin. Mbl. Augenheilk. 75 (1925) 459.
RICHTER, S.: Skleraperforation bei Rosaceakeratitis. Klin. Mbl. Augenheilk. 146 (1965) 422–424.

SPRAFKE, H.: Akne vulgaris. In: Der Augenarzt, edited by K. Velhagen. Verlag für Kunst und Wissenschaft, Leipzig.
TOMKINSON, J. G.: Unusual distribution and morphological modification of comedo. Brit. Med. J. 1931/I, 666.
TRIEBENSTEIN, O.: Die Rosazaeerkrankungen des Auges. Klin. Mbl. Augenheilk. 68 (1922) 11–35.

Diseases of the Hair, Especially Alopecia Areata

Diseases of the hair associated with ocular changes were already mentioned above and include those such as anhidrosis hypotrichotica and mandibulofacial dysostosis, described by LUDWIG and KORTING as resembling the Vogt-Koyanagi syndrome and later classified by ULLRICH as dyscephaly with congenital cataract associated with hypotrichosis. Other syndromes already mentioned are the Vogt-Koyanagy syndrome itself (LEWIS and ESPLIN; MEYER-SCHWICKERATH and MERTENS), incontinentia pigmenti, and the endogenous eczema (Hertoghe-sign). However, alopecia areata (area Celsi, pelade) (Fig. 194), the most important nonscarring *circumscribed* hair loss, will be discussed.

In the *physiological hair cycle* we distinguish a
1. Growing phase or *anagen*,
2. Transitional phase or transformation phase of the lower part of the follicle and the papilla or *katagen*, and a
3. Resting phase or *telogen*.

In alopecia areata we are dealing, according to VAN SCOTT, with effluvium at the anagen phase; BRAUN-FALCO and RASSNER, however, believe that a dystrophic or mixed condition

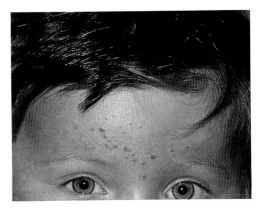

189   Infantile acne.

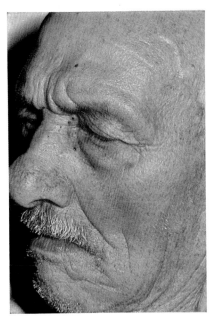

191   Periorbital comedones in the aged.

190   Indurated acne vulgaris.

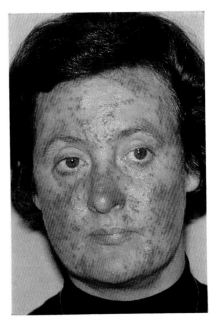

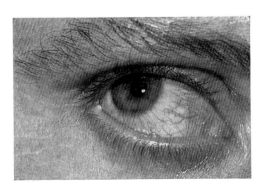

193   Rosacea keratitis.

192 Rosacea.

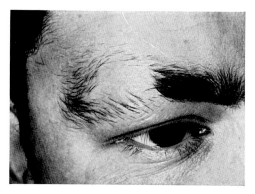

194   Alopecia areata in the area of the eyebrows.

of the hair root is responsible. In more than half of the disciform hair losses the back of the scalp is affected (as so-called *ophiasis*), which prognostically is less favorable. About 25 percent of the cases show frontovertical involvement. In alopecia totalis seu maligna after confluence of affected areas, the entire scalp is without hair. If in addition, hair of the beard, breasts, axillae, and genitals falls out, we are dealing with alopecia universalis. In alopecia areata hairs fall out with the root and regrow soft like silk and with little pigment. In the hair-free interval the affected area assumes an ivorylike color (RICHTER) due to diminution of pigment and hyperplasia of the sebaceous glands. Both processes occur at the same time. Occasionally, a very fine pitting of the nails (KLINGMÜLLER) is present. Prognostically, the course of alopecia areata is definitely more unfavorable if it has manifested itself before puberty (more frequent recurrences). Familial occurrence or association with other neurovegetative disturbances or anomalies (e.g., mongolism) (KORTING and HOLZMANN) are not rare. IKEDA recently stressed for some patients with alopecia areata the disposition to essential hypertension, certain endocrine-autonomous disorders, and atopy; we, however, could not confirm atopy in our patients. Frequently also focal-inflammatory influences must be considered (JACQUET: "épines péladogènes").

Ophthalmologically, a few early reports men-

tioned coincidence with strabismus (TOMKINSON) or cataract (PAPASTRATIGAKIS)* or pupillary disturbances (HERMAN). GOTTRON had already in 1943 reported on a coincidental Horner syndrome in unilateral alopecia areata. Recently, LANGHOF and LEMKE found this combination with the Horner syndrome in 53 out of 63 patients with alopecia areata (e.g., in 82 percent), more on the right than the left side. We could confirm this combination in this high percentage only in rare instances. Langhof and Lemke also found, in 21 patients of their studied group, *increased* tortuosities of the retinal vessels, which these authors consider as pointing "to a possible disturbance of the anlage in the vegetative master center, the diencephalon and the hypophyseal system." HAYNES and PARRY could in addition observe in 61 of 130 patients with alopecia areata refractory anomalies, especially astigmatism, findings which could not be confirmed by FIVAZ and KLINGMÜLLER. Finally, ŠALAMON and STOJAKOVIÇ recently described congenital cataracts, horizontal nystagmus, and tapetoretinal changes in a female patient with alopecia areata and other multiple congenital anomalies.

In contrast to alopecia areata, pseudopelade (alopecia atrophicans Brocq) presents atrophic, hairless, and smooth areas, which can be compared to footsteps in the snow.

In alopecia mucinosa (PINKUS), in contrast to alopecia areata, a circumscribed alopecia develops clinically in the area of follicular papules with keratinous plugs, or, in other words, under a spinulismlike aspect by follicular and seboglandular mucophanerosis (BRAUN-FALCO). In my own observations this happens not too rarely in the eyebrow region (see CABRÉ and KORTING). I believe that in these cases the development of follicular mucinosis should be considered in connection with a

---

* On the other hand, SUMMERLY, WATSON, and MONCKTON found on slit lamp examination that point-like lenticular opacities occur in patients with alopecia areata about as often as in normal persons.

systemic basic disturbance, mostly of reticulo-granulomatous character (e.g., mycosis fungoides) (KORTING) (Fig. 195).

In syphilis, especially in the secondary stage, diffuse or disseminated hair loss may be present. The latter (alopecia areolaris) (RONDELET) presents a picture of "moth-eaten hair" (PINKUS). In the periorbital region hair loss of the eyebrows and lashes may be present. Not rarely, they occur together. LÖHE mentioned that next to normally long cilia "sec-tions with short hairs thinned near the end occur so that the usually regular line of equally long eyelashes assumes a zig-zag line." Alopecia specifica of the eyelids has been called "*signe d'omnibus*" by the French because in the old Parisian horse-drawn busses with opposing long benches (SCHÖNFELD) this alopecia permitted a passenger to make the correct diagnosis of the person sitting opposite him. (Le "signe d'omnibus" usually refers to the loss of lateral eyebrows.—Translators.)

References

BRAUN-FALCO, O.: Mucophanerosis intrafollicularis et seboglandularis. Derm. Wschr. 136 (1957) 1289–1303.

CABRÉ, J., and KORTING, G. W.: Zum symptomatischen Charakter der "Mucinosis follicularis"; ihr Vorkommen beim Lupus erythematodes chronicus. Derm. Wschr. 149 (1964) 513–518.

FIVATZ, L.: Untersuchungen zur Ätiologie der Alopecia areata. Dermatologica (Basel) 108 (1954) 352–360.

GOTTRON, H. A.: Einseitige Alopecia areata. Ref. Zbl. Haut- u. Geschl.-Kr. 69 (1942) 57.

HAYNES, H. A., and PARRY, Th. L.: Alopecia areata associated with refractive errors. Arch. Derm. Syph. (Chic.) 59 (1949) 340–343.

HERMAN, E.: Immobilité pupillaire à la lumière dans un cas d'alopécie en aires d'origines endrocrinosympathique. Encéphale 21 (1926) 64. Ref. Zbl. Haut- u. Geschl.-Kr. 20 (1926) 61.

IKEDA, T.: A new classification of alopecia areata. Dermatologica (Basel) 131 (1965) 421–445.

KLINGMÜLLER, G.: Alopecia areata. Mit besonderer Berücksichtigung der Therapie. Hautarzt 9 (1958) 97–108.

KLINGMÜLLER, G., and REEH, W.: Nagelgrübchen und deren familiäre Häufungen bei der Alopecia areata. Arch. klin. exp. Derm. 201 (1955) 574–580.

KORTING, G. W.: Mykosis fungoides. Derm. Wschr. 135 (1957) 569–570.

KORTING, G. W., and BREHM, G.: Vakuoläre Follikeldegeneration bei einer Lichénification géante. Derm. Wschr. 144 (1961) 1261–1268.

LANGHOF, H., and LEMKE, L.: Ophthalmologische Befunde bei Alopecia areata. Derm. Wschr. 146 (1962) 585–590.

LEWIS, G. M., and ESPLIN, B. M.: Vogt-Koyanagi syndrome. Arch. Derm. Syph. (Chic.) 59 (1949) 526–530.

LÖHE, H.: Wesen und Verlauf der Frühsyphilis im einzelnen. In: Die Haut- und Geschlechtskrankheiten, Bd. IV, edited by L. Arzt and K. Zieler. Urban & Schwarzenberg, Wien, 1934.

LUDWIG, A., and KORTING, G.: Vogt-Koyanagi-ähnliches Syndrom und mandibulofaciale Dysostis (Franceschetti-Zwahlen). Arch. Derm. Syph. (Berl.) 190 (1950) 307–316.

MEYER-SCHWICKERATH, G., and MERTENS, H. G.: Zum Krankheitsbild der Uveo-Encephalitis. Ärztl. Wschr. 9 (1954) 904–910.

PAPASTRATIGAKIS, C.: Un nouveau syndrome dystrophique juvénile, alopécie totale associée à la caractère et à des altérations onguéales. Paris Méd. 12 (1922). Ref. Zbl. Haut- u. Geschl.-Kr. (1923) 485.

PINKUS, H.: Alopecia mucinosa. Arch. Derm. Syph. (Chic.) 76 (1957) 419–426.

PINKUS H., and SCHÖNFELD, R. J.: Weiteres zur Alopecia mucinosa. Hautarzt 10 (1959) 400–403.

RICHTER, R.: Die Haare. In: Hdb. der Haut- und Geschlechtskrankheiten, Bd. I/3, edited by J. Jadassohn. Springer, Berlin, 1963.

ŠALAMON, T., and STOJAKOVIC, M.: Über einen Fall von Alopecia areata maligna bei einer Patientin mit Retinitis pigmentosa (tapeto-retinale Degeneration) und anderen multiplen kongenitalen Anomalien. Z. Haut- u. Geschl.-Kr. 43 (1968) 267–272.

SCHÖNFELD, W.: Alopecia specifica. In: Dermatologie für Augenärzte, edited by W. Schönfeld. Thieme, Stuttgart, 1947.

SUMMERLY, R., WATSON, D. M., and MONCKTON, P. W.: Alopecia areata u. Cataract. Arch. Derm. Syph. (Chicago) 93, 411–412, 1966.

TOMKINSON, J. G.: Alopecia areata and strabismus. A family group of cases. Brit. J. Ophthal. 6 (1922) 505.

TOMKINSON, J. G.: Alopecia areata and strabismus. Familial group. Brit. J. Ophthal. 8 (1924) 518.

Behçet's Disease

In contrast to stomatitis aphthosa or gingivostomatitis herpetica, the first manifestation of the herpes simplex virus, the habitual and recurrent aphthae (first recognized by von Mikulicz and Flusser) is represented by single lesions appearing in crops. They appear as blisters but rapidly develop into fibrinous ulcers surrounded by hyperemia. They favor the anterior part of the oral cavity. Stomatitis, fetor, or salivation is absent. There is a certain familial background. The patients have a labile gastrointestinal tract or are neurotic (which seems doubtful — Translators) (Fig. 196).

The uveo-aphthous syndrome, described by GILBERT, an ophthalmologist from Hamburg, and the Turkish dermatologist BEHÇET (1937), comprises on the skin and mucous membrane very tender, single or multiple, long-lasting aphthous lesions of the oral mucosa or sharply circumscribed, ulcerated aphthae of the genitals (Fig. 197). The disorder is accompanied by pain, fever, profuse sweats, myalgias, arthralgias, and so on, and the patient appears to be severely ill. On the skin also erythema, papulo-pustular, or furunclelike lesions or inflammatory reactions simulating erythema nodosum or ulcerous pyodermas appear. Their development is due to the characteristic unspecific hyperreactivity (NAZZARO) to any occasional cause. On the other hand, a Behçet skin reaction may solve questionable cases (W. JADASSOHN, FRANCESCHETTI, and HUNZIKER).

Histopathologically, most diverse pictures of a vascular "inflammation" are present. Aphthous lesions are due to a necrotizing angiitis. Other cases show a productive substrate comparable to thrombophlebitis migrans. This is also the macroscopic impression.

In addition to the cutaneous and mucosal lesions a *recurrent hypopyoniritis* (Fig. 198) occurs, which according to BIETTI and BRUNA affects 88 per cent of young patients, predominantly males, between 15 and 35 years of age. This happens mostly in certain geographic areas or in certain races (Japan, Turkey, Greece, and Italy [observations in the U.S.A. and other countries show that there are no geographical or racial predilections — Translators]). It occurs usually relatively late, i.e., mostly after the onset of the cutaneogenital changes and with certain peaks in spring and fall. It always affects both eyes, eventually. A few instances have started with spontaneous hemophthalmia. Other ocular manifestations are corneal ulcerations, vascular changes (see FRANARIER and GENEVET), bleedings into the vitreous, tapetoretinal degenerations of the retina, especially if the CNS is involved (e.g., meningoencephalitis) in the so-called "neuro-Behçet," and paresis of the ocular muscles (for details see SCHRECK).

The ocular pathology of Behçet's disease is based, according to SHIKANO, on an "exudative," and according to BORAS and SEBESTYÉN on a "fibroplastic inflammation" followed by pronounced connective tissue proliferation and scar formation. Shikano sees in this development a certain parallel to the so-called collagen diseases, but other authors such as UJIHARA consider the recurrences of Behçet's disease repeated sensibilizations resembling the Arthus phenomenon. Finally, this disorder is absolutely not contagious although Behçet himself considered a viral origin and Gilbert later discussed the possibility of a "malignant" leptospirosis.

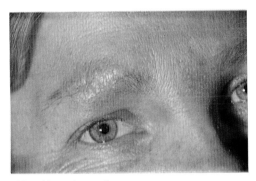

195 Alopecia mucinosa.

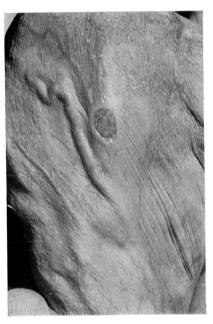

197 Behçet's syndrome. Solitary aphthous lesion on the scrotum.

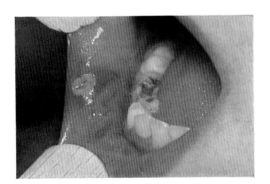

196 Behçet's syndrome. Solitary aphthous lesion in the mouth.

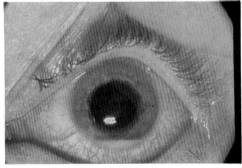

198 Behçet's syndrome. Hypopyonuveitis (Collection of Prof. Nover, Mainz).

References

BEHÇET, H.: Über rezidivierende aphthöse durch ein Virus verursachte Geschwüre am Mund, am Auge und an den Genitalien. Derm. Wschr. 105 (1937) 1152–1157.

BEHÇET, H.: Considérations sur les lésions aptheuses de la bouche et des parties génitales ainsi que sur les manifestations oculaires d'origine probablement parasitaire et observations concernant leur foyer d'infection. Bull. Soc. franç. Derm. Syph. 45 (1938) 420–433.

BEHÇET, H.: Kurze Mitteilung über Fokalsepsis mit aphthösen Erscheinungen an Mund, Genitalien und Veränderungen an den Augen, als wahrscheinliche Folge einer durch Virus bedingten Allgemeininfektion. Derm. Wschr. 107 (1938) 1037–1040.

BEHÇET, H.: Einige Bemerkungen zu meinen Beobachtungen über den Tri-Symptomenkomplex. Med. Welt (Berl.) 13 (1939) 1222–1227.

BEHÇET, H.: Some observations on the clinical picture of the so-called triple symptom-complex. Dermatologica (Basel) 81 (1940) 73–83.

BIETTI, G. B., and BRUNA, F.: An ophthalmic report on Behçet's disease. In: Int. Sympos. on Behçet's disease, Rome, 1964, edited by M. Monacelli and P. Nazzaro. Karger, Basel, 1966.

BOROS, B., SEBESTYÉN: Über das klinische und histologische Bild des Behçet-Syndroms. Klin. Mbl. Augenheilk. 145 (1964) 386–393.

FARNARIER, G., and GENEVET, A.: A propos d'un cas de syndrome de Behçet au cours d'une grande aphthose à prédominance vasculaire. Bull. Soc. Ophthal. Fr. 64 (1964) 660–664. Ref. Zbl. ges. Ophthal. 94 (1965) 141.

GILBERT, W.: Pathologisch-anatomische Befunde bei Iridocyclitis. Arch. Augenheilk. 87 (1921) 27–34.

GILBERT, W.: Über chronische Verlaufsform der metastatischen Ophthalmie ("Ophthalmia lenta"). Arch. Augenheilk. 96 (1925) 119–130.

GILBERT, W.: "Ophthalmia lenta." Klin Mbl. Augenheilk. 113 (1948) 1–8.

GILBERT, W.: Die Ophthalmia lenta als maligne Leptospirose des Auges. Zbl. ges. Ophthal. 55 (1951) 261–264.

IWAMOTO, T.: Electron microscopic studies on Behçet's syndrome. I. Iris tissues from patients of Behçet's syndrome. Acta Soc. Ophthal. Jap. 64 (1960) 528–540. Ref. Zbl. ges. Ophthal. 80 (1960) 271.

JADASSOHN, W., FRANCESCHETTI, A., and HUNZIKER, X.: La réaction cutanée à la "behçetine." Bull. Soc. Ophthal. Fr. 73 (1960) 380–384.

JADASSOHN, W., FRANCESCHETTI, A., and HUNZIKER, X.: Zur "Behçetine-Reaktion." Hautarzt 12 (1961) 64–65.

MONACELLI, M., and NAZZARO, P.: Behçet's disease. Int. Sympos. Rome, 1964. Karger, Basel, 1966.

NAZZARO, P.: La reattività aspecifica della cute e delle mucose visibili nella malatti de Behçet. Minerva derm. (Torino) 39 (1964) 192–195. Ref. Zbl. ges. Ophthal. 94 (1965) 305.

SCHRECK, E.: Veränderungen des Sehorgans bei Haut- und Geschlechtskrankheiten. In: Dermatologie und Venerologie, Bd. IV, edited by H. A. Gottron and W. Schönfeld. Thieme, Stuttgart, 1960.

SHIKANO, S.: Ocular pathology of Behçet's syndrome. Int. Sympos. on Behçet's disease, Rome, 1964. Karger, Basel, 1966.

UJIHARA, H.: Experimental studies on ocular Arthus phenomenon. Related to the ocular localisation of Behçet's disease lesions. Acta Soc. Ophthal. 94 (1965) 43. Ref. Zbl. ges. Ophthal. 94 (1965) 430.

Reiter's Disease

The urethro-oculo-synovial syndrome (FIESSINGER and LEROY; REITER), which distinctly favors men over women, seems to many authors related to Behçet's disease (see FALCK and SCHMIDT) but belongs, according to HORNSTEIN, "rather to the group of rheumatoid disorders." SCHUERMANN and HAUSER and HAUSER alone described the sequence of the chief signs: urethritis, conjunctivitis, and arthritis, and beyond these manifestations mucosal signs such as the so-called Reiter-balanitis (REICH) (Fig. 200), or occasionally, the involvement of the endo- and myocardium, glandular organs (tear glands, prostate, testicle, and epididymis) as well as the peripheral and central nervous systems. The joints present polyarthritis more often than monarthritis or arthralgias. The urethral secretion consists of either clear or purulent — occasionally, however, even hemorrhagic — mucus. The *cutaneous changes* of Reiter's disease are best described as erythematous to papular, or even exudative-psoriasiform lesions so that quite frequently the diagnostic separation from pustular psoriasis or psoriasis arthropathica becomes necessary. Furthermore, it seems possible that some cases of formerly (Vidal, 1893) so-called "keratosis blennorrhagica" did not belong to a neisserian infection but rather to Reiter's disease.

In contrast to Behçet's disease, which involves the mesodermal uvea, the *ocular* changes (see summaries by SCHRECK or CSONKA) affect the ectoderm of the eye entirely, i. e., primarily the lids, conjunctiva, and corneal epithelium, so that conjunctivitis (Fig. 201), episcleritis, keratitis, and only later, occasionally, (recurrent) iridocyclitis, are important. Reports on interstitial keratitis, corneal ulcers, or cataracts have been rare. Hauser mentions as a special finding a secondary glaucoma as a sequel to recurrent iritis. According to CIMBAL "scleroconjunctivitis" (with vague prodromata), which is followed in two to three days by a massive congestion in the conjunctival and episcleral vascular rete and is occasionally associated with small hemorrhages on the corneal limbus, is characteristic. Rather frequently definite chemosis develops under these circumstances.

Opinions on the etiology of Reiter's disease are divided: one often hears discussions on predisposing factors such as bacterial infections associated with intestinal diseases, viral infections, mycoplasmas, hyperergic or polyetiologic reactions, among others.

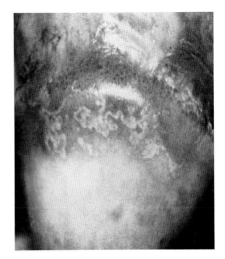

200 Reiter's disease. Parakeratotic balanitis.

References

CIMBAL, O.: Das Reitersche Syndrom als Nachbarkrankheit der Ruhr. v. Graefes Arch. Ophthal. 145 (1942) 142.

CSONKA, G.: Reiter's Syndrome. Ergebn. innerer Med. Kinderheilk. 23 (1965) 125–189.

FALCK, I., and SCHMIDT, G.: Die Beziehungen des Morbus Reiter und des Morbus Behçet zum rheumatischen Formenkreis. Med. Klin. 56 (1961) 1744–1748.

FIESSINGER, LEROY: Bull. Soc. méd. hôp. Paris 40 (1916) 2030.

HAUSER, W.: Zur Diagnostik der Reiterschen Krankheit. Med. Welt (Stuttg.) 1964, 2404–2409.

HORNSTEIN, O.: Hautmanifestationen rheumatischer Krankheiten. In: Hdb. der Haut- und Geschlechtskrankheiten, Bd. II/2, edited by J. Jadassohn. Springer, Berlin, 1965.

REICH, H.: Balanitis circinata bei Reiterscher Krankheit. Arch. Derm. Syph. (Berl.) 194 (1952) 1–29.

REITER, H.: Über eine bisher unbekannte Spirochäteninfektion (Spirochaetosis arthritica). Dtsch. med. Wschr. 42 (1916) 1535–1536.

REITER, H.: Die Reitersche Krankheit. Dtsch. med. Wschr. 82 (1957) 1336–1337.

SCHRECK, E.: Veränderungen des Sehorgans bei Haut- und Geschlechtskrankheiten. In: Dermatologie und Venerologie, Bd. -IV, edited by H. A. Gottron and W. Schönfeld. Thieme, Stuttgart, 1960.

SCHUERMANN, H., and HAUSER, W.: Reitersche Krankheit. Med. Klin. 44 (1949) 1269–1274.

Sjögren's Syndrome

Sjögren's syndrome, which in contrast to Reiter's disease favors females over males, is also called keratoconjunctivitis sicca and is, therefore, characterized by dryness and keratosis of the mucous membranes. The cardinal clinical signs are xerostomia with diminishend secretion of the salivary glands. This process is occasionally accompanied by a characteristic swelling of the parotid gland (dacryosialoadenopathia [*sialo* denotes a relationship to the salivary glands — Translators] atrophicans,

KLEY), a dry and glossy tongue with papillary atrophy and rhinolaryngotracheitis sicca, achylia gastrica, vulvitis, and others. On the skin the atrophy of the sweat and sebaceous glands leads to an ichthyosiform aspect, which occasionally assumes eczematization or pellagroid- or sclerodermalike features. Involvement of the joints in the Gougerot-Sjögren syndrome corresponds to rheumatoid arthritis. In several cases, however, a connection with other aspects of rheumatoid disease, a picture resembling Schönlein's purpura or Felty's syndrome, was observed. Besides, occasionally, the disease is associated with dysproteinemia, hepatosplenomegaly, increase in lymphoid reticulum cell elements in the bone marrow, and others (see SEIFERT and GEILER).

Ophthalmologically, WISSMANN, an ophthalmologist from Wiesbaden, reported as early as 1932 on the classical picture of "filamentary keratitis" dependent on endocrine disturbances, particularly in women after the fourth decade of life. One year later, the Swedish ophthalmologist Sjögren classified the signs of the Sicca syndrome (lack of tears, conjunctivitis sicca, and so on) as a syndrome.

Concerning the *etiology*, the questions whether connections to chronic rheumatoid diseases (SCHULZE and MIEHLKE), a relative "reticulosis," or a diencephalic disturbance exist, are often repeated. In addition, precipitating autoantibodies against tissue of the salivary glands were found in Sjögren's disease (ANDERSON, GRAY, BECK, and KINNEAR).

References

ANDERSON, J. R., GRAY, K. G., BECK, J. S., and KINNEAR, W. F.: Precipitating autoantibodies in Sjögren's disease. Lancet II (1961) 456.

KLEY, W.: Klinik und Differentialdiagnostik der Speicheldrüsenschwellungen (Sialoadenitiden und Sialoadenosen). Münch. med. Wschr. 101 (1959) 997–1001.

SCHULZE, G., and MIEHLKE, K.: Die rheumatischen Speicheldrüsenentzündungen. Z. Rheumaforsch. 19 (1960) 166–172.

SEIFERT, G., and GEILER, G.: Speicheldrüsen und Rheumatismus. Dtsch. med. Wschr. 82 (1957) 1415–1417.

SJÖGREN, H. S. C.: Zur Kenntnis der Keratoconjunctivitis sicca (Keratitis filiformis bei Hypofunktion der Tränendrüsen). Acta ophthal. (Kbh.) 2 (1933) 151.

WISSMANN, R.: Keratitis filiformis, als Teilsymptom innersekretorischer Störungen. Dtsch. med. Wschr. 58 (1932) 1525–1527.

Venereal Diseases

Few other diseases within the speciality of dermatology have shown such fundamental changes during the last thirty years as the venereal diseases; our former opinions regarding possibilities of treatment, prognosis, rate of morbidity, even clinical and pathologic-anatomic manifestations have had to be revised. These changes do not mean merely certain abridgement in the systematic teaching of dermatology brought about by improved therapy, which prevents some reactions or subsequent stages of these diseases from appearing at all. This is shown by the use of antibiotics and sometimes the improvement from the antihyperergic effect of cortisone for attenuation of the Jarisch-Herxheimer and the Krause reactions, and in the field of ophthalmology by the influence on specific parenchymatous keratitis. Through increased clinical experience, classic syphilology in some places had to undergo necessary revisions, especially in differential diagnosis; some disease manifestations originally considered syphilitic are now called pseudolues, owing partly to improved serologic diagnosis (NELSON test) (for details see KORTING, 1954). Similarly, the former classic complications of anterior and posterior gonorrhea of the male are now almost always called "postgonorrheal catarrh" or "pseudogonorrhea." Of the four venereal diseases covered by the German law against venereal diseases (1953), syphilis and gonorrhea are still important, especially with their increased frequency during the last eight years. But at present no official statistics of venereal diseases exist any more in Germany, with the result that figures of their incidence are not exact. However, one can be sure that the age at which infection takes place has moved toward the group of adolescents of 14 to 18 ("teenager-gonorrhea"); also anal primary syphilitic lesions or primary rectal gonorrhea due to homosexual intercourse are occasionally seen at present (these are rather prevalent in the large cities of the United States — Translators).

Syphilis (lues)

The causative organism of syphilis is the spirocheta pallida or the treponema pallidum discovered March 3, 1905 by FRITZ SCHAUDINN jointly with E. HOFFMANN. About three weeks after the infection, which according to Paracelsus (1493–1541) is transmitted, as a general principle, "coitu, partu seu tactu" (through coitus, childbirth, or physical touch), the syphilitic primary lesion emerges; the diagnosis is made from the clinical picture (it is not always the textbook Hunterian chancre —Translators), and is corroborated by the demonstration of the causative organism by darkfield examination, the best method (LANDSTEINER and MUCHA). The initial infiltrated lesion has a tendency to undergo transformation into an ulcer with subsequent formation of a scar. Once in a while, the induration can be absent, i.e., resulting in FOLLMAN's (1931) "balanitis specifica luica" or a condition resembling "balanitis erosiva" (see KRESSBACH). Already RICORD stated "as a shadow follows the body so the regional hard nontender swelling of the lymph nodes follows the initial sclerosis"; aspiration of fluid of these nodes may show spirochetes. Ordinarily, the syphilitic primary lesion is solitary; during the last few years we have repeatedly observed multiple, mostly smaller, or bipolar scleroses. The initial lesion is still found primarily in the genital region. The occurrence of *extragenital chancres* differs according to various statistics from 3 to 15 percent, so that, generally, a proportion of about 10 to 100 can be assumed (see SCHWARZKOPF). Extragenital chancres of the eye will be discussed later.

In the *secondary stage* of acquired syphilis, around the sixth to ninth week after infection, hematogenous spread of the causative organism takes place. This results in the appearance of at first rather macular, later papular, exanthemas followed by successively larger but fewer lesions ("roséole en retour"; Figs. 201 and 202). Other significant signs are opaline (opal-glass-like) patches, specific angina, alopecia diffusa or areolaris, "signe d'omnibus" (see Chapter 27 — Translators), leukoderma (collar of Venus), condylomata lata, involvement of various internal organs such as nephritis and myositis, and changes of the cerebrospinal fluid with cephalea (headaches).

In the *tertiary stage*, the stage of late syphilis (Ricord was the first to divide the disease into stages), beginning around the third to fifth year after infection, a few asymmetrically localized, destructive lesions appear, which regress with either superficially or deeply located scar formation. When tertiary syphilids are superficial, they increase in size through contact with other lesions (tuberoserpiginous syphilid) or appear as gummatous syphilomas (FRACASTORO: "Gumma," Fig. 203), which originate in the deep cutis and form hard nodes, melting slowly. These syphilids show very few spirochetes and frequently are hardly infectious; a general lymphadenopathy is practically absent; my own observations show that such adenopathy is not regularly present any longer in florid secondary syphilis.

The expression "syphilis d'emblée" or decapitated syphilis (syphilis without a primary lesion — Translators) refers mainly to transfusion syphilis. A precipitated tertiarism, still in the secondary period, with severe constitutional signs, ecthymatous-rupioid skin lesions, in most cases negative serology, and absence of lymphadenopathy, is called lues maligna. The term neurosyphilis refers to the disease entities tabes dorsalis, general paresis and cerebrospinal syphilis; lately these three diseases have undergone considerable atypical changes in their clinical appearance. They should be called metalues only, when one wants to emphasize their partly immunopathological genesis, which goes beyond the immediate answer of the organism to the spirochete.

Syphilis of the child is seldom acquired (examples: syphilis of the "innocent" or by criminal assault). Practically all cases of syphilis in children are congenital and result from diaplacental transmission of the treponemes to the fetus. ("Without an infected mother there is no syphilitic child," MATZENAUER, 1903.) (Rarely, syphilis is acquired *intra partum*, when the child passes through the birth canal of a mother who has florid syphilis acquired late in pregnancy — Translators.) As in transfusion syphilis, a true primary lesion per se is missing in "lues connata" (or congenita). Such syphilis progresses as a generalized infection; it rarely becomes evident before the fourth or fifth fetal month. KASSOWITZ established the rule that the more florid the syphilis of the mother, the more certain the infection of the fetus. Signs of congenital syphilis differ considerably in the newborn, in infants, and in the school child (syphilis congenita tarda). The syphilitic newborn shows facial discoloration of a dirty brown or waxy white hue, small or large macular lesions of the skin which are similar to the acquired secondary stage of syphilis, and Hochsinger infiltrates, for instance as specific coryza (rhinitis), palmoplantar bullae, and pemphigoid lesions with an abundance of spirochetes. At birth, sites of predilection of maculopapular syphilids are face and extremities, less frequently the trunk. Later, perioral scars (Parrot's rhagades) permit a diagnosis of congenital syphilis. Circumscribed lesions, such as condylomata lata, occur in the infant; rarely at a later stage when only a few cutaneous lesions are present. Hutchinson's triad is the main sign: Keratitis parenchymatosa, barrel-shaped and notched upper incisors of the second dentition (PASINI found spiro-

chetes in these teeth — Translators), and labyrinth-type deafness. The syphilitic infant shows osseous changes such as osteochondritis dissecans (WEGENER, 1870), periostitis syphilitica (saber shins), and less frequently, osteomyelitis fibrosa rareficans (PICK). Wegener's osteochondritis may be responsible for Parrot's pseudoparalysis of the extremities. In older children (and adolescents), however, the signs of congenital syphilis are frontal or parietal bosses ("front olympique," FOURNIER), the fifth finger sign (DUBOIS) and thickening of the sternoclavicular joint (HIGOUMENAKIS); however, these signs are socalled "weak signs" of congenital syphilis.

At present the Nelson or TPI test is the most reliable serologic reaction because it shows the presence of antibodies against virulent treponemas. However, in cases of early syphilis one has to keep in mind that the immobilizing factors of the Nelson test react later than the reagins of the classic serological reactions. The Nelson test is almost always positive in syphilis "non sanata," in patients with tertiary, latent, or congenital syphilis, even after intensive treatment. On the other hand, negativity of the Nelson test can be judged as prognostically favorable (our own observations have been compiled by MULERT and TUPATH-BARNISKE).

Ophthalmologically, location of *primary lesions on the eyelids* (transmitted through kissing), or on the inner canthus of the eye of physicians through bronchoscopy (GOTTRON, HORNBERGER) have been reported, but while the statistics of extragenital primary lesions mention this location as extremely rare, it still occurs as in former times (BARANOV; SCHWARZKOPF: 1 per cent of all extragenital primary lesions occur in the area of the eye). GREITHER extensively discussed primary lesions of the eye with reference to the old literature and his own cases (further case reports by RODIN and PILLAT). SCHRECK reported that the majority of hard chancres occur on the lid margin and from there extend somewhat to the conjunctiva.

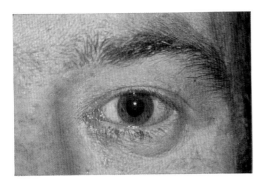

199   Reiter's disease. Conjunctivitis.

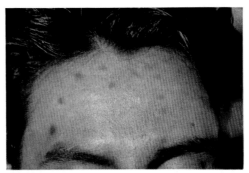

202   Secondary syphilis with hyperpigmented lesions.

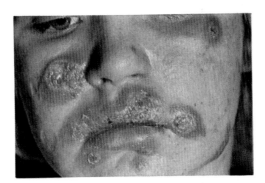

203   Malignant syphilis.

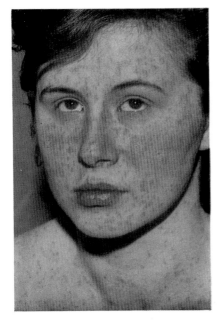

201   Secondary syphilis. Marked roseola.

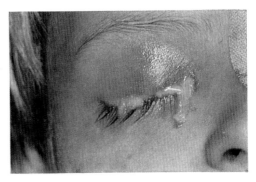

204   Gonoblennorrhea.

Lesions on the eye of the secondary or tertiary periods have become rare because syphilis is diagnosed earlier and treatment has become more effective; there are reports of specific exanthemas involving the eyelids with papules, and condylomas, and of tertiary lesions such as barleycorn- or hailstone-sized gummas, tarsitis, or periostitis. Eyelid changes due to congenital syphilis such as specific angular rhagades similar to Parrot's perioral furrows have been rarely seen during the last decades. The specific alteration of the tear glands (dacryoadenitis) or the bilateral enlargement of the parotid gland due to congenital syphilis also have become rarities; corresponding to the MIKULICZ salivary gland syndrome, this condition of the parotid gland has been called NABARRO's sign.

Parenchymatous keratitis due to congenital syphilis (only exceptionally caused by acquired syphilis) is the most frequent syphilitic manifestation of the cornea; ordinarily it occurs bilaterally (see CREUTZBURG about unilaterality) or it affects within a short time one eye and then the other, usually taking its classical course with only rare exceptions. IGERSHEIMER distinguished five different forms of keratitis but subsequent reviewers have left only three of these as important: one beginning in the periphery at the limbus, the central-axial one, and a third one appearing as a deep fine keratitis (PILLAT). The typical dark blue to gray central clouding of the cornea progresses from the center, accompanied by blepharospasm and abnormal lacrimation and followed by brushlike vascularization starting from the limbus. Iris, ciliary body (the pupil loses its roundness), and peripheral chorioid take part in the inflammatory process probably more often than could, on account of the clouding of the cornea, be ascertained with the ophthalmoscope. For further details see ophthalmological special texts (IGERSHEIMER, LENZ, SCHRECK, and PILLAT). Mention may be made of the typical pepper and salt fundus of syphilitic chorioretinitis; parenchymatous keratitis manifests itself in the majority of cases

between the ages of 6 and 20, females predominating (HERZAU, WERNER, and HOSSMANN).

There exist strictly ophthalmological observations on extremely rare diseases of the cornea due to acquired syphilis (primary lesion; parenchymatous, punctate, deep keratitis; gummas, and so on) and on the equally rare specific diseases of the sclera, iris, and ciliary body (e.g., KRÜCKMANN: roseola of the iris). Strangely enough, a primary specific syphilitic cataract has not been observed so far (PILLAT); the *retina* may become affected through either direct or secondary vascular alteration. Syphilitic changes of the retina sometimes resemble pigmentary retinal degeneration, but in contradistinction to this heredodegenerative disease hemeralopia (night blindness) is not present as a rule.

More important also for the dermatologist is the (preponderantly bilateral) tabetic involvement of the optic nerve, usually accompanied by pupillary signs. The changes of the optic disc are caused by a gangliopetal (directed toward the ganglion — Translators) degeneration with concentric contraction of the visual fields, initially of the red-green color fields. According to IGERSHEIMER, tabes dorsalis shows optic atrophy in 20 per cent, and general paresis in 10 per cent of patients; SCHRECK states that during the last few decades tabetic optic atrophies have become more frequent as compared to osteoarthropathies. Congenital syphilis, however, may show optic atrophy rarely; pathological pupillary changes may occur in the tertiary stage (PILLAT). The phenomenon of the Argyll R. Robertson pupil ("On an interesting series of eye symptoms in a case of spinal disease with remarks on the action of belladonna on the iris, etc.," 1869) has remained most important; clinically, this pupillary reaction permits a diagnosis in late or neurosyphilis in 60 to 90 per cent of cases of tabes, in 50 per cent of general paresis, and in about 10 per cent of cerebrospinal syphilis.

This phenomenon of reflectory fixed pupils may occur rarely also in chronic alcoholism, multiple sclerosis, diabetes mellitus, trauma, and other conditions; is frequently accompanied by anisocoria, deviation from circular shape, and optic atrophy, and is almost always bilateral:

1. The direct and indirect reactions to light are missing.
2. The accommodation and convergence reflex is increased.
3. The pupillary reaction to sensible, sensory, or psychic stimuli is absent or diminished.
4. There is relative or absolute miosis.

However, preponderantly in cerebrospinal syphilis, we find absolute or total areflexia of the pupil. Adie's syndrome is a pseudo Argyll Robertson phenomenon (unilateral myotonic pupil, which is usually larger than the other pupil — Translators) with absent tendon reflexes of the extremities on a nonsyphilitic basis; KYRIELEIS postulated a vegetative disturbance of the reflex action. Cogan's syndrome mentioned above, is a syntropy of a nonsyphilitic, interstitial keratitis with loss of function of the stato-acoustic nerve.

Gonorrhea

The discovery of the causative agent of this disease was made in the year 1879 by the then 23-year-old Albert Neisser in Breslau; the title of his publication was "A form of micrococcus specific for gonorrhea." His classmate Paul Ehrlich then suggested the term "gonococcus," which is still widely used, and today's nomenclature identifies it as "Neisseria gonorrhea"; it belongs to the Neisseria group and is gram-negative, as are also however, all pseudogonococci. A few days after an infection the organisms stained with methylene blue present coffeebean-shaped diplococci, arranged (through phagocytosis) like a swarm of bees within a cell. They are all of the same size (0.6 to 1.0 micron). In legal cases culture plates inoculated with pus

are imperative; this was first accomplished by LEISTIKOW (1882) and BUMM (1885). After a short incubation period (one to eight days) the male presents urethral discharge. The female shows urethral and cervical and later also perhaps rectal involvement due to infection with the vaginal secretion; in the female child and in the senile woman vulvovaginitis gonorrheica may develop after a short incubation period (one to eight days).

Gonorrhea of the eye has to be diagnosed bacteriologically in every case since there are many differential diagnostic possibilities (including blennorrhea due to a filtrable virus, pneumococcic blennorrhea, swimming pool conjunctivitis, and so on). Gonoblennorrhea has undergone a change in its frequency and also in its clinical picture. The incubation period for gonorrheic disease of the eye is the same as for gonorrhea of the genitals. It is essential to separate the different ways of infection; either exogenous (gonorrheic conjunctivitis of the newborn—acquired while the child passes through the genital tract of a gonorrheic mother), or hematogenous as a true gonorrheic metastasis (gonoblennorrhea of the adult, gonorrheic iridocyclitis—MACKENZIE; KRÜCKMANN). Even today protective glasses are recommended when examining a case of gonorrheic blennorrhea. The course and the intensity of gonoblennorrhea increase with age because, as stated by PILLAT, "the power of resistance of the avascular cornea against the toxins of the gonococci and against the proteolysis of the pus cells diminishes with age." Gonorrheic infection of the eye is dangerous because the pathologic process may involve the cornea (infiltrates, ulcers, perforation into the interior, anterior polar cataract, adherent leukoma, and blindness).

Chancroid (ulcus molle)

The causative organism of the soft chancre is the streptobacillus, so called on account of its peculiar chain formation; it was recognized in smears by DUCREY (1889) and later in histologic sections by UNNA (1892); today its name is "hemophilus ducreyi." Chancroids occur mostly as multiple ulcers after an incubation period of 2 to 14 days; the ulcer itself presents as a sharply circumscribed, not indurated, lesion, covered with yellow gray pus showing undermined borders. In 1852 BASSEREAU recognized autoinoculations, so characteristic for the soft chancre. The bacilli are identified after drying of the smear in air, fixation for 15 minutes in 96 per cent alcohol, and staining with Löffler's methylene blue (one to two minutes) or with the Pappenheim-Unna methyl green-pyronin stain (15 minutes). A double infection with the Ducrey bacillus and the spirocheta pallida results in the formation of a mixed ulcer (ulcus mixtum) described by Rollet in 1866; he mentioned the possibility of the simultaneous presence in the same place of a syphilitic chancre (ulcus durum), scabies, "gonorrhea," and vaccine. In about one-third of the cases of chancroid the regional lymph nodes are bilaterally involved and considerably enlarged ("bubo").

Ophthalmologically, SCHRECK refers to rare observations of extragenital chancroids on lids and conjunctiva; reports of such cases go far back to old dermatologic references (GRUNER). The most common extragenital location of the chancroid is the site on the fingers.

Lymphogranuloma inguinale venereum (lymphopathia venereum)

Lymphopathia venerea was known as "strumous" (LANG) or "climatic" bubo; in 1913, in Lyon, the dermatologist Nicolas, the pathologist Favre, and the surgeon Durand established this condition as an independent disease entity of venereal character; Rost had advanced the same idea a year previously. In 1935–36, the causative organism was described by the Japanese MIYAGAWA

et al. in smears stained according to Giemsa's method as a relatively large, granulocorpuscular element, and in 1937 its presence in victoria-blue stains was ascertained by RUGE; before, PRIETO probably had seen this organism in 1927.

The Frei (1925) test plays an important role in the diagnosis of this disease. About 20 days after the development of swollen inguinal lymph nodes, which are frequently divided by Poupart's ligament, the Frei test becomes positive. (This is an intradermal test with injection of 0.1 cc. of a sterilized suspension of bubo material or an emulsion of brain obtained from mice or monkeys who had been infected with lymphogranuloma virus. This technique has been successfully used also in other skin diseases (i.e., Mitsuda test in leprosy, Nickerson-Kveim test in sarcoidosis, or Debré-Mollaret test in cat-scratch disease).

The primary lesion of lymphogranuloma venereum appears about 14 days to weeks after the infection; it is inconspicuous and easily overlooked and ranges from millet- to rice-corn size. The lymphogranuloma bubo is highly characteristic with a central fistula; a late manifestation is the anorectal syndrome (Huguier's disease, "esthiomène") with severe stenosis of the rectum occasionally.

Besides general signs and symptoms such as fatigue, inflammation of the joints, hyper- and dysproteinemias, and polymorphous skin eruptions, one observes inflammatory mucosal changes also on the conjunctiva. Such conjunctivitis and blepharitis can occur together with enlargement of the preauricular lymph nodes in a case of; Parinard syndrome. In any case, either exogenous or hematogenous specific ocular changes can occur with lymphopathia venerea. Some cases of conjunctivitis may be due to *photosensitivity* (SONCK). In a detailed study, Sonck collected 23 cases of ocular affections in lymphogranuloma venereum; he found four cases of recurrent iritis, and seven patients presented mostly phlyctenular conjunctivitis; in one case herpetiform cutaneous erosion occurred;

other patients showed subconjunctival ecchymoses. All patients with ocular involvement presented simultaneous manifestations of other extragenital organs (hydrarthrosis, erythema nodosum, eruptions of light sensitivity and similar manifestations). Special mention has to be made of changes of the *ocular fundus*; according to KITAGARA and FUNAKAWA they chiefly consist of peripapillary edema, dilatations, and tortuosities of the retinal vessels. There are no apparent connections of such fundus changes to other manifestations of lymphogranuloma inguinale. (Further findings such as hemorrhages of the fundus, spotty exudations, and similar ocular changes in animal experiments as well were reported by HENSCHLER-GREIFELT and SCHUERMANN.)

References

ADIE, W. J.: Pseudo Argyll Robertson pupils with absent tendon reflexes; a benign disorder simulating tabes dorsalis. Brit. Med. J. I (1931) 928–930.

BARANOV, A.: Charakteristik der Primäreffekte der Abteilung für Syphilis des GVI. Sovetsk. Vstn. Venerol. Derm. 3 (1934) 245–256. Ref. Zbl. Haut- u. Geschl.-Kr. 48 (1934) 588.

BRUNER, E.: Zur Kenntnis des Ulcus molle extragenitale. Ein Fall von Ulcus molle am Fuße. Derm. Wschr. 54 (1912) 277–286.

COGAN, D. G.: Syndrome of non syphilitic interstitial keratitis and vestibulo auditory symptoms. Arch. Ophthal. 33 (1945) 144–149.

COGAN, D. G.: Non syphilitic interstitial keratitis with vestibulo auditory symptoms. Report of additional cases. Arch. Ophthal. 42 (1949) 42–49.

CREUTZBURG: Beitrag zur Frage der einseitigen Keratitis parenchymatosa e lue connata. Ophth. Kongr. DDR 1952.

GREITHER, A.: Primäraffekte an den Augenlidern. Z. Haut- u. Geschl.-Kr. 7 (1949) 331–332.

HENSCHLER-GREIFELT, A., and SCHUERMANN, H.: Klinik des Lymphogranuloma inguinale. In: Hdb. der Haut- und Geschlechtskrankheiten, Bd. VI/1, edited by J. Jadassohn. Springer, Berlin, 1964.

HERZAU, B., and HOSSMANN, E.: Über Keratitis parenchymatosa. Klin. Mbl. Augenheilk. 88 (1931) 464–477.

HORNBERGER, W.: Hilusdrüsenbeteiligung bei Lues. Fortschr. Röntgenstr. 73 (1950) 553.

IGERSHEIMER, J.: Syphilis und Auge. In: Hdb. der Haut- und Geschlechtskrankheiten, Bd. XVII/2, edited by J. Jadassohn. Springer, Berlin, 1928.

IGERSHEIMER, J.: Syphilis und Auge. In: Kurzes Hdb. der Ophthalmologie, Bd. VII.

IGERSHEIMER, J.: Gibt es irgendwelche ocularen Befunde, welche man definitiv als ausschließlich luetisch klassifizieren kann? Amer. J. Ophthal. 31 (1948) 50, 351, 615.

KORTING, G. W.: Zur Differentialdiagnose und Spezifität einiger luischer Krankheitszeichen. Medizinische 39 (1954) 704–709.

KRESBACH, H.: Zum gegenwärtigen Bild der Frühsyphilis. Z. Haut- u. Geschl.-Kr. 43 (1968) 109–118.

KYRIELEIS, W.: Versuche über Strychninwirkung bei der sogenannten Pupillotonie. Arch. Ophthal. 123 (1923) 1.

LENZ, G.: Syphilis des Auges. In: Haut- und Geschlechtskrankheiten, Bd. IV, edited by L. Arzt and K. Zieler. Urban & Schwarzenberg, Wien, 1934.

LÖHE, H.: Wesen und Verlauf der Frühsyphilis im einzelnen. In: Haut- und Geschlechtskrankheiten, Bd. IV, edited by L. Arzt and K. Zieler, especially p. 256 ff. Urban & Schwarzenberg, Wien, 1934.

MULERT, H., and TUPATH-BARNISKE, R.: Die Lues-Therapie der Universitäts-Hautklinik Mainz in den Jahren 1953–1964 unter besonderer Beachtung der Penicillin-Therapie. Ärztebl. Rhld.-Pfalz 21 (1968) 71–85.

NABARRO, D.: Some aspects of congenital syphilis. Brit. J. Vener. Dis. 25 (1949) 133.

NABARRO, D.: Congenital syphilis. Arnold, London, 1954.

OEHME, J.: Lues (Syphilis) des Kindes. In: Hdb. der Kinderheilkunde, Bd. V, edited by H. Opitz and F. Schmid, 800–887. Springer, Berlin, 1963.

PILLAT, A.: Syphilis und Auge. In: Hdb. der Haut- und Geschlechtskrankheiten, Bd. VI/2A, edited by J. Jadassohn, 926–1052, Springer, Berlin, 1962.

RODIN, A.: Primäraffect unter dem linken Oberlide. Z. Haut- u. Geschl.-Kr. 7 (1949) 21.

SCHRECK, E.: Veränderungen des Sehorgans bei Haut- und Geschlechtskrankheiten. In: Dermatologie und Venerologie, Bd. IV, edited by H. A. Gottron and W. Schönfeld. Thieme, Stuttgart, 1960.

SCHWARZKOPF, A.: Über extragenitale Primäraffekte. Arch. Derm. Syph. (Berl.) 162 (1930) 189–196.

SONCK, C. E.: Die Augenaffektionen bei Lymphogranuloma inguinale. Acta Derm.-Venereol. (Stockh.) 23 (1943) 512–546.

Ocular Changes as Side Effect of Several Drugs Used in Dermatological Practice

The risk of drug effects on the eye as side effects of drugs used in dermatological practice, although small, cannot be considered negligible, since clinical and experimental experience demands caution. During the salvarsan era ocular changes (paresis of ocular muscles, nystagmus, iridocyclitis, damage to the optic nerve, and blocking of the central arteries, as well as loss of hair around the eye, arsenical melanosis of the lids, and conjunctiva) from arsenic were rather common. Nowadays, however, with the negative attitude of dermatologists toward arsenical therapy (for details see KORTING, HOLZMANN, and DENK), such damaging reports have almost disappeared. However, as late as 1967, NOVER and KOINIS could observe retinal bleeding, retinitis, and optic atrophy in a 39-year-old man who had taken by mouth an arsenical preparation for three years on account of psoriasis.

The place of the formerly used arsenic in certain chronic dermatoses (psoriasis vulgaris and lichen planus) has now been taken over by cortisone, which is frequently even given when not strictly indicated. It is known that it may cause ocular damage in acute infections, especially herpetic corneal conditions; it may retard the healing of corneal wounds, and in several reports, when given internally (COVELL) it also may increase the intraocular pressure (reported already in 1960 by KÜCHLE). In a personal observation, in which the suspicion of a connection between a cataract and the prolonged use of corticosteroids seems well founded, a colleague had for years taken triamcinolone for psoriasis vulgaris. He developed a cataract. It seems important to quote OGLESBY, BLACK, V. SALLMANN, and BUNIM's observation of subcapsular cataract-formation in 30 of 72 patients treated with cortisone. These findings occurred in patients who had taken at least 16 mg. of prednisone per day for over a year. Since, however, cataracts do not constitute a manifestation of Cushing's syndrome it must be mentioned that a number of other authors definitely refuse to see a connection between cortisone and cataract (ABRAHAMSON et al.; GORDON et al.; TOOGOOD et al.; and PFAHL et al.).

Other ophthalmological side effects of various drugs used in dermatological practice (see review of KÜCHLE) are reactions due to irritation or hypersensitivity, primarily of the conjunctiva and lids; on rare occasions retinal changes, but more frequently refractive changes (e.g., with the sulfonamides), as well as diplopia, abnormal color vision, and damage to the optic nerve, are observed, as for instance with the antibiotic chloramphenicol, although the retinal damage could not be reproduced experimentally (MEIER-RUGE). There are additional reports on xanthopsias (objects appear yellow — Translators) and transitory myopias following streptomycin, fine corneal opacities following PAS (para-amino-salicylic acid), damage to the optic nerve following isoniazid, and many others. With cytostaticas, among which "Bayer E39" was used early by ophthalmologists for intrafocal treatment of lid basaliomas (PILLAT), an increase in intraocular pressure was observed following mustard gas. Recently, LISCHKA described multiple epithelial corneal defects following methotrexate. It is not surprising that during treatment with anticoagulants, bleeding into the interior of the eye may occur. It should be remembered that this kind of therapy is especially indicated in those ocular diseases (thromboses of the central vein and so on) which are themselves characterized by hemorrhagic tendencies. Compared to these rather rare side effects, which, however, may become serious in an individual case, there are mainly two extensively used drugs, the phenothiazines and chloroquine, whose ocular side effects will be discussed in greater detail.

If the usual dose of the phenothiazines is decisively exceeded, ocular changes may occur, which, however, at the earliest stage of symptoms and signs, are completely reversible (MEYER-RUGE). With chloropromazine opacities of the cornea and lens (CAIRNS, CAPOORE, and GREGORY; CAMERON) were observed and also bluish discolorations of the face and dorsa of the hands and (because of acute photosensitization) redness and swelling of the exposed skin. Changes of the fundus, however, are totally absent.

Chloroquine, which followed another antimalarial (atabrine), was also used in lupus erythematosus (PROKOPTSCHOUK, 1940; PAGE, 1951; GOLDMAN, COLE, and PRESTON; and GRUPPER) all over the world. Various photodermatoses (rosacea, among others) were also treated with this drug. Cutaneous side effects were described as lichenoid or purpuric eruptions or as the more important dyschromias. They represented either pigmentations of the skin, palate, tongue, and nailbed, or depigmentations, especially of the scalp (BARR; LIPPARD and KAUER). Exacerbations of psoriasis were also reported. In general, these side effects were elicited if the daily dose of 250 mg. of chloroquine had been maintained for more than four to six months (STEWART, MILES, and EARNSHAW).

Ophthalmologically, apparently in connection with the pigmentary deposits in the skin resulting from continued use of chloroquine (HOLLWICH and LANGHOF; TUFFANELLI, ABRAHAM, and DUBOIS), irreversible changes develop occasionally. Patients complain of a sensation of blinding and, as we ourselves observed at an early date, disturbances of accommodation. More important, however, are corneal opacities and changes of the retinal vessels (CALKINS, PAU, and BAUMER), which are caused, according to Hollwich and Langhof, partly by the vasoconstrictive effect of this drug and partly by the individual disposition of the patient. The corneal disturbances are presumably due to fine deposits of the break-down products of such antimalarials

in the epithelium. Retinal changes (HOBBS, SORSBY, and FREEMANN) are associated with peripheral contraction of the visual field, nyctalopia, and progressive loss of vision. In the fundus, edema, narrowing of the arterioles, macular degeneration, and, occasionally, massive deposits of pigment occur, with, however, no hemorrhages or exudations. Already fifty years ago such iatrogenic retinal changes from quinine derivatives were reported by JESS. Recently, Meier-Ruge has extensively studied chloroquine-retinopathy. We have learned from this detailed monograph, among other facts, that chloroquine is stored in the eye excessively, about eighty times more than in the liver. By fixation to melanin it can apparently greatly change the normal melanin metabolism (IPPEN). Moreover, chloroquine seems to inhibit the protein metabolism and the enzymatous activities of the melanocytes in the chorioid. Chloroquine-retinopathy is experimentally reproducible in rabbits and cats within a short time (MEIER-RUGE); even if chloroquine is discontinued, progression of the retinal changes continues. Meier-Ruge, in discussing the pathogenesis, suggests the formation of chloroquine-protein associates, "which document themselves possibly in the gray PAS-positive pigment with yellow fluorescence of its own in the pigmented epithelium of the paraffin slides." Knowledge of these retinopathies in the course of chloroquine medication seems to indicate that the eyes of every patient under this therapy have first to be examined by an ophthalmologist before treatment is begun and thereafter about every three months (see also THIES). Of course, a strong indication for the use of the drug and also for the duration of the treatment must exist, if these drugs are to be given at all.

Dimethyl-sulfoxide (DMSO) has been known since 1867 when ALEXANDER SAYTZEFF published its composition. It is a waste product of paper manufacturing and has lately aroused great attention by clinicians on account of its unusual qualities, which are, among other things physiochemical, penetrating, histamine-releasing, and anti-inflammatory (for details see WEYER; LAUDAHN and GERTICH). The sale of the drug has, however, been temporarily interrupted because DMSO causes lenticular opacities, which, when DMSO was orally or transcutaneously applied, could be reproduced in young dogs, pigs, and rabbits, depending on dose and duration of use (RUBIN and BARNETT). In these experiments the effects seemed dependent on unusually high doses, which could not in the least be compared to previous doses used in humans (KLEBERGER; SMITH, MASON, and EPSTEIN; LAUDAHN). In general, ophthalmologic follow-ups on large series of patients treated with DMSO resulted in a denial of lenticular toxicity of DMSO, even if the drug had been applied for months transcutaneously (GORDON; HOFFMANN and MEYER-ROHN).

References

ABRAHAMSON, I. A., Jr., and ABRAHAMSON, I. A.: Cataracta complicata and corticosteroids: the question of a possible relationship between subcapsular cataracts and corticosteroids. Eye, Ear, Nose, Throat Monthly 40 (1961) 266–268.

BARR, J. F.: Subungual pigmentation following prolonged atabrine therapy. U. S. nav. med. Bull. 43 (1944) 929.

CAIRNS, R. J., CAPOORE, H. S., and GREGORY, I. D. R.: Oculocutaneous changes after years on high doses of chlorpromazine. Lancet I (1965) 239–241. Ref. Zbl. ges. Ophthal. 95 (1965/66) 143.

CAMERON, M. E.: Ocular melanosis with special reference to chlorpromazine. Brit. J. Ophthal. 51 (1967) 295–305.

COVELL, L. L.: Glaucoma induced by systemic steroid therapy. Amer. J. Ophthal. 45 (1958) 108.

GOLDMAN, L., COLE, D. P., and PRESTON, R. H.: Chloroquine diphosphate in treatment of discoid lupus erythematosus. J. Amer. Med. Ass. 152 (1953) 1428–1429.

GORDON, D. M.: Dimethyl sulfoxide in ophthalmology, with especial reference to possible toxic effects Ann. N. Y. Acad. Sci. 141 (1967) 392–401.

GORDON, D. M., KAMMERER, W. H., and FREYBERG, R. H.: Examination for posterior subcapsular cataracts: a preliminary report of results in 45 rheumatoid patients treated with corticosteroids. J. Amer. Med. Ass. 175 (1961) 127–129.

GRUPPER, Chr.: Lupus érythémateux et antipaludiques de synthèse. Expérience personelle avec la nivaquine. Ses avantages sur la quinacrine à propos de 36 cas. Bull. Soc. franc. Derm. Syph. 60 (1953) 423–428.

HOBBS, H. E., SORSBY, A., and FREEDMAN, A.: Retinopathy following chloroquine therapy. Lancet II (1959) 478–480.

HOFFMANN, D., and MEYER-ROHN, J.: Ophthalmologische Untersuchungen bei langdauernder Anwendung von DMSO. In: DMSO-Symposion in Wien 1966. Saladruck, Berlin, 1966.

HOLLWICH, F., and LANGHOF, H.: Augenveränderungen nach Chloroquin. Med. Klin. 58 (1963) 1625–1626.

IPPEN, H.: Zur Problematik der Pigmentforschung. Dtsch. med. Wschr. 89 (1964) 798.

JESS, A.: Eine neue experimentelle Retinitis (Chinolin-Retinitis). Actes XII. Congr. Internat. Ophthal. (St. Petersburg) 1 (1914) 109.

JESS, A.: Die Gefahren der Chemotherapie für das Auge, insbesondere über eine das Sehorgan schwerschädigende Komponente des Chinins und seiner Derivate. v. Graefes Arch. Ophthal. 104 (1921) 48–74.

JESS, A.: Die Geschichte einer vergessenen retinotoxischen Substanz. Klin. Mbl. f. Augenheilk. 152 (1968) 649–654.

KIERLAND, R. R., SHEARD, C., MASON, H. L., and LOBITZ,

W. C.: Fluorescence of nails from quinaricine hydrochloride. J. Amer. Med. Ass. 131 (1946) 809–810.

KLEBERGER, K.-E.: An ophthalmological evaluation of DMSO. Ann. N. Y. Acad. Sci. 141 (1967) 381–385.

KORTING, G. W., HOLZMANN, H., and DENK, R.: Nach wie vor aktuell: Arsen-Spätschäden. Ärztebl. Rhld.-Pfalz, 1965.

KÜCHLE, H. J.: Ophthalmologie. In: Klinik und Therapie der Nebenwirkungen, edited by H. P. Kuemmerle, A. Senn, P. Rentchnick, and N. Gossens. Thieme, Stuttgart, 1960.

LAUDAHN, G.: Discussion of the foregoing papers. Ann. N. Y. Acad. Sci. 141 (1967) 402.

LAUDAHN, G., and GERTICH, K.: Dimethyl-Sulfoxid DMSO. DMSO-Symposium in Wien 1966. Saladruck, Berlin, 1966.

LEAKE, Ch. D.: Biological actions of dimethyl sulfoxide. Ann. N. Y. Acad. Sci. 141 (1967).

LIPPARD, V. W., and KAUER, G. L.: Pigmentation of the palate and subungual tissues associated with suppressive quinacrine hydrochloride therapy. Amer. J. Trop. Med. 25 (1945) 469–471.

LISCHKA, G.: Auffallend rasche Wirkung des Methotrexats bei Psoriasis eines 82jährigen Patienten mit gleichzeitigen Nebenwirkungen am Auge. Hautarzt 19 (1968) 473.

MEIER-RUGE, W.: Zur Pathologie arzneimitteltoxischer Nebenwirkungen am Auge. Med. Welt (Stuttg.) 18 (1967) 1601–1609, 1611.

MEIER-RUGE, W.: Medikamentöse Retinopathie. In: Zwanglose Abhandlungen aus dem Gebiet der normalen und pathologischen Anatomie, Heft 18, edited by W. Bargmann and W. Dörr. Thieme, Stuttgart 1967.

NOVER, A., and KOINIS, G.: Arsenschädigung der Netzhaut. Klin. Mbl. Augenheilk. 150 (1967) 535–537.

OGLESBY, R. B., BLACK, R. L., v. SALLMANN, L., and BUNIM, J. J.: Cataracts in rheumatoid arthritis patients treated with corticosteroids: description and differential diagnosis. Arch. Ophthal. 66 (1961) 519–532.

OGLESBY, R. B., BLACK, R. L., v. SALLMANN, L., and BUNIM, J. J.: Cataracts in patients with rheumatic diseases treated with corticosteroids: further observations. Arch. Ophthal. 66 (1961) 625–630.

PAGE, F.: Traitement du lupus érythémateux par la nivaquine. Lancet 1951, 6687.

PFAHL, S. B., Jr., MAKLEY, T., ROTHERMISH, N. O., and McCOY, F. W.: Relationship of steroid therapy and cataracts in patients with rheumatoid arthritis. Amer. J. Ophthal. 52 (1961) 831–833.

PILLAT, A.: Die Wirkung des Cytostaticum Bayer E 39 bei malignen Geschwülsten der Lider. Wien. klin. Wschr. 70 (1958) 383–386.

PROKOPTSCHOUK, A. J.: Traitement du lupus érythémateux par l'acriquine. Vestn. Vener. Derm. (1940) 23.

RUBIN, L. F., and BARRNETT, K. C.: Ocular effects of oral and dermal application of dimethyl sulfoxide in animals. Ann. N. Y. Acad. Sci. 141 (1967) 333—345.

SAYTZEFF, A.: Über die Einwirkung von Salpetersäure auf Schwefelmethyl und Schwefeläthyl. Ann. Chem. 144 (1867) 148.

SMITH, E. R., MASON, M. M., and EPSTEIN E.: The influence of dimethyl sulfoxide on the dog with emphasis on the ophthalmologic examination. Ann. N. Y. Acad. Sci. 141 (1967) 386—391.

SMITH, J. L.: Chloroquine macula degeneration. Arch. Ophthal. 68 (1962) 186—190.

STEWART, T. W., MILES, D. W., and EARNSHAW, E. R.: Pigmentation and retinopathy due to chloroquine. Acta. Derm..-Venereol. (Stockh.) 48 (1968) 47—52.

THIES, W.: Die Chloroquin-Behandlung der Hautkrankheiten. In: Fortschritte praktischer Dermatologie und Venerologie, Bd. IV, edited by A. Marchionini. Springer, Berlin, 1962.

TOOGOOD, J. H., DYSON, C., THOMPSON, C. A., and MULARCHYK, E. J.: Posterior subcapsular cataracts as a complication of adrenocorticalsteroid therapy. Canad. Med. Ass. J. 86 (1962) 52—56.

TUFFANELLI, D., ABRAHAM, R. A., and DUBOIS, E. I.: Pigmentation from antimalarial Therapy. Arch. Derm. Syph. (Chic.) 88 (1963) 419—426.

WULF, K., and ULLERICH, K.: Zur Frage der Augenhintergrundsveränderungen bei der Behandlung tuberkulöser Hautaffektionen mit hohen Dosen Vitamin D 2. Derm. Wschr. 126 (1952) 1235—1236.

Index

Note: Numbers in *italics* indicate pages containing illustrations.

Acantholysis, 81
Acanthomas, hyperkeratotic, 46
Acanthosis nigricans, 78
Acarus scabiei, 55
Acne conglobata, 189
Acne vulgaris, 26
Acrocyanosis, 68
Acrodermatitis chronica atrophicans, 132
Acroleukopathia, 100
Acroparesthesia, 32, 136
Actinic dermatoses, 90
Actinomycosis, 52
Actinomycosis israelii, 54
Adalin purpura, 108
Adenoma sebaceum, 140, *141*
Adie's syndrome, 203
Ainhum, 73
Albinism, 98, *99*
Albinism solum bulbi, 98
Albinism solum fundi, 98
Albinoidism, 98
Alder, granulation anomaly of, 66
Aleppo boil, 36
Alkaptonuria, 122
Allergens, affecting periorbital area, 7
  inhaled, as cause of urticaria, 16
Allergies, contact, 16
  to cobalt, *19*
  to cosmetics, 7
  to light, 91
  to penicillin, 16
Alopecia areata, 190, *192*
Alopecia areolaris, 193
Alopecia atrophicans Brocq, 190
Alopecia mucinosa, 192, *195*
Alopecia specifica, 193
Alopecia totalis seu maligna, 190
Amyloidosis, 116
  familial, 116
  lichen, 116
Anetoderma, 67

Anetodermia erythematosa,
  Pellizzari type, 132
  Schweninger-Buzzi type, 132
Aneurin, lack of, 134
Aneurysma verminosum hominis, 103
Angiitis, of retinal vessels, 102
Angiitis obliterans, 102
Angioid streaks, with pseudoxanthoma elasticum, 76, *77*
Angiokeratoma corporis diffusum, 121, *121*
Angiolopathias, 102
Angiolupoid, 29
Angioma, 149
  definition of, 106
  multiple progressive (Darier), 150
  nodular, *151*
Angioma racemosum, 152
Angiomatosis, encephalo-oculo-cutanea, 150
Angiomatosis miliaris, 121
Angiomatosis retinae, 150
Angiomatosis trigemino-cerebralis, 150
Angiomatous phakomatoses, 142, 150
Angioneuropathias, 102
Angioneurotic edema, *19*
Angiopathies, 106
Angiophakomatoses, 149
Anhidrosis hypotrichotica, 60
Anthrax, of skin, 22
Antigen, Mollaret-Debré, 48
  Tularin, 48
Aphthae, 194
Aphthoid, of Pospischill-Feyrter, 44
Aplasia, of skin, *63*
Aplasia cutis congenita circumscripta, 61
Arachnodactylia, 67
Arcus lipoides, 114
Arcus senilis, 114
Area bullosa, 40
Area Celsi, 190
Area migrans, 40
Argyria, 122
Ariboflavinosis, 134

Arteriolitis "allergica" cutis, 104
Arteritis, hyperergic, 45
Arteritis temporalis, 104
Ascaris, 58
Ascher syndrome, 32, 34
Asthma, as symptom of endogenous eczema, 8
Asteroid bodies, 29
Ataxia telangiectatica, 152
Atopic cataract, *13*
Atopic dermatitis, 42
Atrophoderma vermiculatum, 132
Atrophy, 131
  macular, 132
  sclerosing, of skin, 62
  senile-degenerative, 131
Aurantiasis, 122
Auspitz's phenomenon, 4
Auto-toxic reactions, 18
Avitaminoses, 133, 136

Bacillus anthracis, 22
Bacillus pyocyaneus, 20
Balanitis circumscripta preblastomatosa, 170
Balanitis erosiva, 198
Balanitis specifica luica, 198
Basal cell epithelioma, 155
Basal cell nevus, 142
Basal cell nevus–jaw cyst syndrome, 144
Basaliomas, 142
Bedbugs, 58
Behçet's disease, 194, *195*
Berloque dermatitis, 92
Bindegewebsnaevi, 146
Biopsy, of skin, 31
Bitot spots, 133
Blastomycosis, 52
Blepharitis, mycotic, 51
Blepharitis atrophicans progressiva cum chalasi, 32
Blepharitis granulomatosa, 32, *33*
Blepharochalasis, 32, *33*
Blepharoconjunctivitis, eczematous, 28, 55

Blisters, fever, 44
Boeck's disease, contrasted to sarcoidosis, 29
Boils, Aleppo, 36
  Oriental, 36
Bones, changes in, with familial dermochondrocorneal dystrophy, 65
Bonnet-Dechaume-Blanc syndrome, 150
Bonnevie-Ullrich-Turner syndrome, 69
Botryomycosis, 150
Botulism, 23
Bourneville-Pringle phakomatosis, 140
Bowen tumor, 160, *161*, *163*
Brain, congenital defects of, 39
Brooke's epithelioma, 142
Bruch's membrane, calcification of, 78
Bubo, 204
Bullous pemphigoid, 82
Bullous staphylodermia, superficial, 20
Burns, 87, *89*

Café-au-lait spots, 139
Calcification, of postauricular lymph nodes, in tuberculosis, 25
Calcinosis, dystrophic, definition of, 121
Calcinosis cutis, 121
Callus, light, 90
  milker's, 42
Cancer, metastatic cutaneous, 155
  squamous cell, 155, *159*
Cancroid, 155
Candida albicans, 51
Candida conjunctivitis tarsi et bulbi, 51
Candidiasis, 51
  granulomatous, *53*
Capillaritis, hyperergic, 45
Carcinoma, corium, 155
  of sebaceous gland, 158
  of sweat gland, 158
  prickle cell, 155
Cardio-vaso-renal syndrome, 121
Carotinosis, 122
Cat-scratch disease, 23, 46
  histologic appearance of, *48*
Cataract, 60
  atopic, 12, *13*

Cataract, congenital, 39
  in endogenous eczema, 12
  mongoloid, 68
  syndermatotic, 72
  with scleroderma, 127
  x-ray, 12
  zonular, 66
Catarrh, of conjunctiva, 58
Caterpillars, 58
Cephalosporium, 54
Chagas' disease, 16
Chalazodermia, 32, 35
Chancre, Hunterian, 198
Chancrelike pyoderma, 20, 21
Chancroid, 204
"Chart of stars," 45
Chediak-Higashi syndrome, 98
Cheilitis, 16
Cheilitis granulomatosa, 32
Cheilitis granulomatosa of Miescher, 34
Cheilosis, 134
Chemicals, injury from, 88
Chicken pox, 45
Chilblain lupus, 128
Chlamydobacteriaceae, 54
Chloasma, 95
Chloroquine, as cause of ocular damage, 207
Chorioretinitis, 36
Choroiditis, 28
Christmas disease, 109
Ciliary muscle, in botulism, 23
  in diphtheria, 23
Circulatory disorders, 102
Coagulation necroses, *89*
Coagulopathies, 109
Cobalt, allergic reaction to, *19*
Cogan syndrome, 102, 203
Cold sores, 44
Collagenoses, 126
Coloboma, 60, 65
Comedones, lack of, 26
  palpebral, 189
  periorbital, *115*, 189, *191*
Compound nevi, 167
Condylomas, pointed, 46
Conjunctivae, calcification, 29
Conjunctivitis, massive, 23
  phlyctenular, 28
  pseudomembranous, 23
  unilateral, 48
Connective tissue, diseases of, 126
Contact dermatitis, subacute, eye drops as cause of, *9*

Coombs test, 108
Coproporphyria, congenital erythropoietic, 112
Corium carcinoma, 155
Cornea, alteration of, in tuberculosis of skin, 25
  calcification of, 29
  candidiasis of, 51
  dystrophy of, 60
  ulcers of, 45
  vaccinia lesion of, 42
Cornea verticillata, 122
Corticosteroids, as cause of exanthems, 2
  as cause of ocular damage, 206
Corynebacterium diphtheriae, 22
Cosmetics, allergic reactions to, 7
Cottonwool spots, 127
"Coverings crease," of outer eye, 65
Cryptococcosis, 52
Curtius syndrome, 61
Cutaneous horn, 160, *161*
Cutis laxa, 64, *65*, 68, 75
Cutis marmorata, 68
Cutis nuchae rhomboidalis, 131
Cysts, 149
  horny, with hereditary epidermolyses, 85
  keratin, *147*
Cytoid bodies, 128, 130

Dacryocystitis, sporotrichotic, 52
Dacryosialoadenopathia atrophicans, 197
Dactylolysis spontanea, 73
Darier's disease, 74, *77*
"Delled wart," 43
Dependent tumor, 155
Dermanyssus avium, 55
Dermatite lichenoide purpurique et pigmentée, 108
Dermatite papulo-squameuse atrophiante, 102
Dermatitis, atopic, 42
  berloque, 92
  contact, 7, 8
  dysseborrheic, 8
  in periorbital region, 7
  "scrotal," 134
  seborrheic, 9, 11
Dermatitis bullosa striata pratensis, 92
Dermatitis exfoliativa, 20

Dermatitis exfoliativa neo-
  natorum, 81
Dermatitis herpetiformis, 81, 82,
  83, 84
Dermatitis macular atrophicans,
  132
Dermatitis solaris acuta, 90
Dermatofibroma lenticulare,
  146
Dermatofibrosarcoma pro-
  tuberans, 166
Dermatolysie palpébrale, 32
Dermatomycoses, 50
Dermatomyositis, 126, 129, 130
Dermatoses, 54
  actinic, 90
  bullous, 81
  papulosquamous, 4
  subcorneal pustular, 85
Dermochondrocorneal dys-
  trophy, familial, 65
Descemet's membrane,
  wrinkling of, 12
Detachment, of retina, 14
Diabetes insipidus, in
  sarcoidosis, 29
Diatheses, hemorrhagic, 106
Diencephaloretinal degenera-
  tion, 67
Dimethyl-sylfoxide, as cause of
  ocular damage, 207
Diphtheria, of skin, 22
Diphtherial keratoconjunctivi-
  tis, 42
Diphtheritis blepharitis, 23
Disease. See name of specific
  disease.
Diseases due to light, 90
Dislocation, spontaneous, of
  lens, 64
Dolichostenomely, 67
Drug eruptions, toxic, 2, 3
Drug-induced reactions, 2, 7,
  18, 206
Dubreuilh's melanosis, 170
Ducts, lacrimal,
  inflammation in measles, 1
Dysautonomia, familial, 68
Dyskeratosis congenita, 62
Dyskeratosis intraepithelialis
  benigna hereditaria, 75
Dysmorphia, mandibulo-oculo-
  facialis, 68
Dysostosis, maxillofacial, 68
Dysostosis mandibulofacialis,
  65
Dysostosis multiplex, 66

Dysplasias, congenital, of skin,
  61
  Crouzon's, 66
  ectodermal, 60, 60, 63
  neuroectodermal, 139
  renofacial, 66
Dystonia, vegetative, 11
Dystrophic calcinosis, definition
  of, 121
Dystrophy, familial dermochondrocorneal, 65
  of cornea, 60
  of elastic tissue, 75

Ecchymosis, of bulbar conjunctivitis, 111
Eccrine sweat glands, multiple
  abscesses of, 22
Ecthyma gangrenosum, 22
Ectodermal dysplasia, 63
Ectopia lentis, 66
Eczema, 7
  airborne contacts causing, 8
  chronic, 42
  endogenous, 8
  flexural, 8, 13
  histologic manifestations of, 9
  in periorbital region, 7
  seborrheic, 8, 9
Eczema herpeticum, 41, 42, 44
Eczema solare, 91
Eczema vaccinatum, 41, 42
Eddowe-symptom complex, 67
Edema, angioneurotic, 16, 19
  in erysipelas, 22
  interstitial, of cornea, 12
  of periorbital tissue, 32
  Quincke's, 16
Ehlers-Danlos syndrome, 64, 65
Elastorrhexis systematisata, 76
Elastosis perforans serpiginosa,
  75
Elephantiasis, 16
  subacute, 19
Elephantiasis palpebrarum, 22
Embryopathy, 39
Emotional influences, in lichen
  ruber planus, 18
Emphysema, of skin, 16
Encephalitis, postvaccinial, 40
Endangiitis obliterans, 102
Endogenous eczema, 8
Endophyte, 155
Entropion, congenital, 65
Eosinophilic granuloma, 182
Eosiniphilic granuloma faciale,
  185, 186

Eosinophilic granuloma, 178
Ephelides (freckles), 95, 139
  of Siemens, 139
Ephelidosis maligna, 90
Epicanthus, 68
Epidermolysis acuta toxica, 20
Epidermolysis bullosa, 18
Epidermolysis bullosa heredi-
  taria, 85
Epidermolysis bullosa heredi-
  taria dystrophica dominans,
  85
Epidermolysis bullosa heredita-
  ria dystrophica recessiva, 86
Epidermolysis bullosa hyper-
  plastica, 85
Epidermophyton floccosum, 50
Epithelioid cell granuloma, 28
Epithelioma, basal cell, 142, 155,
  157
  Brooke's, 142
  calcifying, 158
  keratinizing squamous cell,
    155
  Malherbe, 158, 159
  "pagetoid," 155
  squamous cell, 155
Epithelioma adenoides cysti-
  cum, 142
Erysipelas, 21, 22
Erysipeloid erythemas, 40
Erythema, acute, 1
  erysipeloid, 40
Erythema induratum, 28
Erythema migrans, 57
Erythema multiforme, 1, 2, 3
Erythema nodosum, 2
Erythroplasia tumor, 160
Espundia, 36
Euryopia, 66
Euthyroid myxedema, 118
Exanthems, toxic-allergic, 2, 3
  toxic-allergic bullous epi-
    dermolytic, variant of, 20
Exophthalmos, in sarcoidosis,
  29
  malignant endocrine, 118
  resulting from urticaria, 16
Eyeball, protrusion of, 118
Eyebrows, changes in, in endo-
  genous eczema, 11
  lack of, in Thomson's syn-
    drome, 62
  tumor of, 158
Eye drops, as cause of subacute
  contact dermatitis, 9
Eyeglass frames, reaction to, 7, 8

Eyelashes, loss of, 62
  poliosis of, *99*
Eyelid axis, mongoloid, 60
Eyelids, allergic reactions to
    cosmetics, 7
  basal cell nevi of, 144
  blepharitis of, in infants, 8
  edema of, 23
  hemangioma of, 149
  in hordeolum, 20
  in pyodermas, 20
  in sarcoidosis, 29
  in urticaria, 16
  tumor of, 158
  vaccinial lesions, 40, *41*
  with lice, *57*
Eyelid xanthelasma, 113

Fabry's disease, 121
Face, leonine, 14
  telangiectases of, 102
Favus, 50
Fetus, injury to, 39
Fever blisters, 44
Fibromas, 146
Fibrome en pastille, 146
Fibrosis, nodular subepidermal,
    146
Filariasis, 58
Flies, as cause of disease, 58
Fogo selvagem, 82
Follicles, keratoses of, 74
Follicularis, 48
Folliculitis, 20
Foreign bodies, in eye, 88
Freckles, 139
  axillary, 139
  melanotic, 168, *169*
  summer, 95
Freezing injuries, 87
Frei test, 31, 204
Fruit, as cause of urticaria, 16
Fundus, hyperpigmentation
    with melanin, 62
Fundus oculi, alteration of, 25,
    205
Fungi, pathogenic, 50
Furuncles, 20

Gastrophilus equi, 58
Genodermatosis, spinulous
    keratotic, 74
German measles, 39
Giant cell arteritis (arteritis
    temporalis), 104
Glaucoma, 66

Glomus tumor, 150
Goldenhar syndrome, 148
Gonoblennorrhea, *201*, 203
Gonorrhea, 198, 203
Gougerot-Sjögren syndrome,
    197
Gout, cutaneous, 120
  eosinophilic, 178
  epithelioid cell, 28
  pyogenic, 150
  sarcoidlike foreign body, *27*
  tuberculous, 26
Granuloma annulare, *117*
Granuloma telangiectaticum
    benignum, 150
Granulomatosis, recurrent
    edematous, 32
  rhingenous, 103
  Wegener's, 103
Griseofulvin, as cause of
    exanthems, 2
Grönblad-Strandberg syn-
    drome, 78
Growth, stunted, in Weiner's
    syndrome, 62

Hair, diseases of, 190
Hair growth, changes in, 11
Hallermann-Streiffe syndrome,
    68
Halogens, as cause of exan-
    thems, 2
Hand-Schüller-Christian
    disease, 178
Hartnup's syndrome, 112
Hayfever, as sign of endogen-
    ous eczema, 8
Heerfordt syndrome, 29
Hemangioma, 149
  cavernous, *151*
  sclerosing, *147*
Hemangiomatosis, viscero-
    cutaneous (Bean), 150
Hematoma, *111*
Hemiatrophy, facial, 131
  Romberg's, 131
Hemophilia A and B, 109
Henoch's purpura, 108
Hermans-Herzberg phakoma-
    toses, 142
Herpes febrilis, 44
Herpes sepsis, 44
Herpes simplex, 44, *47*
Herpes zoster, 44, *47*
Herpetica, aphthous vulvo-
    vaginitis, 44

Hertoghe phenomenon, 11, *13*
Hidradenoma, 144
Hippel-Lindau syndrome, 142,
    150
Histiocytoma, *147*
Histiocytosis X, 178
Hives, 16
Hochsinger infiltrates, 200
Hodgkin's disease, 180
Hordeolum (sty), 20
Hutchinson's triad, 200
Hyalinosis cutis et mucosae,
    116, *119*
Hydralazine, as cause of
    exanthems, 2
Hydroa aestivalis, 91, 110
Hydroa vacciniforme, 91, 110
Hydroadenitis profunda, 22
Hydrocephalus, inflammatory,
    23
Hydrocystomas, 149
Hyperacusis, 32
Hypercholesterinemic
    xanthomatoses, 113
Hyperglobulinemica, Walden-
    ström's purpura, 108
Hyperkeratosis, diffuse, 72
Hyperlipemic xanthomatosis,
    114
Hyperphosphatemia, familial,
    78
Hyperpigmentation, with con-
    genital anomalies, 65
  with melanin, of fundus, 62
Hyperplasia, of skin, circum-
    scribed lymphocytic, 182
  of skin, circumscribed
    plasmocytic, 182
  pseudoepitheliomatous, 166
Hypertelorismus, 66, 68
Hyphomycetes, 50
Hypodontia, 60
Hypohidrosis, 60
Hypopyon, with keratitis, 44
Hypopyoniritis, recurrent, 194
Hyporegulation, in endogenous
    eczema, 10
Hypothelia, 68
Hypotrichosis, 60

Ichthyosis congenita, 72
Ichthyosis vulgaris, 72
Imbecility, with congenital
    anomalies, 65
Impetigo contagiosa due to
    staphylococci, *21*
Incontinentia pigmenti, 96, *99*

Infectious causes, of lichen ruber planus, 18
Infectious diseases, of skin, 20
Inflammation of retina, 30
INH, disturbances due to, 2, 135
Injuries, physical and chemical, to skin
Iridocyclitis, 26, 28, 31, 44
  exudative, 58
  in sarcoidosis, 29
Iridocyclitis granulomatosa, 36
Iris, color of, 14
  tuberculids of, 25
Iritis, 26, 28
  serous, with corneal vaccinia, 42
  with sporotrichosis, 52
Ixodes ricinus, as cause of disease, 55

Jarisch-Herxheimer reaction, 198
Jellinek sign, 95
Junction nevi, 167
Juvenile melanoma, 168, *169*

Kaposi's sarcoma, 166
Kasabach-Merritt syndrome, 109, 150
Keloids, 148, *148*
  formation of, with hereditary epidermolyses, 85
Keratitides, 43
Keratitis, interstitial, 22
  nonsyphilitic interstitial, 102
  parenchymatous, 28
  vascular, as manifestation of candidiasis, 51
Keratitis dendritica, 44
Keratitis disciformis, 44
Keratitis filiformis, 44
Keratitis punctata, 44
Keratitis serpiginosa, 44
Keratitis superficialis punctata, 52
Keratitis vesicularis, 44
Keratoacanthoma, 164, *165*
Keratoconjunctivitis, 23, 42
Keratoconjunctivitis herpetica, 42, 44
Keratoconjunctivitis scrofulosa, after measles or whooping cough, 25
Keratoconjunctivitis sicca, 197
Keratoconus, 66
Keratomalacia, 133

Keratoses, acneiform follicular, 61, 74
  congenital, 73
  diffuse, 72
  follicular, 74
  seborrheic, *161, 162, 163*
  senile, 160, *161, 162*
Keratosis blennorrhagica, 196
Keratosis palmoplantaris diffusa, 73
Keratosis palmoplantaris insuliformis seu striata, 73
Keratosis palmoplantaris mutilans, 73
Keratosis palmoplantaris papulosa, 73
Keratosis pilaris rubra, 74
Keratosis punctata, 74
Keratosis superficialis, 74
Kidney changes, in sarcoidosis, 29
Klippel-Trenaunay syndrome, 142
Klippel-Trenaunay-Weber phakomatosis, 152
Koplik spots, in measles, 1
Krabbe syndrome, 150
Krause reaction, 198
Kveim-Nickerson test, 31

Langhans-type giant cells, 28
Larva migrans, 58
Laser rays, reactions to, 92, 93
Lawford syndrome, 152
Leiomyoma, 148
Leishmaniasis cutis, 36
Leishmanids, 36
Lens, spontaneous dislocation of, 64
Lentigines, 139
Lentiginosis centrofacialis (Touraine), 139
Lentiginosis profusa (Darier), 139
Lentigo, *141*
Leonine face, 14
Lepidopteriasis, 56
"Lepotrix," 54
Leprosy, 34
  lepromatous, 35
Letterer-Siwe disease, 178
Leukemia, lymphatic, 177
  monocytic, 177
  myeloid, 177
Leukism, 98
Leukoderma acquisitum centrifugum, 100

Leukokeratoses, of conjunctiva, 62
  of oral mucosa, 62
Leukopathias, 98
Leukoplakia, 62
Leukoses, 177
Lice, as cause of disease, 23, 55
  of eyelids, *57*
Lichen amyloidoses, 116
Lichen myxedematosus, 118, *119*
Lichen nitidus, 18
Lichen ruber, 18, *19*
Lichen scorbuticus, 136, *137*
Lichen scrofulosorum, 25, 26, 74
Lichen simplex, 10
Lichen simplex chronicus, 14
Lichen trichophyticus, 74
Lichenoid form of sarcoidosis, *33*
Light, as cause of disease, 90
  callus, 90
  eruption, chronic polymorphic, 91
  urticaria, 91
    familial protoporphyrinimic, 91
"Light-shrunk skin," 90
Lingua plicata, 16, 32
Lingua scrotalis, 32
Lipid proteinosis, 116
Lipoidoses, 113
Lipomas, 148
Lips, alteration of, in Ascher syndrome, 32
  swelling of, 32
Livedo reticularis, 102
Liver mask, 95, *97*
Loefgren syndrome, 29
Louis-Bar syndrome, 102, 142, 152
Lues, 198
  congenital, 22
Lues connata, 200
Lues maligna, 198
"Lumpy jaw," in actinomycosis, 52
Lupus erythematodes chronicus cum exacerbatione acuta, 128
Lupus erythematosus, 36, 126, 128, *129*
  chilblain, 128
  visceral, 126
Lupus miliaris disseminatus faciei, 25, *27*
  histologic appearance, *28*

Lupus pernio, 29, *33*
Lupus vulgaris, 25, 26, *27*, 36
Lyell's syndrome, 20, 81
Lymph, in development of eczema, 7
Lymphadenitis, nonbacterial regional, 46
Lymphadenitis nuchalis, 55
Lymphadenopathy, of neck nodes, 23
Lymphadenosis, 177
Lymphadenosis benigna cutis, 182, *185*, 186
Lymphadenosis benigna orbitae plasmoma, 187
Lymphangiomas, 149
Lymphangitis, with vaccinial disease, 40
    with varioliform lesions, 42
Lymphatic leukemia, 177
Lymphocytes, in development of eczema, 7
Lymphogranuloma inguinale, 46
Lymphogranuloma inguinale venereum, 204
Lymphogranulomatosis, of skin, 180, *182*
Lymphopathia venereum, 204

Macroglobulinemia, 179, *181*
Macula, as sign of primary tuberculosis, 25
Macular atrophy, 132
Maculoanesthetic leprosy, 34
Madarosis, 22, 42
Mafucci syndrome, 150
Mal de meleda, 73
Malherbe, epithelioma, 158, *159*
Mandibulofacial dysostosis, *67*
Marchesani syndrome, 66
Masque biliare, 95
Maxillofacial dysostosis, 68
McManus reaction, *53*
Measles, 1
    German, 39
Meekeren-Ehlers-Danlos syndrome, 64
Melanoblastose neuro-cutanée (Touraine), 142
Melanocytosis, ocular and dermal, 96
    oculodermal (Fitzpatrick), 142
Melanogenesis, 95
Melanophakomatoses, 142
    juvenile, 168, *169*

Melanophakomatoses, malignant, 170, *173*
    metastasizing, 172
Melanoplakia, 170
Melanosis circumscripta preblastomatosa, 168, *169*
Melanosis circumscripta precancerosa, 168
Melkersson-Rosenthal syndrome, 16, 32, *33*, 34
Meningoencephalitis herpetica, 44
Metabolic diseases, 109
Metalues, 200
Microabscesses, 85
Microophthalmia, 60
Microsporon audouini, 50
Microsporosis, 50
Miescher, cheilitis granulamotosa of, 34
Migraine attacks, 32
Mikulicz syndrome, 30
Milia, *151*
"Milk rash," 8, *13*
Milker's calluses, 42
Milker's nodules, 42
Mitsuda reaction, in leprosy, 34
Mitsuda test, 31
Miyagawanelloses, 46
Mollaret-Debré antigen, 48
Möller-Barlow disease, 136
Molluscum bodies, 43
Molluscum contagiosum, *41*, 42, *43*
Mongolism, 68
Mongoloid eyelid axis, 60, 68
Moorens ulcer, 156
Morphea, 126
Morpionosis, 56
Mucormycosis, 54
Mucous membrane, benign pemphigoid of, 84
Muscle, ciliary, in diphtheria, 23
Muscle paralysis, 23
Muscle sphincter pupillae, in botulism, 23
Mycids, 50
Mycosis fungoides, 181, *183*
Myeloid leukemia, 177
Myelosis, 177
Myocardium, injury of the, 23
Myxedema, 16, 118
    circumscribed pretibial, 118
    euthyroid, 118

Nabarro's sign, 202
Necrobiosis lipoidica, 114
Necrosis, multiple, of skin in thromboangiitis, 102
Nelson test, 198, 200
Nephrosis, paraproteinemic, 179
Nerve, optic. See *Optic nerve.*
Neuritis, retrobulbar, 32
Neurodermatitis circumscripta, 14
Neuroectodermal dysplasias, 139
Neurofibromatosis, 139, *141*
Neuronaevus blue Masson, 167
Neurosyphilis, 198
Nevoid syndromes, 139
Nevoxanthoendothelioma, 114, *117*, 146
Nevus (i), 139
    basal cell, 142
        –jaw cyst syndrome, 144
    blue, 167, *168*, *169*
    cellular blue, 168
    compound, 167
    connective tissue, 140
    fascicular, 168
    "familiare chromatophore," 98
    hard, 144
    junction, 167
    nevus cell, *165*, 167
    of corium, 167
    of dermis, 167
    of Ota, 96, *97*
    of Sutton, 100
    organic, 144
    spindle cell, 168
Nevus caeruleus, 167, *168*
Nevus depigmentosus (albinism), 100
Nevus elasticus, 146
Nevus epitheliomatosus multiplex, 142
Nevus flammei, 149
Nevus fuscocaeruleus ophthalomaxillaris of Ota, 96, 142
"Nevus pellitus," 167
Nevus sebaceus, 144, *145*
Nevus spili, 139
Niacin, definition of, 135
Nicotinic acid, deficiency of, 135
Niemann-Pick disease, 120
Nikolski sign, 20, 81
Nodet sign, 81
Nodules, milker's, 42
Nystagmus, 60

Ochronosis, 122
Oligophrenia, phenylpyruvic, 122
Onchocerca caecutiens, 58
Onchocerca volvulus, 58
Ophiasis, 192
Ophthalmia nodosa, 58
Ophthalmomyiasis, 58
Optic nerves, alteration of, 30
  atrophy of, 78
  infiltration of, in sarcoidosis, 29
Orbit, involvement in actinomycosis, 54
Oriental boil, 36
Osler's disease, 106, *107*
Osteitis cystica multiplex, 29
Osteogenesis imperfecta, 67, 75
Osteomyelitis, of maxilla, 16
Ostitis, sporotrichotic, 52
Ota, nevus of, 96, *97*
Oxyuriasis, 58

Pachydermoperiostosis, 16
Pachyonychia congenita, 61
"Pagetoid" epithelioma, 155
Paget's disease, 160
Palmar keratoses, 73
Pannus formation, peripheral, 74
Pantothenic acid, deficiency of, 135
Papilloma, basal cell, 166
Papulose atrophiante maligne, 102
Papulosquamous dermatoses.
  See *Dermatoses, papulosquamous.*
Parahemophilia, 109
Parakeratosis variegata (parapsoriasis lichenoides), 6
Parapemphigus, 82, *83*
Paraproteinemic nephrosis, 179
Parapsoriasis, 6
Parapsoriasis en plaque, 6
Pareiitis granulomatosa, 32
Parinaud's syndrome, 48, 52, 204
Parotitis, 16
Parrot's rhagades, 200
Patch test, to determine allergen, 8
Pediculoides ventricosus, 55
Pediculosis, 55
Pediculosis pubis, 56
Pelade, 190

Pellagra, 135, *137*
  prosperity, 135
Pemphigoid, benign mucous membrane, 84
  bullous, 82
Pemphigus, 81
  Brazilian, 82
  familial, 84
Pemphigus chronicus, 81
Pemphigus erythematosus, 82
Pemphigus febrilis, 20
Pemphigus foliaceus, 81, 82
Pemphigus vegetans, 81
Pemphigus vulgaris, 81, *82, 83*
Pencil, indelible, tissue necroses of, 87
Penicillin, allergic reaction to, 16
Periarteritis nodosa, 103
Perifolliculitis, 20
Periorbital area, allergens affecting, 7
Periorbital comedones, *115*
Periorbital tissue, edema of, 32
Periostitis, sporotrichotic, 52
Perlèche, 134
Pernio, springtime, 91
Peutz-Jeghers syndrome, 139, 142, *143*
Pfaundler's disease, 66
Pflastersteinnaevi, 146
Phakomatosis, 139
  angiomatous, 142, 150
  Bourneville-Pringle, 140
  epitheliomatous, 142
  Hermans-Herzberg, 142
  Klippel-Trenaunay-Weber, 152
Phenistix test, 122
Phenolphthalein, as cause of exantheme, 2
Phenomenon, Hertoghe, 11
Phenylketonuria, 122
Phenylpyruvic oligophrenia, 122
Phlyctenular reactions, 26
Phosphatid storage disease, 121
Photodermatosis, 91
  familial protoporphyinemic (urticaria due to light), 112
Photodynamic reactions, 92
Phthiriasis, 56
Phrynoderma, 133, 136
Pigmentation, disturbances in, 95
Pilomatrixoma, 158
Pityriasis lichenoides chronica, 6
Pityriasis rubra pilaris, *5*

Pityriasis simplex, signs of, 8
Plantar keratoses, 73
Plants, as cause of contact dermatitis, 8
Plasmocytoma, 179
Plexuschoriordeus, 23
"Plica polonica," 55
Plummer-Vinson syndrome, 135
Poikiloderma, Rothmund's, 60
Poikilodermatous changes, in dyskeratosis congenita, 62
Poliosis, of eyelashes, *99*
Poliosis circumscripta, 98
Poltauf-Sternberg's disease, 180
Polykeratosis, congenital, 61
Polymyositis, 130
Porphyria, 110
  acute, 113
  erythropoietic, 112
Porphyria congenita, 112
Porphyria cutanea tarda, 112, 113, *115*
Porphyria hepatica acuta intermittens, 112
Porphyria variegata, 112
Pospischill-Feyrter, aphthoid of, 44
Postgonorrheal catarrh, 198
Postvaccinial encephalitis, 40
Precancers, 160
Prickle cell carcinoma, 155
Pringle's disease, 140, *141, 143*, 146, 148
Progeria, 62
Prognathy, 67
Prosperity pellagra, 135
Proteinosis, 116
Prurigo, 16, 58
  summer, 110
Prurigo lymphatica, 177
Pseudoacanthosis nigricans, 78
Pseudoalkaptonuria, 122
Pseudocancers, 164
Pseudomonas aeruginosa, 20
"Pseudo-Hertoghe" sign, *13*
Pseudolues, 198
Pseudopelade, 190
"Pseudotrichinosis" (dermatomyositis), 16
Pseudoxanthoma elasticum, 75, 76, *77*
Psoriasis vulgaris, 4, *5*
PTC deficiency, 109
Pterygia, 68
Pterygium syndrome, 69

Ptosis, 32
  in leprosy, 36
Pupil, Argyll R. Robertson, in
  syphilis, 202
Puppet string phenomenon, 136
Purpura, Adalin, 108
  eczematidlike, 108
  Henoch's, 108
  itching, 108
  "orthostatica," 108
  progressive pigmentary, 108
  Schönlein's, 108
  thrombotic thrombocyto-
    penic, 109
Purpura arciformis, 108
Purpura Majocchi, 108
Pustulosis varicelliformis, 42
Pyemotids, 55
Pyocyaneus infection of skin,
  20, 22
Pyoderma, 20, 21
Pyogenic granuloma, 150
Pyrazolon, derivatives of, as
  cause of exantheme, 2
Pyridoxine, deficiency of, 135
Pyridoxinosis, 135
Quincke's edema, 16

Rabbit fever (tularemia), 23
Radiation, reactions to, 90, 92,
  93
Radiodermatitis, 159
Reaction, McManus, 53
  Mitsuda, 34
"Red eye" disease, 75
Refsum syndrome, 72
Reiter's disease, 196, 201
Reiter-balanitis, 196
Renal changes, in sarcoidosis, 29
Rendu-Osler disease, 108
Reticulogranulomatosis, 180
Reticuloses, 177, 178
Retina, cottonwool spots, 127
Retinal detachment, 14
Retinal inflammation, 30
Retinitis, with sporotrichosis,
  52
Rhinophyma, 190
Riboflavin, deficiency of, 134
Riehl syndrome, 62
Riley-Day syndrome, 68
Ritter's disease, 20
Rodents, as cause of tularemia,
  23
Roentgen rays, reactions to, 12,
  92

Romberg's hemiatrophy, 131
Rosacea, 189, 191
Rosacea erythematosa, 189
Rosacea keratitis, 190, 191
Rosacea lupoid, 26
Rosacea papulosa, 189
Rothmund syndrome, 62
Rothmund's poikiloderma, 60
Rubella, 39
Rud syndrome, 72
Rumpel-Leed test, 109

Sarcoidosis, 28
  lichenoid form of, 33
Sarcoma, 166
  Kaposi's, 166
  multiple idiopathic hemor-
    rhagic, 166
Sarcoptes hominis, 55
Savin syndrome, 72
Scabies, 55
  Norwegian, 56
Scalds, 87
Scale, facial, 6
Schamberg's disease, 108
Schaumann bodies, 29
Schistosomiasis, 16
Schönlein's purpura, 108
Scleredema, adultorum, 127
Scleroderma, 126, 129
  circumscribed, 126
  edematous, 127
Scleromyxedema, 118
Sclerosis, tuberous cerebral, 140
Scotomas, 32
Scrofuloderma, 25, 26
Scurvy, 136
Sebaceous gland, carcinoma of,
  158
  diseases of acne vulgaris, 189,
  191
"Seborrheic appearance"
  in psoriasis vulgaris, 4
Senear-Usher syndrome, 82
Senile-degenerative atrophy of
  skin, 131
Serpiginous vaccinia, 40
Sézary's syndrome, 178
Shingles, 44
Siemens-Schäfer syndrome, 62
"Signe d'omnibus," 103
Sjögren's syndrome, 31, 197
Sjögren-Larsson syndrome, 72
Skeletal malformations, 68
  skin and eye changes with, 66

Skin, anthrax of, 21, 22
  atrophy of, 131
  chronic infectious diseases of,
    25
  congenital abnormalities of,
    60
  diphtheria of, 22
  farmer's, 131
  infectious diseases of, 20
  "light-shrunk," 90
  lymphogranulomatosis of,
    180, 182
  malignant tumors of, 155
  mycotic infections of, 50
  physical and chemical injuries
    to, 87
  sailor's 131
  toxoplasmosis of, 23
  tumors of, 139
  virus diseases of, 39
Smallpox, 41
Sores, cold, 44
Spanlang-Tappeiner syndrome,
  73
Sphincter pupillae muscle, in
  botulism, 23
Sphingolipidoses, 120
Spider web appearance, in lichen
  ruber planus, 18
Spiegler's tumors, 142
Spinalioma, 155
"Spinolusism," 74
Spongiosis, 9
Sporotrichosis, 52
Spots, Bitot, 133
  café-au-lait, 139
  Koplik, in measles, 1
Springtime pernio, 91
Squamous cell epithelioma, 155
Staphylococcal impetigo
  contagiosa, 20
Staphylococcal pemphigoid, 20
Staphylococcus aureus, 20
Staphylodermia, superficial
  bullous, 20
Stevens-Johnson syndrome, 1
Stomatitis, 44
Stomatitis angularis, 134
Storage diseases, 110
Strabismus, 60
Streptococcus pyogenes, 20
"Streptotrix," 54
Strophulus, 16
"Strumous," 204
Sturge-Weber syndrome, 142,
  150, 151
Sty (hordeolum), 20

Sulfonamides, as cause of
  exanthems, 2
Summer freckles, 95
Summer prurigo, 110
Sunburn, 90, 92
Sutton, nevus of, 100
Sweat glands, abscesses of, 22
  carcinoma of, 158
Sycosis parasitaria, 51
Sycosis parasitaria ciliaris, 51
Sycosis vulgaris, 20
Syndrome, Adie's, 203
  Ascher, 32, 34
  Bonnet-Dechaume-Blanc, 150
  Bonnevie-Ullrich-Turner, 69
  Chediak-Higashi, 98
  Cogan, 102, 203
  Curtius, 61
  Goldenhar, 148
  Gougerot-Sjögren, 197
  Grönblad-Strandberg, 78
  Hallermann-Streife, 68
  Hartnup's, 112
  Heerfordt, 29
  Hippel-Lindau, 142
  Kasabach-Merritt, 109, 150
  Klippel-Trenaunay, 142
  Krabbe, 150
  Lawford, 152
  Loefgren, 29
  Louis-Bar, 102, 142, 152
  Lyell, 20, 81
  Mafucci, 150
  Marchesani, 66
  Meekeren-Ehlers-Danlos, 64
  Melkersson-Rosenthal, 32,
    33, 34
  Mikulicz, 30
  nevoid, 139
  Parinaud's, 48, 52, 204
  Peutz-Jeghers, 139, 142, 143
  Plummer-Vinson, 135
  pterygium, 69
  Refsum, 72
  Riehl, 62
  Riley-Day, 68
  Rothmund, 62
  Rud, 72
  Savin, 72
  Senear-Usher, 82
  Sézary's, 178
  Siemens-Schäfer, 62
  Sjögren's, 31, 197
  Sjögren-Larsson, 72
  Spanlang-Tappeiner, 73
  Stevens-Johnson, 1
  Sturge-Weber, 142, 150, 151

  Thibierge-Weissenbach, 76
  Thomson's, 62
  Treacher-Collins, 67
  Vogt-Koyanagi, 61
  Vogt-Koyanagi-Harada, 98,
    99
  Ward, 144
  Weiner's, 62
  Wildervanck, 148
Synechiae, 68
Syphilid, tuberoserpiginous, of
  Lewis, 26
Syphilis, 198, 201
  cerebrospinal, 203
  congenital, 23
  primary, 20
Syphilis congenita tarda, 200
Syringoma, 144, 145

Tabes dorsalis, 202
Tear duct, 26
  obliteration of, 62
Tear sac, 26
Tears, reduction in secretion,
  60
Telangiectases, of face, 102
Test, Coombs, 108
  for actinomycosis, 52
  for sarcoidosis, 31
  for smallpox, 40
  Frei, 31, 204
  Kveim-Nickerson, 31
  Mitsuda, 31
  Nelson, 198, 200
  patch, to determine allergen, 8
  phemstix, 122
  Rumpel-Leed, 109
  TPI, 200
  tuberculin, 25, 31
Tetradactyly crease, 68
Tetrastichiasis, with congenital
  anomalies, 65
Thibierge-Weissenbach
  syndrome, 76
Thomson's syndrome, 62
Thromboangiitis cutaneo-
  intestinalis disseminata, 102
Thrombocytopathies, 108
Thrombocytopenia, 108
Ticks, causing tularemia, 23
Tinea corporis, 50
Tinea inguinalis, 50
Tinea interdigitalis, 50
Toadskin, 133
Torulosis, 52

Toxic-allergic bullous epi-
  dermolytic exanthem, variant
  of, 20
Toxic allergic exanthems, 2, 3
Toxoplasmosis of skin, 23
TPI test, 200
Transfusion syphilis, 198
Trauma, 16
Treacher-Collins syndrome,
  67
Trichiasis, 22
Trichinosis, 16, 58
Trichiuris, 58
Trichomatrioma, 158
Trichophyton mentagrophytes,
  50
Trichophyton rubrum, 50
Trichophyton schönleini, 50
Trichophytosis, 51
  superficial, 53
Trichophytosis ciliaris, 51
Trisomy 21, 68
Trombiculus autumnalis, 55
Trophic ulcer, of nose, 89
Tuberculids, 25
  papulonecrotic, 28
Tuberculin, cutaneous sensitiv-
  ity to, 26
  tests, 25, 31
Tuberculoid leprosy, 34
Tuberculosa verrucosa cutis, 25
Tuberculosis, atypical form of,
  29
Tuberculosis colliquativa cutis,
  25
Tuberculosis cutis, 25
Tuberculosis fungosa serpigi-
  nosa, 26
Tuberculosis miliaris ulcerosa
  cutis et mucosae, 25
Tuberculous granulomata, 26
Tuberoserpiginous syphilid of
  Lewis, 26
Tuberous cerebral sclerosis, 140
Tularemia, 23
  oculoglandular form of, 20
Tularin antigen, 48
Tumor, Bowen, 160, 161, 163
  dependent, 155
  erythroplasia, 160
  eyelid, 58, 158
  glomus, 150
  Koenen's, 140
  of skin, 139
    malignant, 155
  pigmented, 167
  Spiegler's, 142

Ulcer, Mooren's, 156
  of cornea, 45
  trophic, of nose, *89*
  vaccinial, 40
Ulcus durum, 204
Ulcus mixtum, 204
Ulcus molle, 204
Ulcus serpens with hypopyon,
  51
Ulerythema ophryogenes, 74, *77*
Uranitis granulomatosa, 32
Urethro-oculo-synovial
  syndrome, 196
Urticaria, 16, *19*, 58
  light, 91
Urticaria pigmentosa, 178, *183*
Uveitis, "granulomatous," 23
  in sarcoidosis, 29
Uveo-aphthous syndrome, 194
Uveoparotitis, 29

Vaccination, side effects of, 40
"Vagabond skin," caused by
  lice, 55
Varicella (chicken pox), 44, 45,
  47
Vascular disorders, 102
Vasculitides, relation to tuber-
  culous lesions, 25
Vasomotor disorders, 102

Venereal disease, 198
Vermiasis, 58
Verruca seborrheica, 166
Verruca senilis, 160, 166
Verrucae vulgaris, 46, *47*
Viral diseases of skin, ocular
  manifestations of, 39 (table)
Virus, varicella, 44
Vision, acuity of, 45
Vitamin A, deficiency of, 133
Vitamin B$_2$ (riboflavin),
  deficiency of, 134
Vitamin B$_6$, deficiency of, 135
Vitamin C, definition of, 136
Vitamin D, 2
Vitiligo, *99*, 100
Vitreous, bleeding into, 29
Vogt-Koyanagi syndrome, 61
Vogt-Koyanagi-Harada
  syndrome, 98
Von Recklinghausen's disease,
  139, 140, *141*
Vulvitis "pellagrosa," 134

Waldenström's purpura hyperglobulinemica, 108
Waldenström's disease, 179, *181*
Ward syndrome, 144
Warts, 46, *47*
  "delled," 43
  seborrheic, 166

Wegener's granulomatosis, 103
Weiner's syndrome, 62
Werlhof disease, 108, *111*
Wickham's striae, 18
"Wild fire," 82
Wildervanck syndrome, 148
Wilson's disease, 114
Worms, as cause of urticaria, 16

Xanthelasma, 113
Xanthelasma palpebrarum, 113,
  115
Xanthelasma sebaceum, 113
Xanthoma, *115*
Xanthomatoses, hyper-
  cholesterinemic, 113
  hyperlipemic, 114
Xeroderma pigmentosum, *89*,
  90
Xerophthalmia, 133
X-rays, reactions to, 12, 92

Yeast septicemia, 51

Ziehl-Neelsen technique, in
  leprosy, 34
Zona, 44
Zoonoses, 55
Zoster ophthalmicus, 45
Zoster oticus, 45